Ecological Studies, Vol. 158

Analysis and Synthesis

Ecological Studies

Volumes published since 1996 are listed at the end of this book.

Springer
Tokyo
Berlin
Heidelberg
New York
Barcelona
Hong Kong
London
Milan
Paris

T. Nakashizuka Y. Matsumoto (Eds.)

Diversity and Interaction in a Temperate Forest Community

Ogawa Forest Reserve of Japan

With 109 Figures, 4 in Color, and 40 Tables

Springer

Dr. Tohru Nakashizuka
Professor, Research Institute for Humanity and Nature
Kitashirakawa-Oiwake, Sakyo, Kyoto 606-8502, Japan

Dr. Yoosuke Matsumoto
Head of Global Forest Conservation Laboratory
Forestry and Forest Products Research Institute (FFPRI)
P.O. Box 16, Tsukuba Norin Kenkyu Danchi-nai
Tsukuba, Ibaraki 305-8687, Japan

ISSN 0070-8356
ISBN 978-4-431-68000-0 ISBN 978-4-431-67879-3 (eBook)
DOI 10.1007/978-4-431-67879-3

Library of Congress Cataloging-in-Publication Data applied for.

Printed on acid-free paper

Softcover reprint of the hardcover 1st edition 2002

Foreword

The studies in the Ogawa Forest Reserve (OFR) were initiated by a group of plant ecologists and gradually expanded into a comprehensive project covering various aspects of biology, soil science, and silviculture. The project was integrated as part of the Forest Ecosystem Team under the BIO-COSMOS Program funded by the Ministry of Agriculture, Forestry and Fisheries. As the coordinators of the Forest Ecosystem Team, we are pleased that reports of the long-term studies carried out in the OFR are being published in this first volume on Japanese ecosystems in the Ecological Studies series.

Scientists and researchers have made numerous contributions to the field of forest ecology during more than 10 years of studies in the OFR. Two reasons can be cited for the success of the project: scientists from various disciplines concentrated on a single target forest ecosystem, and the research continued over a relatively long term. It is now recognized that ecological processes include complicated mechanisms supported by interactions among organisms and large temporal variations. The researchers in the OFR project were motivated by their interest in the history of ecosystems and the interactions of diverse creatures in the forest. Integrating their research interests yielded answers concerning the mechanisms by which biodiversity is maintained and how it functions in the forest ecosystem. It also offered insights into sustainable management approaches, one of the most important issues related to forest ecosystems worldwide. We hope that the studies in the OFR will continue and expand to meet the increasing need for greater knowledge of forest ecosystems and their management.

Much remains unknown concerning biodiversity and its function in ecosystems. We hope that this book will act as a stimulus for further research in this field by introducing intensive studies conducted in Japan to an international audience.

August 22, 2001
SHOBU SAKURAI
YASUSHI MORIKAWA

Preface

Studies on mechanisms for maintaining biological diversity in an ecosystem are of primary importance in both basic ecology and applied ecological sciences. The causality of the variation in species diversity among ecosystems, maintaining mechanisms of diversity, and the role of biodiversity in ecosystems are the central issues of modern ecology. A general decrease in biodiversity has been recognized as one of the most critical global environmental problems. In silviculture and forest management, biodiversity is one of the key criteria for sustainable management practices.

This book was compiled after 13 years of intensive investigations on the mechanisms by which tree diversity is maintained in a forest community. The Ogawa Forest Reserve (OFR) is a remnant old-growth forest in central Japan and has one of the highest tree species diversities of temperate deciduous forests worldwide. Researchers with various backgrounds, including plant ecology, tree physiology, entomology, mycology, pedology, forest meteorology, and silviculture, carried out studies in the OFR to elucidate the structure and dynamics of the forest community. Among the wide range of ecological results, those on tree demographies and their variation, interactions with other organisms, and reactions to disturbance are particularly important. The investigations covered more than 20 tree species in this community, and the findings have been integrated at the community level.

Although a period of 13 years is not sufficient to understand the entire structure of the forest community, the studies targeting the same community published here have greatly extended our knowledge. The results indicate the rules of ecological assembly in this community, along with several key interactive processes that control its organization and dynamics. The results also have numerous implications for sustainable forest management practices. We believe that the OFR research will make further contributions as the various studies continue in the future.

The OFR studies were funded by various sources. In particular, the BIO-COSMOS Program of the Ministry of Agriculture, Forestry and Fisheries supported the research for 10 years. The Ministry of Education, Science, Sports and Culture (08454250, 10354013, and 09NP1501) and the Ministry of Environment (Grant B52.3.2, 1996–1998) also provided several grants for aspects of the project. This publication was supported by the Japan Society for the Promotion of Science (Grant 135308). The Forestry and Forest Products Research Institute provided all facilities and on-site support necessary for the project. The Tokyo Regional Forestry Office shared our enthusiasm for the research and gave approval for activities to be carried out in the OFR. On behalf of all the scientists who participated in this project, we would like to express our sincere appreciation to those organizations.

Special thanks are also due to Drs. Shobu Sakurai and Yasushi Morikawa, who coordinated the entire program of the Forest Ecosystem Team of BIO-COSMOS, for their understanding and encouragement. We also thank Dr. Satoko Kawarasaki and Nobuya Koike, MSc, who assisted in editing this volume.

TOHRU NAKASHIZUKA
YOOSUKE MATSUMOTO

Contents

Part 5: Eco-physiological Studies on Tree Growth

Part 6: Genetic Studies of Tree Populations

Part 7: Interaction Between Plants and Other Organisms

Contributors

ABE, SHIN
Hokkaido Research Center, Forestry and Forest Products Research Institute (FFPRI), 7 Hitsujigaoka, Toyohira, Sapporo, Hokkaido 062-8516, Japan

AIZAWA, SHUHEI
Tohoku Research Center, Forestry and Forest Products Research Institute (FFPRI), 72 Shimokuriyagawa, Morioka, Iwate 020-0123, Japan

IIDA, SHIGEO
Forestry and Forest Products Research Institute (FFPRI), P.O. Box 16, Tsukuba Norin Kenkyu Danchi-nai, Tsukuba, Ibaraki 305-8687, Japan

ISAGI, YUJI
Faculty of Integrated Arts and Sciences, Hiroshima University, 1-7-1 Kagamiyama, Higashi-Hiroshima, Hiroshima 739-8521, Japan

ISHIDA, ATSUSHI
Forestry and Forest Products Research Institute (FFPRI), P.O. Box 16, Tsukuba Norin Kenkyu Danchi-nai, Tsukuba, Ibaraki 305-8687, Japan

ISHIZUKA, MORIYOSHI
Forestry and Forest Products Research Institute (FFPRI), P.O. Box 16, Tsukuba Norin Kenkyu Danchi-nai, Tsukuba, Ibaraki 305-8687, Japan

KANAZASHI, TATSUO
Tohoku Research Center, Forestry and Forest Products Research Institute (FFPRI), 72 Shimokuriyagawa, Morioka, Iwate 020-0123, Japan

KAWANO, SHOICHI
Department of Botany, Graduate School of Science, Kitashirakawa-Oiwake, Sakyo, Kyoto University, Kyoto, 606-8502, Japan
Present address: 303-204 Greentown Makishima, 51-1 Makishima-cho, Motoyashiki, Uji, Kyoto 611-0041, Japan

KIMURA, KYOKO
The Master's Program in Environmental Sciences, Tsukuba University, Tennodai, Tsukuba, Ibaraki 305-8572, Japan
Present address: Kuchierabu-jima, Kamiyaku-cho, Kumage-gun, Kagoshima 891-4208, Japan

KITAMURA, KEIKO
Forestry and Forest Products Research Institute (FFPRI), P.O. Box 16, Tsukuba Norin Kenkyu Danchi-nai, Tsukuba, Ibaraki 305-8687, Japan

KOMINAMI, YOHSUKE
Kyusyu Research Center, Forestry and Forest Products Research Institute (FFPRI), 4-11-16 Kurokami, Kumamoto 860-0862, Japan

KUBO, TAKUYA
Frontier Research System for Global Change, 3173-25 Showa-machi, Kanazawa, Yokohama, Kanagawa 236-0001, Japan

MARUYAMA, YUTAKA
Hokkaido Research Center, Forestry and Forest Products Research Institute (FFPRI), 7 Hitsujigaoka, Toyohira, Sapporo, Hokkaido 062-8516, Japan

MASAKI, TAKASHI
Tohoku Research Center, Forestry and Forest Products Research Institute (FFPRI), 72 Shimokuriyagawa, Morioka, Iwate 020-0123, Japan

MATSUMOTO, YOOSUKE
Forestry and Forest Products Research Institute (FFPRI), P.O. Box 16, Tsukuba Norin Kenkyu Danchi-nai, Tsukuba, Ibaraki 305-8687, Japan

MATSUURA, YOJIRO
Forestry and Forest Products Research Institute (FFPRI), P.O. Box 16, Tsukuba Norin Kenkyu Danchi-nai, Tsukuba, Ibaraki 305-8687, Japan

MIZOGUCHI, YASUKO
Forestry and Forest Products Research Institute (FFPRI), P.O. Box 16, Tsukuba Norin Kenkyu Danchi-nai, Tsukuba, Ibaraki 305-8687, Japan

MORIKAWA, YASUSHI
Department of Basic Human Science, Waseda University, 2-579-15 Mikajima, Tokorozawa, Saitama 359-1164, Japan

MORISAWA, TAKESHI
Kiso Experimental Station, Forestry and Forest Products Research Institute (FFPRI), 5473 Kisofukushima, Kiso, Nagano 397-0001, Japan

NAKASHIZUKA, TOHRU
Research Institute for Humanity and Nature, Kitashirakawa-Oiwake, Sakyo, Kyoto 606-8502, Japan

NIIYAMA, KAORU
Forestry and Forest Products Research Institute (FFPRI), P.O. Box 16, Tsukuba Norin Kenkyu Danchi-nai, Tsukuba, Ibaraki 305-8687, Japan

NOGUCHI, KO
Department of Biology, Graduate School of Science, Osaka University, 1-16 Machikaneyama-cho, Toyonaka, Osaka 560-0043, Japan

OCHIAI, YUKIHITO
Japan International Research Center for Agricultural Sciences (JIRCAS), 1-2 Ohwashi, Tsukuba, Ibaraki 305-8686, Japan

OHKOCHI, ISAMU
Forestry and Forest Products Research Institute (FFPRI), P.O. Box 16, Tsukuba Norin Kenkyu Danchi-nai, Tsukuba, Ibaraki 305-8687, Japan

OHTANI, YOSHIKAZU
Forestry and Forest Products Research Institute (FFPRI), P.O. Box 16, Tsukuba Norin Kenkyu Danchi-nai, Tsukuba, Ibaraki 305-8687, Japan

OKABE, HIROAKI
Forestry and Forest Products Research Institute (FFPRI), P.O. Box 16, Tsukuba Norin Kenkyu Danchi-nai, Tsukuba, Ibaraki 305-8687, Japan

PARK, YONG-MOK
Department of Life Science, College of Natural Science and Engineering, Chongju University, 36 Naedok-Dong, Chongju 360-764, Korea

SAKURAI, SHOBU
Forestry and Forest Products Research Institute (FFPRI), P.O. Box 16, Tsukuba Norin Kenkyu Danchi-nai, Tsukuba, Ibaraki 305-8687, Japan

SHIBATA, MITSUE
Forestry and Forest Products Research Institute (FFPRI), P.O. Box 16, Tsukuba Norin Kenkyu Danchi-nai, Tsukuba, Ibaraki 305-8687, Japan

SONE, KOSEI
Department of Biology, Graduate School of Science, Osaka University, 1-16 Machikaneyama-cho, Toyonaka, Osaka 560-0043, Japan

SUZUKI, WAJIROU
Forestry and Forest Products Research Institute (FFPRI), P.O. Box 16, Tsukuba Norin Kenkyu Danchi-nai, Tsukuba, Ibaraki 305-8687, Japan

TAKAHASHI, MASAMICHI
Forestry and Forest Products Research Institute (FFPRI), P.O. Box 16, Tsukuba Norin Kenkyu Danchi-nai, Tsukuba, Ibaraki 305-8687, Japan

TANAKA, HIROSHI
Forestry and Forest Products Research Institute (FFPRI), P.O. Box 16, Tsukuba Norin Kenkyu Danchi-nai, Tsukuba, Ibaraki 305-8687, Japan

TERASHIMA, ICHIRO
Department of Biology, Graduate School of Science, Osaka University, 1-16 Machikaneyama-cho, Toyonaka, Osaka 560-0043, Japan

UEDA, AKIRA
Kansai Research Center, Forestry and Forest Products Research Institute (FFPRI), Momoyama-machi, Fushimi, Kyoto 612-0855, Japan

UEMURA, AKIRA
Forestry and Forest Products Research Institute (FFPRI), P.O. Box 16, Tsukuba Norin Kenkyu Danchi-nai, Tsukuba, Ibaraki 305-8687, Japan

UTSUGI, HAJIME
Hokkaido Research Center, Forestry and Forest Products Research Institute (FFPRI), 7 Hitsujigaoka, Toyohira, Sapporo, Hokkaido 062-8516, Japan

YOSHINAGA, SHUICHIRO
Forestry and Forest Products Research Institute (FFPRI), P.O. Box 16, Tsukuba Norin Kenkyu Danchi-nai, Tsukuba, Ibaraki 305-8687, Japan

Part 1

Introduction

1 Studies on the Coexistence of Tree Species in Forest Communities

TOHRU NAKASHIZUKA and HIROSHI TANAKA

1.1 Diversity and Species Coexistence in Forest Ecosystems

What mechanisms enable different species to coexist in a community? What are the ecological assembly rules that determine the composition and organization of a community? These are among the key questions of ecology. The answers to these questions and the elucidation of these mechanisms for maintaining species diversity in a community will help us develop techniques to protect species diversity in forest management practices. The studies in this book focus on the diversity and coexistence mechanisms of tree species, while incorporating some aspects of the interaction between trees and other organisms. Trees are the main primary producers in forest ecosystems and determine the fundamental characteristics of structure, productivity and dynamics of the ecosystem. These characteristics of the forest ecosystem generate the necessary conditions for maintaining the diversity of the other organisms, and the interactive processes between trees and other organisms enable different species of trees to coexist.

1.2 Coexistence Mechanisms of Tree Species

Hubbell and Foster (1986) proposed two major states of species coexistence: equilibrium (resulting from niche partitioning and/or trade-offs among species) and nonequilibrium (by chance). The composition of each tree community reflects both these states, although the relative contribution of each differs among communities (Brokaw and Busing 2000). The studies in this book mainly discuss species coexistence from the perspective of an equilibrium state between species, because the composition of temperate forests is apparently less chance-determined than that of more species-rich tropical forests (Nakashizuka 2001).

Ecological Studies, Vol. 158
Nakashizuka, Matsumoto (eds.) Diversity and Interaction in a Temperate Forest Community: Ogawa Forest Reserve of Japan

Belonging to a single trophic level, tree species basically compete for the same resources of light, water, and nutrients. If the spatial, temporal, biotic, or abiotic resources or habitat are not partitioned, one species having a competitive advantage may monopolize the resources (Pacala and Tilman 1994). On the other hand, if resources are allocated by trade-offs of some kind, different species can coexist under the spatially or temporally heterogeneous conditions created by these trade-offs. The mechanisms of this coexistence can be studied by analyzing the dissimilarities or trade-offs in ecophysiology, the resource allocation pattern, and demographic parameters throughout the life cycle of the trees.

Many theories and hypotheses have been proposed to explain species coexistence mechanisms in a community (Tokeshi 1999). Even if we limit the discussion to forest communities, a variety of mechanisms that may enable species coexistence have been described for many aspects of the tree life cycle. Since the tree life cycle is usually very long and may encompass a vast range of body sizes, different mechanisms or ontogenetic variations, each specific to a different life stage, may be required to make coexistence possible. A trade-off or dissimilarity can operate as a coexistence mechanism both between the life history traits of species at the same life stage and between those of species at different life stages. Keddy (1992) summarized the coexistence mechanisms as the combination of particular sets of community conditions (environmental heterogeneity, presence of a disturbance, and the biotic environment) and the responses of tree species to those conditions.

1.2.1 Community Conditions

We propose that four major factors enhance the coexistence of tree species: (1) horizontal heterogeneity in resource distribution; (2) forest architecture or stratification (vertical heterogeneity of the light resource); (3) presence of disturbance; and (4) interaction between trees and other organisms (Nakashizuka 1996, 2001). Horizontal heterogeneity refers to topographic variations, the spatial distribution of nutrients, and water availability in a forest community. Many traditional theories on niche separation along environmental gradients focused on this factor. Vertical heterogeneity of the light resource results from the interaction through growth and competition among individual trees (Kohyama 1994), and this factor is prominent in forest communities. Disturbances have been reported to enable tree coexistence in various ways (White and Pickett 1985). Trees have adapted both to colonize safe sites created by disturbances and to attain resistance or resilience against disturbances. All tree species interact with other organisms through processes such as pollination, seed dispersal, herbivory, and nutrient uptake. Specific interactions (i.e., specific to a particular species) and/or nonspecific interactions (i.e., between any particular combination of species) between trees and other organisms may give a competitive advantage or disadvantage to one species in a particular species pool that has been historically determined. The key question is which factor contributes most to species coexistence or species diversity in a given community?

1.2.2 Tree Response to Community Conditions

The responses of a tree species to a set of community conditions determine whether or not the species can maintain its population in a community. A tree species can respond mainly at three levels: (1) at the organ level by an ecophysiological response; (2) at the individual level by changing resource allocation patterns; and (3) at the population level through demographic variations. Differences in photosynthetic ability and resistance to environmental stress, and their variation under differing environmental conditions, affect the growth and survival of trees. The allocations to leaves vs. roots and to vegetative vs. reproductive organs are important strategic options in resource utilization. Demographic variations under biotic and abiotic conditions are critical for population dynamics. The key question is which traits at these three levels are particularly important for species coexistence in the community?

1.3 Methodology for Studying Mechanisms of Coexistence

To approach the questions mentioned above, studies on tree coexistence should (1) elucidate the whole tree life cycle, (2) concentrate on a single community, and (3) be based on long-term observations.

1.3.1 Studies on the Whole Life Cycle

To understand tree species coexistence, it is important to investigate the whole life history and demography of each tree species and how they vary under the community conditions created by the presence of coexisting (or co-occurring) tree species. A mechanism operating at a particular life stage may play an important role in the maintenance of a tree population in a community, and this critical life stage may be different among species. The demography at the seed stage may be particularly important for one species, although the competition at the adult stage might be more important for another species. Thus, a niche partitioning or trade-off at the seed stage may not always be important for all species in a community. One of the key analyses that must be performed is the identification of this critical stage. Moreover, the contribution of life history traits to the maintenance of a population in the community should be evaluated quantitatively.

1.3.2 Concentration of Studies on a Single Community

A complete picture of the coexistence mechanisms controlling community organization cannot be obtained from comparisons of tree life history traits studied separately at different sites because it is necessary to know how these traits reflect the response of the tree species to a particular set of community conditions. Even a species with a fixed life history trait may play different roles in forest communities with different disturbance regimes or a different species pool. Thus, studies on co-occurring species should be conducted simultaneously on sympatric species. After determining the demographic parameters throughout the tree life cycle in a community, differences and variations in these parameters should be analyzed and related to those conditions present, such as the disturbance regime and species pool. To evaluate the species response and variations among responses, further comprehensive analyses combining ecophysiology, resource allocation, and demography should be conducted for component species of a particular species pool under common environmental and disturbance regimes.

1.3.3 Long-Term Studies

Ecological phenomena that require long-term studies if they are to be understood include slow processes, rare events or episodic phenomena, processes with high annual variability, subtle processes, and complex phenomena (Franklin 1989). To understand forest community dynamics, the analysis of these types of phenomena is indispensable. Reproductive patterns of long-lived plants (for example, the masting behavior of trees and forest-floor bamboo), and disturbances with long return intervals such as wildfires, floods, droughts, landslides, and outbreaks of pests or pathogens are examples of rare or episodic events that may occur in forest ecosystems. They may be critical for the forest community organization because of the great longevity of forest trees, which may extend over several hundred years. Subtle or highly variable processes such as changes in forest productivity or tree growth and mortality can also be detected only through long-term studies of the ecosystem. Interactions between trees and other organisms may include indirect effects and time lags. Thus, the evaluation of complicated or slow-acting mechanisms with large temporal variations is critical for understanding tree coexistence and diversity.

1.4 Studies in Ogawa Forest Reserve

Taking into account the considerations discussed above, we organized studies concentrating on a single target community and carried out baseline investigations over a longer time period than that used by earlier investigations. The basic design of the studies is introduced together with the climatic and topographic conditions and disturbance regime in Part 2. With this as background, ommunity-level vegetation studies (Part 3) are further analyzed by integrating population (Part 4) and ecophysiological studies of the component tree species (Part 5). Intensive genetic analyses of certain species (Part 6) and studies on interactions between trees and other organisms are also included (Part 7). All results are discussed again along with their implications for sustainable management and the conservation of biodiversity of forest ecosystems in the final part (Part 8).

References

Brokaw N, Busing R (2000) Niche versus chance and tree diversity in forest gaps. Trends Ecol Evol 15:183–188

Franklin JF (1989) Importance and justification of long-term studies in ecology. In: Likens GE (ed) Long-term studies in ecology: approaches and alternatives. Springer, Berlin, Heidelberg, New York, pp 3–19

Hubbell SP, Foster RB (1986) Canopy gaps and the dynamics of a neotropical forest. In: Crawley MJ (ed) Plant ecology. Blackwell, Oxford, pp 77–96

Keddy PA (1992) Assembly and response rules: two goals for predictive community ecology. J Veg Sci 3:57–164

Kohyama T (1994) Size-structure-based models of forest dynamics to interpret population- and community-level mechanisms. J Plant Res 107:107–116

Nakashizuka T (1996) Factors maintaining forest tree diversity. In: Turner IM, Diong CH, Lim SSL, Ng PKL (eds) Biodiversity and the dynamics of ecosystems. The international network for DIVERSITAS in Western Pacific and Asia (DIWPA), pp 51–62

Nakashizuka T (2001) Species coexistence research in temperate, mixed deciduous forests. Trends Ecol Evol 16:205–210

Pacala S, Tilman D (1994) Limiting similarity in mechanistic and spatial models of plant competition in heterogeneous environments. Am Nat 143:222–257

Tokeshi M (1999) Species coexistence. Blackwell, Oxford

White PS, Pickett STA (1985) Natural disturbance and patch dynamics. In: Pickett STA, White PS (eds) The ecology of natural disturbance and patch dynamics. Academic Press, New York, pp 3–13

Part 2

What Is Ogawa Forest Reserve?

2 Climate in Ogawa Forest Reserve

Yasuko Mizoguchi, Takeshi Morisawa, and Yoshikazu Ohtani

2.1 Introduction

In the middle of the latitudes in which the Japanese islands are situated, the climate is characterized by severe winters and summer monsoons. This is because they face the east coast of the Asian continent and cover a wide range of elevations. Strongly influenced by the monsoons, the climatic difference between the areas facing the Pacific Ocean and those facing the Sea of Japan is especially dramatic. The most remarkable is the winter climate. The former has dry, sunny weather, while the latter has a great deal of snow. From late spring to early summer, the northeastern Pacific coast of the islands is influenced by the chilly northeastern winds, which often bring cloudy weather or fog.

Ogawa Forest Reserve (OFR), which is classified as being in the temperate zone, is located on the undulating plateau at the southern edge of the Abukuma mountainous region (36°56′N, 140°35′E). OFR is approximately 20 km west of the Pacific Ocean and 50 km north of the Kanto Plain, where the elevation ranges from 610 m to 660 m (Fig. 2.1).

In 1993, meteorological observations were started at a temporary observatory at Ogawa, 2 km south of OFR, and continued to the end of 1995. The data collected at Ogawa were used to investigate the climatic characteristics of OFR. In this chapter, the climatic characteristics around OFR are described, and the basic climatic data are listed for future ecological applications.

2.2 Meteorological Observations and Data Processing

At the Ogawa observatory, four meteorological elements were measured, i.e., global solar radiation (a Moll–Gorczynski-type pyranometer, EKO, Tokyo) 8 m above the ground, air temperature (Thermistor thermometer, North High Tech, Sapporo) 2 m up in the radiation shield, air humidity (Ceramic hygrometer, North High Tech,

Ecological Studies, Vol. 158
Nakashizuka, Matsumoto (eds.) Diversity and Interaction in a Temperate Forest Community: Ogawa Forest Reserve of Japan

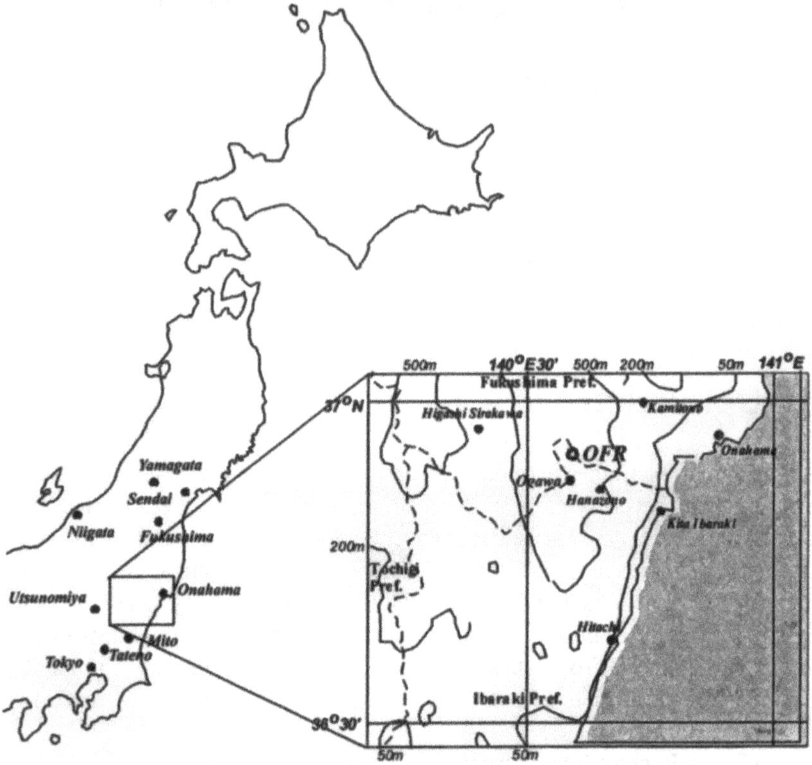

Fig. 2.1. Location of Ogawa Forest Reserve and weather stations nearby

Sapporo) at the same height as the air temperature, and precipitation (Tipping bucket rain gauge, Nakaasa, Tokyo) 1 m above the ground. The data were sampled every 10 min, and hourly sums and means were calculated. In addition, data obtained at other meteorological stations around OFR by the Japan Meteorological Agency (JMA) were used to estimate longer-term climatic data (10 years mean) and to fill any data gaps. The locations and specifications of these stations are shown in Fig. 2.1 and Table 2.1.

The data were processed in the following way. (1) For global solar radiation, a correction for diffuse radiation intercepted by the surrounding topography was attempted, and the daily sums were compared with those of Onahama. Then the monthly mean of the daily global solar radiation was estimated in the same manner as for the air temperature. (2) For the air temperature, the hourly means of the data observed at Ogawa were compared with those observed at Higashishirakawa for 2 years (1994, 1995), and an empirical equation was obtained. Then the monthly mean air temperatures at Ogawa were estimated for a 10-year period (1986–1995). (3) For the vapor pressure, the hourly means were calculated from the relative humidity and air temperature data from Ogawa, and compared with those from Onahama in 1994. The procedure for estimating the monthly mean vapor pressure was similar to that used for the air temperature. Daily, monthly, and annual sets of basic meteorological data were prepared by the methods described above. (4) For the precipitation, data interruption occurred frequently because of data-logger and power-line problems. Therefore, the monthly data at Ogawa were estimated from the weighted averaging of data obtained at Hanazono and Higashishirakawa.

Table 2.1. Locations and observational elements of Ogawa (FFPRI) and surrounding JMA meteorological stations

Station	Organization	Altitude (m)	Location	Observation
Tateno	JMA (weather station)	25	36°03′N 140°08′E	Elements of the surface meteorological observation
Sendai	JMA (weather station)	39	38°16′N 140°54′E	Elements of the surface meteorological observation
Onahama	JMA (weather station)	3	36°57′N 140°54′E	Elements of the surface meteorological observation
Higashi Shirakawa	JMA (AMeDAS)	217	36°57′N 140°24′E	Wind direction, wind velocity, sunshine duration, air temperature, precipitation
Kamitono	JMA (AMeDAS)	125	37°00′N 140°44′E	Wind direction, wind velocity, sunshine duration, air temperature, precipitation
Kitaibaraki	JMA (AMeDAS)	45	36°50′N 140°46′E	Wind direction, wind velocity, sunshine duration, air temperature, precipitation
Hanazono	JMA (AMeDAS)	370	36°52′N 140°38′E	Precipitation
Ogawa	FFPRI	555	36°54′N 140°35′E	Global solar radiation, precipitation, air temperature, air humidity

JMA, Japan Meteorology Agency; AMeDAS, automated meteorological data acquisition system; FFPRI, Forestry and Forest Products Research Institute.

2.3 Climatic Characteristics at Ogawa

The monthly means of the daily global solar radiation, air temperature, and vapor pressure, and the monthly totals of precipitation are shown in Table 2.2, where all the elements are estimated averages for a period of 10 years from 1986 to 1995.

2.3.1 Global Solar Radiation

The mean annual daily global solar radiation was 11.7 MJ m^{-2} day^{-1}. The strongest global solar radiation was in May (15.8 MJ m^{-2} day^{-1}) and the weakest in December (7.2 MJ m^{-2} day^{-1}). In general, the global solar radiation is related to the latitude, although some variations appeared in each location. This is discussed in the following section.

2.3.2 Air Temperature

The mean annual air temperature in Ogawa was 10.7°C. The mean monthly air temperatures in January and February were below 0°C, while in July and August they were above 20°C. Table 2.2 also shows the monthly means of the daily maximum and minimum air temperatures, as well as the diurnal ranges of air temperature. The monthly means of the daily maximum air temperatures were above 20°C from June through September, with the highest being recorded in August. The monthly means of the daily minimum air temperatures were below 0°C from December through

Table 2.2. Climatic elements in Ogawa (1986–1995)

	Solar radiation	Air temperature				Vapor pressure	Precipitation
Month	Mean ($MJ\ m^{-2}\ day^{-1}$)	Mean (°C)	Max. (°C)	Min. (°C)	Range (K)	Mean (hPa)	Total (mm)
January	7.9	–0.9	4.7	–5.3	10.0	5.3	52.0
February	10.1	–0.4	5.2	–4.9	10.1	5.3	81.3
March	11.9	2.8	8.4	–2.0	10.4	6.5	136.2
April	15.2	8.9	15.3	2.7	12.6	9.2	125.4
May	15.8	13.5	19.5	7.8	11.7	12.4	187.7
June	14.2	17.5	22.5	13.3	9.3	16.3	188.5
July	13.7	21.0	25.6	17.4	8.2	20.4	163.1
August	15.4	22.6	27.8	18.9	8.9	23.0	338.2
September	11.4	18.6	23.2	15.0	8.2	19.1	299.1
October	9.5	12.3	17.6	8.1	9.5	13.2	174.2
November	8.0	6.1	12.4	1.2	11.3	9.0	122.5
December	7.2	1.3	7.6	–3.4	11.0	6.4	42.1
Annual	11.7	10.7	15.9	5.8	10.1	12.8	1910.3

Monthly means of global solar radiation, monthly means, daily maximums, daily minimums, and diurnal ranges of air temperature, monthly means of vapor pressure, and monthly totals of precipitation are tabulated. All data are averages for 10 years.

March. The diurnal ranges of air temperatures in April and May were larger than those in any other month.

2.3.3 Vapor Pressure

The vapor pressure was low in winter when the air temperature was low and the weather was usually fine. The lowest pressure, 5.3 hPa, was recorded in January and February. The vapor pressure exceeded 20 hPa in July and August, when the air temperature was high and there was a lot of rain.

2.3.4 Precipitation

The mean annual precipitation in Ogawa was 1,910 mm. Except for January, February, and December, monthly precipitations were more than 100 mm. The largest and smallest monthly precipitation was recorded in August and December, respectively.

2.4 Discussion

Yoshino (1967) has classified the local climate around OFR, including that at the weather stations mentioned above, into three zones according to the distance from the Pacific coast, i.e., the coastal area and the land continuing into the northeast Kanto Plain, the outer boundary of the plain, and the southern edge of the Abukuma mountainous (plateau) area. The common climatic characteristic in these zones is rainy or overcast weather, caused by the prevailing northeastern wind, from late spring to early summer.

Depending on the distance from the coast, the annual precipitation varies between 1,549 mm in Kitaibaraki (the coastal area), 2,166 mm in Hanazono (in the middle of the mountains), 1,910 mm in Ogawa (the plateau following the mountainous area), and 1,377 mm in Higashishirakawa (the basin adjoining the plateau). When the rainfall distribution is affected by the topography, the maximum is recorded in the middle of the mountainous area (Tsukamoto 1992). Among the stations mentioned above, the maximum annual precipitation was recorded in the middle of the mountainous area at Hanazono, and the next highest was recorded on the plateau at Ogawa. This suggests that the rainfall distribution in these areas is indeed orographic, and that the precipitation of Ogawa is also under the influence of the orographic effect.

A hythergraph is a climatic diagram indicating the relationship between the monthly mean air temperature and the monthly precipitation. Figure 2.2 shows hythergraphs of the data obtained from Kamitono, Ogawa, and Kitaibaraki. Judging from the similarity in the shapes of those from Ogawa and Kamitono, the annual changes in air temperature and precipitation largely depend on the distance from the coast and the elevation in these zones.

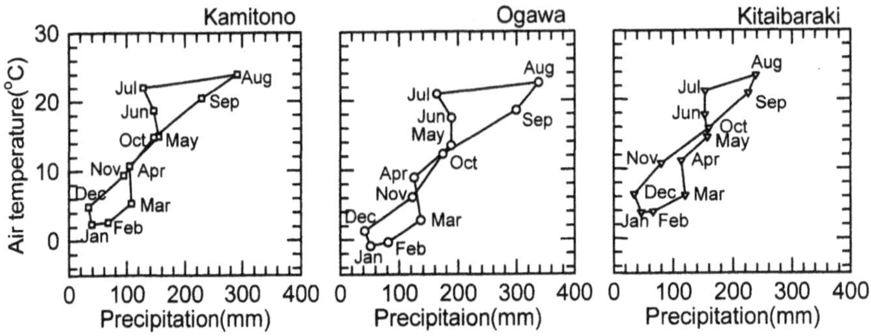

Fig. 2.2. Hythergraph from Ogawa and two nearby meteorological stations. Kitaibaraki is located in the coastal area, and Ogawa and Kamitono are located in the mountainous (plateau) area

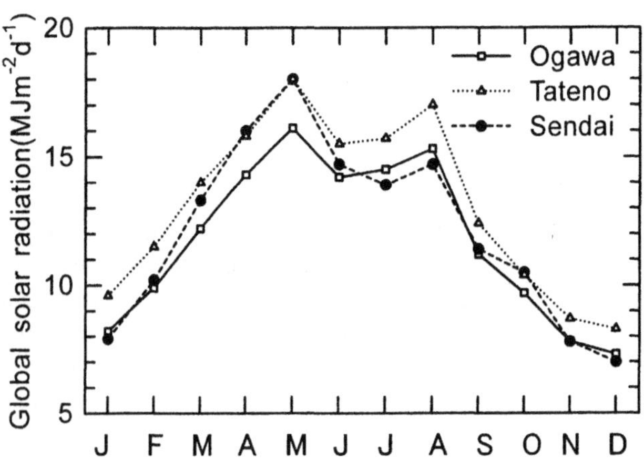

Fig. 2.3. Annual variation of monthly means of daily global solar radiation at Ogawa, Tateno, and Sendai. For Sendai and Tateno, the data are averages of 17 years (1974–1990). For Ogawa, the data are estimated for the same period by considering additional data from Onahama

The topographic influence on the climate in Ogawa has another aspect. Figure 2.3 shows the annual variation in the monthly mean of the global solar radiation in Tateno (National Astronomical Observatory 2000), Ogawa, and Sendai (National Astronomical Observatory 2000). Global solar radiation generally increases with a latitudinal decrease when there is no effect of cloud cover. Although the latitudinal location of Ogawa is between Tateno and Sendai (these locations are shown in Fig. 2.1), the global solar radiation in Ogawa is less than that at the other two stations from spring to early summer. It is suggested that the decrease in global solar radiation is caused by the topographic cloud formation that mainly occurs at this time.

In connection with the cloud formation, the condensation level, which indicates the theoretical lower limit of the clouds, is estimated by the air temperature and the vapor pressure. Figure 2.4 shows the frequency of the number of hours that the

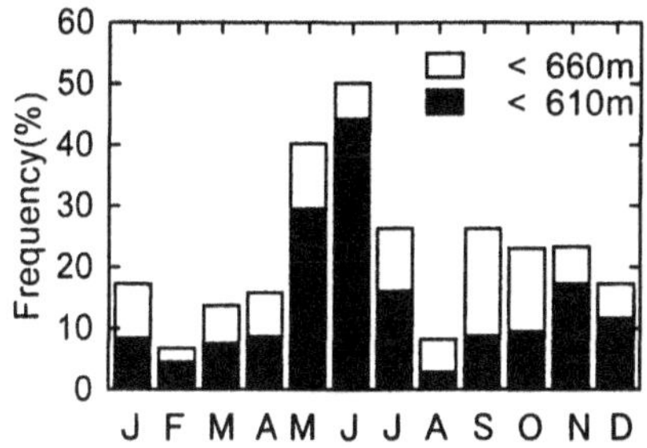

Fig. 2.4. Hourly based frequency of the expected appearance of fog in Ogawa in 1994. The figure shows the frequency of the number of hours when the condensation level is estimated to be lower than the highest (660 m) and lowest (610 m) elevation of Ogawa Forest Reserve

condensation level is lower than the elevation of OFR. The highest frequency occurs in May and June, i.e., approximately 40% and 50% of the time, respectively. Thus, fog is liable to cover OFR in this season.

The term climatic net primary productivity (NPP) was proposed by Uchijima and Seino in 1985. This means the productivity in relation to the net radiation and precipitation. The climatic NPPs of Ogawa and Onahama, which were calculated using this method, are approximately 15 MgDW ha^{-1} $year^{-1}$ and 16 MgDW ha^{-1} $year^{-1}$, respectively. In Ogawa, the global solar radiation is about 13% less than that in Onahama, although the shortfall in climatic NPP is not more than 7%. It has been suggested that because Ogawa has 41% more precipitation, this compensates for its lack of radiation.

2.5 Conclusion

The results of this study can be summarized as follows. The mean annual air temperature and the annual precipitation are 10.7°C and 1,910 mm, respectively. OFR is located in a zone that has more precipitation than the coast and less global solar radiation than the inland zone. The climatic properties of this area are affected by the meteorological phenomena resulting from the prevailing wind in early summer and the topography. The calculated climatic NPP of Ogawa is approximately 15 MgDW ha^{-1} $year^{-1}$. This suggests that the large amount of precipitation results in relatively high productivity in spite of the lack of radiation.

Acknowledgment

The meteorological data from the automated meteorological data acquisition system and those obtained from the weather stations were supplied by the Japan Meteorological Agency.

References

National Astronomical Observatory (ed) (2000) Chronological scientific tables (in Japanese). Maruzen, Tokyo

Tsukamoto Y (ed) (1992) Forest hydrology (in Japanese). Buneido, Tokyo

Uchijima Z, Seino H (1985) Agroclimatic evaluation of net primary productivity of natural vegetation. 1. Chikugo model for evaluating net primary productivity. J Agric Meteorol 40:343–352

Yoshino M (1967) Climatic division of Kanto district, Japan (in Japanese with English abstract). Tohoku Geog 19:165–171

3 Landforms and Soil Characteristics in Ogawa Forest Reserve

SHUICHIRO YOSHINAGA[1], MASAMICHI TAKAHASHI[1], and SHUHEI AIZAWA[2]

3.1 Geology and Classification Systems of Landforms and Soils

We describe the characteristics of the topography and soils as the substrates of the tree community in Ogawa Forest Reserve (OFR) in this chapter. Metamorphic rocks underlie the OFR, and late Quaternary volcanic ash has been widely deposited in the area. These parent materials affect the characteristics and distribution of soils in the OFR. The topography of OFR is gently undulating, as shown on the contour map (Fig. 3.1). A stream runs from north to south through the center of the reserve. The east-facing valley sideslope is steeper than the west-facing slope.

The topography of the OFR is composed of various slopes that can be described as configurations of convexity and concavity. On the basis of detailed observations, the landforms in the OFR have been classified into six units (Fig. 3.2), modifying the landform classification system proposed by Tamura (1981). The landform units are recognized by discontinuities in slope angles and by the forms of adjacent units. The valley-head and the upper and lower sideslopes are subdivided by referring to the slope transverse profile. Soil properties and distributions are described according to this landform classification system.

Soil profiles are described and soils are classified according to the methods described in *Classification of Forest Soil in Japan 1975* (Forest Soil Division 1976). This soil classification system, which was developed to evaluate the growth and distribution of plant communities, is the only system widely used for Japanese forest soils. In this system, the soil types are subdivided by the soil moisture regime dictated by the hydrocatena of the sloping topography. Subtypes from A to F correspond to the soil-moisture gradient from dry to wet. The system is also advantageous for describing soil distributions on narrow and complex landforms in mountainous areas. This system is compared with other classification systems (Soil Survey Staff 1998; FAO-UNESCO, 1988) in Table 3.1.

Ecological Studies, Vol. 158
Nakashizuka, Matsumoto (eds.) Diversity and Interaction in a Temperate Forest Community: Ogawa Forest Reserve of Japan

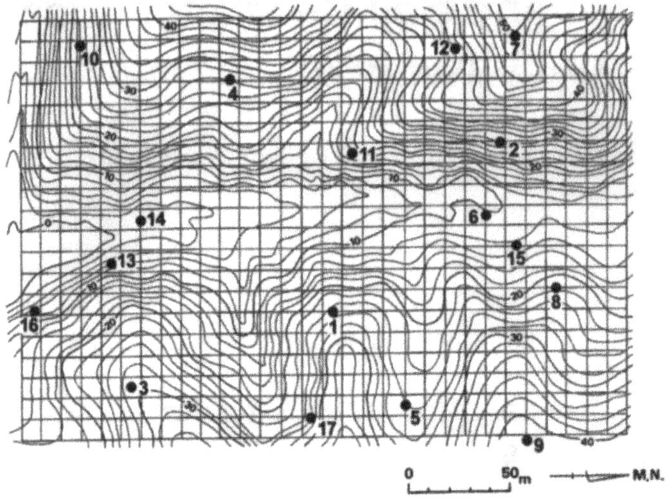

Fig. 3.1. Topography of the permanent 6-ha plot in the Ogawa Forest Reserve (OFR). Contour interval is 2 m. *M.N.*, magnetic north. *Numerals* show pedon locations

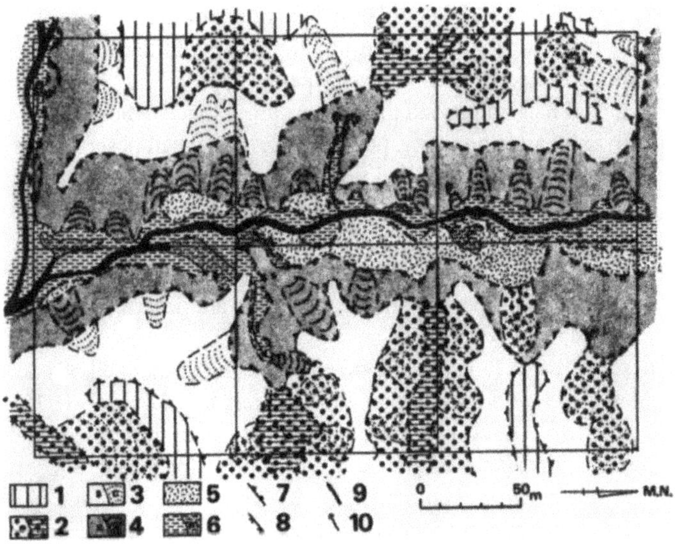

Fig. 3.2. Landform classification in the permanent 6-ha plot of the OFR using the criteria proposed by Tamura (1981). *1*, crestslope; *2*, valley-head (*s*, valley-head slope; *f*, valley-head flat); *3*, upper sideslope (*s*, flat or convex transverse slope profile; *c*, flat or concave transverse slope profile); *4*, lower sideslope (*s*, flat or convex; *c*, flat or concave); *5*, footslope; *6*, bottomland (*m*, marsh); *7*, slope break from gentle to steep; *8*, slope break from steep to gentle; *9*, permanent stream; and *10*, ephemeral spring and channel. *M.N.*, magnetic north

Table 3.1. Correlation of the soil types according to *Classification of forest soil in Japan* (Forest Soil Division 1976) with those of the FAO soil classification system (FAO-UNESCO 1988) and soil taxonomy system of the Soil Survey Staff (1998)

Classification of forest soil in Japan	Soil taxonomy	FAO system	Landform unit
B_B: Dry Brown forest soil	Typic Dystrudepts Dystrudepts Andic Dystrudepts	Dystic Cambisols	Crestslope Upper sideslope
$B_D(d)$: Moderately moist Brown forest soil (drier subtype)	Typic Dystrudepts Andic Dystrudepts Humic Dystrudepts	Dystic Cambisols Umbric Andosols	Upper sideslope
B_D: Moderately moist Brown forest soil	Andic Dystrudepts Typic Dystrudepts Humic Dystrudepts	Humic Cambisols	Lower sideslope Footslope
B_E: Slightly wetted Brown forest soil	Humic Dystrudepts	Humic Cambisols	Valley-head flat
Bl_D: Moderately moist Black soil	Pachic Melanudands	Humic Andosols	Valley-head slope
Bl_E: Slightly wetted Black soil	Pachic Melanudands Humic Pachic Dystrudepts	Humic Andosols	Valley-head flat
$lBl_D(d)$: Moderately moist Light-colored black soil (drier subtype)	Acrudoxic Fulvudands	Umbric Andosols	Crestslope
G: Gley soil	Typic Dystrudepts Lithic Endoaquents	Dystic Cambisols Dystic Gleysols	Bottomland

3.2 Soil Distribution

Soils in the OFR are distributed with a very heterogeneous and mosaic pattern of Brown forest soils (mainly Inceptisols or Cambisols) and Black soils (mainly Andisols or Andosols) (Fig. 3.3). These soil types are associated with specific landform units and with the amount of volcanic ash at the site.

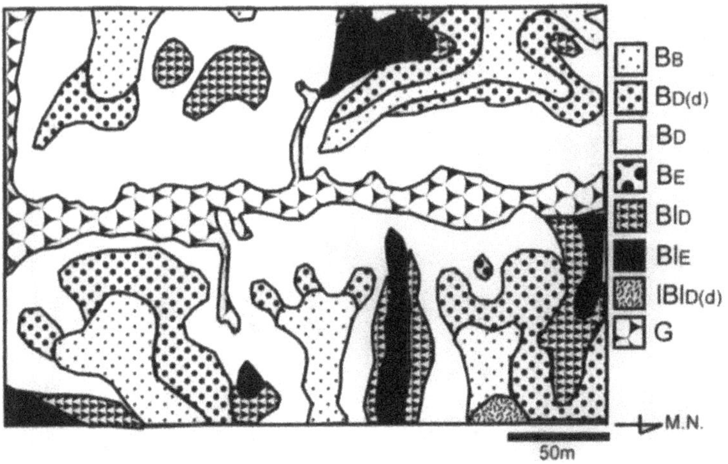

Fig. 3.3. Soil map of the permanent 6-ha plot according to the classification system of the Forest Soil division (1976) (also see Table 3.1)

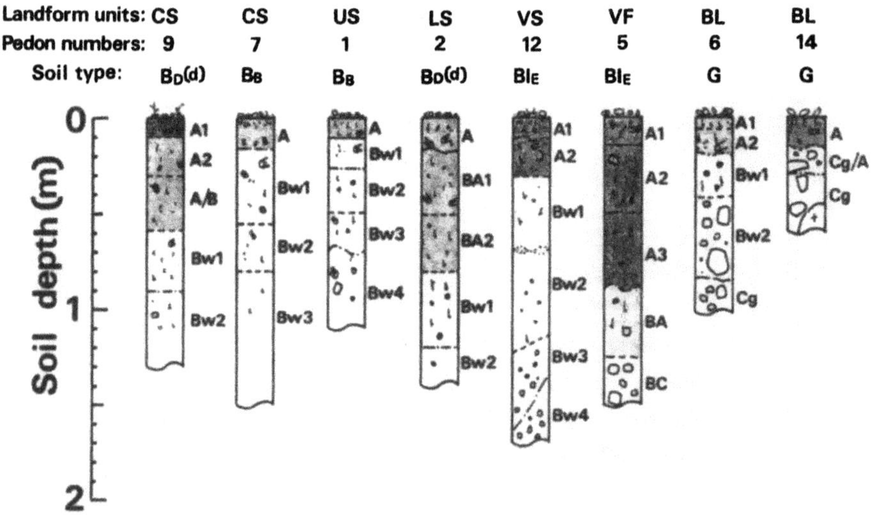

Fig. 3.4. Selected soil profiles in the permanent 6-ha plot. *CS*, crestslope; *US*, upper sideslope; *LS*, lower sideslope; *VS*, valley-head slope; *VF*, valley-head flat; and *BL*, bottomland

Soils on the crestslope and upper sideslope are classified into relatively dry subtypes (B_D(d) type). Soils on narrow crestslopes are classified as dry Brown forest soils (B_B), which are characterized by a shallow (less than 10 cm deep) A horizon (Fig. 3.4), while those on the wide, gentle crestslopes of the west-facing valley sideslopes, are moderately moist Light-colored black soils (drier subtype) (lBl_D(d)) (Pedon 9). The lBl_D(d) type has a relatively thick (but less than 30 cm), light black (Munsell Soil Color Charts, 7.5 YR2/2–10 YR2/3) A horizon above a brown B horizon, with a clear boundary between them. The moderately moist Brown forest soil (drier subtype) (B_D(d)) is found on the gentle upper sideslopes (Pedon 2 and 10). On the valley sides, especially on steep slopes, moderately moist Brown forest soils (B_D) are widely distributed. The B_D type is characterized by a thick, dark brown A horizon above a brown B horizon, with a diffuse boundary.

The A horizons on the slopes reflect the erodibility of volcanic ash materials. The A horizon thickness decreases on the steep and convex crestslopes and upper sideslopes. The color of the A horizons, however, was often blackish, suggesting that the soil had likely formed under the influence of volcanic ash materials.

Moderately moist Black soils (Bl_D) are distributed on the slopes of the valley-head and on the footslopes (Pedon 4). The distinctive feature of the Black soils is the development of very thick (sometimes > 1 m), blackish (Munsell value/chroma smaller than 2/2) A horizons. Slightly wetted Brown forest soils (B_E) and slightly wetted Black soils (Bl_E) are distributed on valley-head flats (Pedons 5, 12, and 15). In the bottomlands along the main stream, Gley soils are distributed. A gley horizon formed in waterlogged soil is present from 70 to 100 cm depth in the terraced bottomlands (Pedon 6), but is found in the surface horizon in the nonterraced bottomlands (Pedon 14).

Because the influence of volcanic ash is suggested even in the Brown forest soils, Andic soil properties as described by the Soil Survey Staff (1998) were exam-

Fig. 3.5. Soil map of the permanent 6-ha plot according to the classification system of the Soil Survey Staff (1998)

ined to evaluate the contribution of volcanic ash to the parent material. Andic soil properties are defined on the basis of organic carbon content, bulk density, amorphous aluminum and iron contents, and ability to retain phosphate. Andic soils are soft, light, porous soils with high water-holding capacity but low available phosphate content. Our examination found that most surface soils in the OFR exhibit Andic properties to some extent. The distribution of Andic soil properties is shown in Fig. 3.5. The surface soil characteristics often satisfied the criteria for Andisols (Soil Survey Staff, 1998) or Andosols (FAO-UNESCO 1988), except with regard to the thickness of the A horizon when the soil was distributed on slopes.

3.3 Soil Chemical Properties

The chemical properties of typical soil profiles are shown in Fig. 3.6. In most areas of the OFR, soils are acidic to slightly acidic (pH 4.7–6.2). The sum of the exchangeable cations is low throughout the profiles, mostly less than 5 $cmol_c kg^{-1}$ in the surface soils, and the dominant exchangeable cation is calcium. Cation exchange capacity (CEC; measured by 1 M NH_4OAc, pH 7) is high, ranging from 20 to 64 $cmol_c kg^{-1}$ in the A horizons and from 7 to 36 $cmol_c kg^{-1}$ in the B horizons, where values are strongly correlated with organic carbon content. Consequently, most of the soils in the OFR have low base saturation, less than 20%.

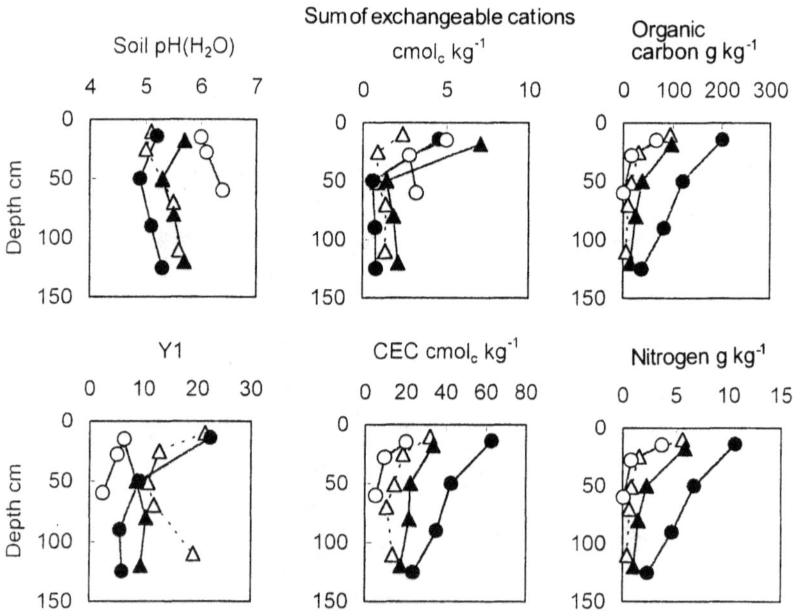

Fig. 3.6. Chemical properties of soils in selected pedons. *Open triangle*, pedon 1 (B_B); *solid triangle*, pedon 2 (B_D); *solid circle*, pedon 5 (lBl_E); and *open circle*, pedon 6 (G)

The exchange acidity, Y_1, which is related to the amount of exchangeable aluminum by the equation

Exchangeable Al ($cmol_c kg^{-1}$) = 0.30 Y_1 + 0.04 (r = 0.97, n = 153; Inoue 1986), varied among pedons, ranging from 1.5 to 53.2 in the surface soils. Amorphous minerals in Andisols usually have low Y_1 (Inoue 1986), but some Andisols in the OFR showed very high Y_1 in the surface soils. Exchangeable Al seems to be distributed in a complex pattern at this site.

Carbon and nitrogen content showed much variation. The Black soils were rich in organic carbon (7.8%–20.1% in the A horizon; 0.7%–3.7% in the B horizon) and nitrogen (0.4%–1.06% in the A horizon; 0.06%–0.23 % in the B horizon). On the other hand, the Brown forest soils, in particular the B_B soils, had less carbon and nitrogen in the A horizon. Therefore, organic matter in mineral soils showed a very heterogeneous distribution in the OFR.

3.4 Soil Physical Properties

The soils in the OFR showed low bulk densities and were moist and well-drained. Dry bulk densities of the surface soils were smaller than 0.6 Mg m^{-3}. In most pedons, except for in the bottomlands, the surface soils were moderately moist, ranging from 0.4 to 0.5 kg kg^{-1}, and water permeability was high (hydraulic conductivity of saturated soils: 10^{-3}–10^{-5} ms^{-1}). In addition, subsurface soils also showed low dry bulk densities, less than 0.8 Mg m^{-3} and moderately high water content ranging from 0.3 to 0.5 kg kg^{-1}. Black soils were clayey and very sticky, but showed the low resistance to penetration by the soil hardness meter (Yamanaka model DIK-5552, Daiki Rika Kogyo, Tokyo, Japan) throughout the profiles, less than 20 mm (< 6.3 kg cm^{-2}).

3.5 Evaluation of the Soils in the OFR for Growth of the Forest Community

The physical properties of the soils in the OFR indicated ideal conditions, such as good drainage, sufficient moisture, and low penetration resistance, for plant growth. These soils exhibited the characteristics of allophanic soils derived from volcanic ash materials (Maeda and Soma 1985). In addition, the deeper horizons, which were derived from the weathering of metamorphic rocks and granite, were also well-drained and soft.

Cation content in the soils, on the other hand, was very low overall. These mineral elements may be supplied from decomposing litter on the surface. The forest floor in the reserve consists mostly of an L layer and a thin F layer, indicating rapid decomposition of litter at this site.

Available nitrogen in the soils is heterogeneous in the OFR. Large pools of organic nitrogen in the Black soils (0.94–1.10 kg N m^{-2}, 0–30 cm depth) indicate that available nitrogen is sufficient for plant growth, because available nitrogen content in mineral soil is correlated with total nitrogen content (Ohta and Kumada 1978). However, the B_B soils on the crestslopes had low mineral nitrogen content, owing to the small nitrogen pool in those soils (0.356–0.759 kg N m^{-2}, 0–30 cm depth). Moreover, poor litter coverage of the forest floor on the crestslopes, probably because litter is blown off by wind during the winter, suggests that the soils of the B_B type contain few nutrients for plant growth.

Because aluminum is toxic to plants (Saigusa et al. 1980; Bertsch 1989), the high Y_1 level may affect the forest community dynamics. The distribution of aluminum in the surface soil may affect the establishment of herbaceous species and tree seedlings. However, exchangeable aluminum (as estimated by Y_1) displayed large variation among pedons and within profiles. Further studies are required to determine the factors controlling the distribution pattern of aluminum at this site. In Chapter 4, we analyze the relationship between the distribution of plant communities and estimated content of exchangeable aluminum in the surface soils by sampling at grid (10 m × 10 m) points.

References

Bertsch PM (1989) Aluminum speciation: methodology and applications. In: Norton SA, Lindberg SE, Page AL (eds) Acid precipitation, vol. 4: soil, aquatic processes, and lake acidification. Springer, New York, pp 63–105

FAO-UNESCO (1988) Soil map of the world. World soil resource report 60, FAO, Rome

Forest Soil Division (1976) Classification of forest soil in Japan 1975 (in Japanese with English summary). Bull Gov For Exp Sta 280:1–28

Inoue K (1986) Chemical properties. In: Wada K (ed) Ando soils in Japan. Kyushu University Press, Fukuoka, Japan, pp 69–98

Maeda T, Soma K (1985) Physical properties. In: Wada K (ed) Ando soils in Japan. Kyushu University Press, Fukuoka, Japan, pp 99–111

Ohta S, Kumada K (1978) Studies on the humus forms of forest soils. VI. Mineralization of nitrogen in brown forest soils. Soil Sci Plant Nutr 24:41–54

Saigusa M, Shoji S, Takahashi T (1980) Plant root growth in acid Andosols from northern Japan: 2. Exchange acidity Y1 as a realistic measure of aluminum toxicity potential. Soil Sci 130:242–250

Soil Survey Staff (1998) Keys to soil taxonomy, 8th ed., United States Department of Agriculture Natural Resources Conservation Service

Tamura T (1981) Multiscale landform classification study in the hills of Japan, part II. Application of the multiscale landform classification system to pure geomorphological studies of the hills of Japan. Sci Reports Tohoku Univ 7th Series (Geog) 31:85–154

4 Forest Vegetation in and Around Ogawa Forest Reserve in Relation to Human Impact

WAJIROU SUZUKI

4.1 Introduction

Ogawa Forest Reserve (OFR) is located on a quasiplain (about 610–660 m above sea level) in the southern part of the Abukuma Mountains, which run from Fukushima Prefecture to Ibaraki Prefecture along the Pacific Coast of central Japan. This area has the typical climate of the Pacific Coast of Japan, which is characterized by cold, dry winters with little snowfall (see Chapter 2; Suzuki 1952), and deciduous broad-leaved forests which have developed naturally and which are mainly composed of *Fagus* and *Quercus* species (Hukushima et al. 1995). The area is located in the lower part of the montane zone, as well as in the transition zone between the warm temperate and cool temperate climate zones (Suzuki 1952; Kashimura 1968; cf. Table 7.3 in Chapter 7). Moreover, in the past, the forests in this area have been subjected to human activities such as burning, cattle grazing, and clear-cutting for fuelwood for at least the last 500 years (see Chapter 7). As a result, almost all the natural forest vegetation has now changed into secondary forest or coniferous plantations (Ibaraki Prefectural Forest Experiment Station 1980).

The OFR (98 ha) is one of few old-growth deciduous forests remaining in the Abukuma Mountains, and this heritage forest has been strictly conserved as a forest reserve since 1969. However, there is some evidence to suggest human impact in and around the OFR. This chapter describes the vegetation and its history in the OFR and in the forests around it, and compares them with those of the old-growth and secondary deciduous broad-leaved forests in Ibaraki Prefecture.

4.2 Land-Use History in the Southern Abukuma Mountains

According to pollen analyses in the Kameyachi bog, which is about 8 km from OFR, the genera *Quercus*, *Castanea*, and *Carpinus* as well as *Fagus* had been growing in this area for about 3,000 years. After the fifteenth century, the amount of pollen from

Ecological Studies, Vol. 158
Nakashizuka, Matsumoto (eds.) Diversity and Interaction in a Temperate Forest Community: Ogawa Forest Reserve of Japan

Pinus and *Betula* species increased gradually, and Gramineae pollen increased rapidly. Simultaneously, carbonized ash produced by forest fires appeared (S. Ikeda, unpublished data). These findings suggest that Gramineae pasture and *Pinus* forest expanded as a result of the forest fires in this area during the past 500 years. In the past 100 years, the amount of pollen from *Cryptomeria japonica* and *Chamaecyparis obtusa* has increased with the increasing size of plantations of these two species.

4.2.1 Ancient Times to the Medieval Period

The southern Abukuma Mountains were first described in the early eighth century in *Hitachi-Fudoki* (723, see Kawano 1981), which was the first book on local natural history. According to this book, the area was uncultivated woodland until the fifth century (Kawano 1981). However, in the Nara era (eighth century), the southern Abukuma Mountains were famous for horse breeding, and many meadows and pastures for horses existed (Taga Livestock Industry Cooperation 1982). Therefore, the area might already have been partly deforested and used as pasture for horses as early as the eighth century.

In 1678, the local government (the local lord of Mito) established a large pasture (Ohno Pasture, 25 km in length from north to south) in order to encourage the development of a horse-breeding industry. With the increasing horse population, horses escaping from the pasture began to cause serious damage to agricultural products. During this era, Ohno Pasture was maintained by burning in early spring, and dwarf bamboo and *Lespedeza* species were planted as cattle food for the winter (Taga Livestock Industry Cooperation 1982).

During the Edo era (1600–1867), the forests in the southern Abukuma Mountains belonged to the Mito local government (lord), and had been strictly managed by the warden. At the same time, the inhabitants of small communities scattered throughout this area carried out small-scale farming, and cultivated meadows for cattle feed and green manure. They also had the common right to obtain fuelwood from the surrounding forests. There is no clear reference to charcoal-burning in the southern Abukuma Mountains, which are in Ibaraki Prefecture, but Hitachi-no-kuni (the old name of Ibaraki Prefecture) used to be a famous charcoal-producing district (Japan Fuel Wood Association 1960). Moreover, the demand for charcoal is thought to have been high in this area in the past, since smithies for farming instruments prospered in Higasi-shirakawa, few kilometers from Ogawa (Hanawa-machi 1986).

4.2.2 Modern Times

During the disorganized period from the collapse of the Tokugawa Government to the establishment of the Meiji Government, vast areas of forest throughout Japan were cut and fired without any control, and became dilapidated. This also happened to the forests in the Ogawa area (Fig. 4.1). In 1886, the Meiji Government established a national forest system to manage the forests belonging to the Tokugawa Government and the local lords, and set up regional and district forest offices through-

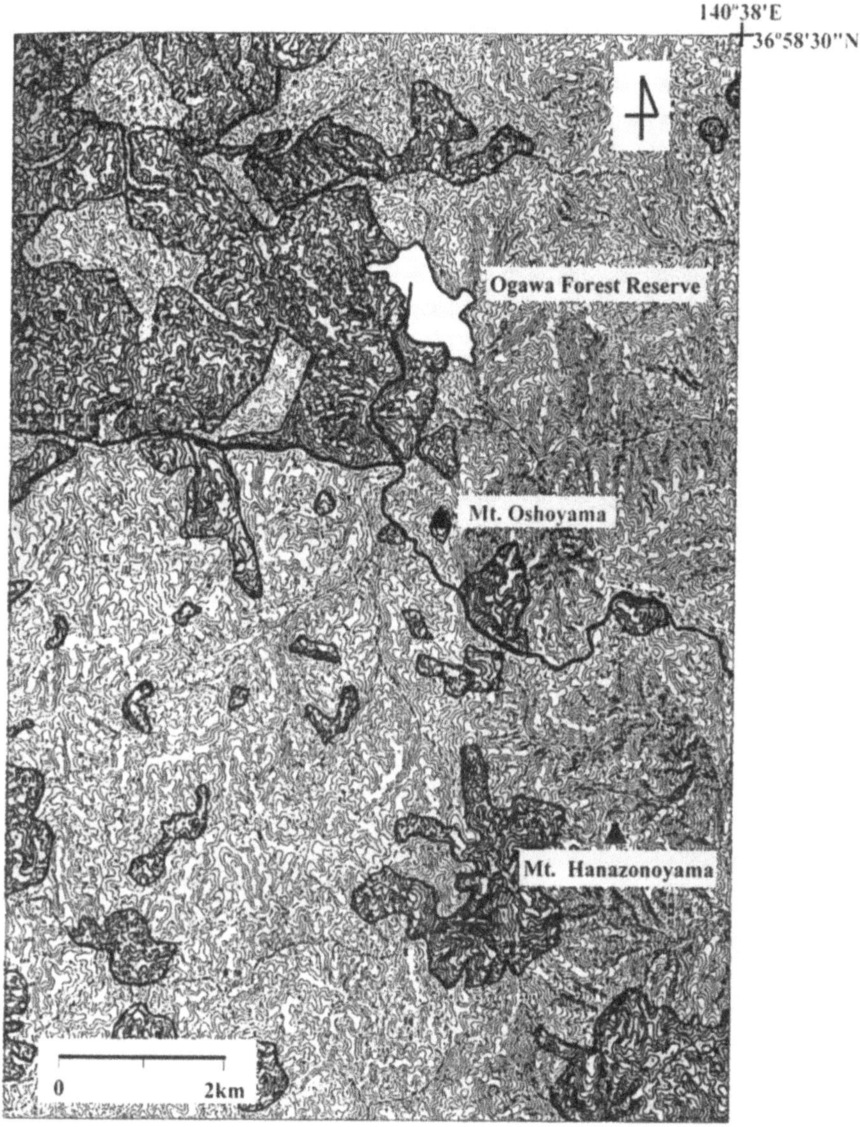

Fig. 4.1. Land-use map of the southern Abukuma Mountains published in 1911. The *darker* areas show the pastures and meadows

out Japan. The national forests in the southern Abukuma Mountains were managed by the Takahagi District Forest Office.

The modern management of Japanese National Forests began after the issue of new laws and regulations on forest practice in 1899. At first, the Forest Agency attempted to reforest the woodlands, which had been degraded by unrestricted land-use such as cutting and burning. The Takahagi District Forest Office carried out a large-scale afforestation project by planting *Chamaecyparis obtusa* and *Cryptomeria japonica* in the pastures and meadows which belonged to the national forests, and these plantations covered 6,509 ha (55% of the total national forest) by 1925. At the same time, with the expansion of coniferous plantations, many confrontations occurred between officers of the national forest and the local inhabitants who had been using the national forest for pasture, meadowland, and fuelwood. In 1916, the Forest Agency established a "custodial forest" system for using the timber from the forests as fuelwood for home consumption, and "national pasture" and "national meadow" systems to compensate the local people for the loss of pasture and meadowland by the expanding coniferous plantations. In the area managed by the Takahagi District Forest Office, 10 pastures (1,039 ha) and 22 meadows (669 ha) were set up, and a large area of pasture (1,138 ha) was sold to a private landowner. A modern management system was then introduced to "national pasture" and "national meadows", and cattle grazing and burning were controlled (Tokyo Regional Forest Office 1925a, b).

After the establishment of the national forest management system, commercial logging started in earnest. However, the forest resources of the Abukuma Mountains, which had historically been subjected to human activities such as burning and cattle grazing, were not protected. Old-growth forests remained only in riparian areas and on steep hillsides where they avoided the effects of forest fires, and middle-aged stands, plantations, and open land formed a complex mosaic landscape in other areas (Tokyo Regional Forest Office 1925a, b). For these reasons, most trees in this area were used as fuelwood for home consumption, except for the commercial logging of large trees of *Castanea, Quercus,* and *Fagus* species. With development of domestic industries after World War I, the demand for charcoal increased dramatically, and during World War II, charcoal-burning and manufacturing equipment for military horses became the main industries in the area.

4.2.3 After World War II

The political regime and the economic system of Japan were democratized after World War II. The land-use system in the Abukuma Mountains also changed during the rapid economic growth in the 1960s. First, horse breeding, which had been a main industry in this area, declined because of the mechanization of farming, and the livestock industry did not need such large pastures and meadows (Taga Livestock Industry Cooperation 1982). As a result, the large pastures and meadows were either converted into coniferous plantations, or were abandoned and developed a secondary growth of deciduous broad-leaved trees by natural succession. Charcoal production, which had reached a peak shortly after World War II, also declined rapidly because of the energy revolution in Japan in the 1960s. On the other hand, with increasing demands for timber at this time, afforestation was promoted, especially in the national forests, by changing deciduous broad-leaved forests into coniferous plantations. At present, all secondary deciduous forests remain in private hands. They produce logs for mushroom production, and wood chips instead of fuelwood and/or charcoal.

4.3 Current Forest Vegetation in the Southern Abukuma Mountains

The species composition of tree layers and forest-floor vegetation were investigated in the 24 old-growth stands which remain in Ibaraki Prefecture, including several forests outside the Abukuma Mountains, and in 26 secondary stands in and around the OFR. These stands were all deciduous temperate forests, located from 300 m to 880 m above sea level. To classify the stand characteristics, cluster analysis (the group average method) was applied to the compositional dissimilarity of both tree-layer and forest-floor vegetation among the stands (Suzuki et al. 1997).

Table 4.1 shows the tree species composition of each forest type, classified by cluster. The stands were classified into six major groups: (1) *Fagus crenata* type, (2) *F. japonica* type, (3) *Carpinus laxiflora* type, (4) mixed type I, mainly composed of *Fagus, Quercus,* and *Carpinus* species, (5) mixed type II, mainly composed of *Quercus, Carpinus,* and *Prunus* species, and (6) *Q. serrata* type. These characterizations are based on the dominance of component species in each stand.

Old-growth stands belonged to the *Fagus crenata*, *F. japonica*, and *Carpinus laxiflora* types, or mixed type I. The *F. crenata* type, which developed on the upland slopes with deep organic soil, was dominated by *F. crenata*, and composed of *Quercus*, *Carpinus,* and *Acer* species. The *F. japonica* type mainly developed on unstable hillsides with rocks and shallow organic soil, and was a low-diversity forest dominated by *F. japonica,* accompanied by *F. crenata*, *Acer,* and *Carpinus* species. The *Carpinus laxiflora* type, which was mainly composed of *F. japonica* and *Magnolia obovata* as well as *C. laxiflora*, developed on and around narrow ridges. On the other hand, mixed type I, which had a high species diversity mainly composed of *Fagus*, *Quercus*, *Carpinus*, and *Acer* species, developed on the hillsides. A few old-growth stands of *Abies firma*, which is the natural forest species in the transition zone between warm temperate and cool temperate climates, were observed on the narrow ridges and steep slopes. The characteristics of each tree species are analyzed in detail in Chapter 6 (Table 6.1).

Secondary stands belonged to the *Quercus serrata* type, the *Carpinus laxiflora* type, or mixed type II. The *Quercus serrata* type is a stand of low species diversity dominated only by *Quercus serrata*. The *Carpinus laxiflora* type was mentioned above in the old-growth stands. Mixed type II was composed of *Q. serrata*, *Q. crispula,* and *Castanea crenata*, but had no *Fagus* species. Secondary growth of *F. crenata* was also observed after clear-cutting of *F. crenata* old-growth forest, while no secondary growth of *F. japonica* was observed. *Betula platyphylla* var. *japonica,* which predominates after clear-cutting or forest fires in the montane zone, was limited to small areas of distribution.

Table 4.2 shows the species composition of each type of undergrowth vegetation classified according to the cluster analyses of old-growth and secondary forests. The forest-floor vegetation of these stands was classified into six major groups: (1) *Sasamorpha* type, (2) *Sasa* type, (3) *Pleioblastus* type, (4) *Hydrangea* type, (5) *Disporum* type, and (6) *Carex* type. The *Sasamorpha*, *Sasa,* and *Pleioblastus* types were common in both the old-growth and the secondary-growth stands, but the *Carex*

Table 4.1. Species composition of each forest type in old-growth and secondary-growth forests in and around the Abukuma Mountains, Ibaraki Prefecture

Forest type		*F. crenata*	Mixed I	Mixed II	*C. laxiflora*	*Q. serrata*	*F. japonica*	Mean value of relative dominance (%)
Number of stands		11	5	6	6	9	5	
Total number of species		46	42	34	37	30	25	
Quercus serata	+	II	III	V	II	V	I	18.5
Fagus japonica	+	II	IV	.	IV	.	V	16.0
Fagus crenata	+	V	IV	.	I	I	II	14.5
Carpinus laxiflora	+	III	III	III	V	III	II	8.4
Qurcus crispula	+	IV	.	IV	.	II	.	3.9
Castanea crenata	+	I	I	IV	.	II	I	3.6
Acer rufinerve	+	III	III	II	II	.	III	1.9
Swida controversa	+	I	I	III	II	.	I	1.9
Magnolia obovata	+	I	III	II	IV	II	.	1.9
Prunus jamasakura	–	I	.	V	.	III	.	1.9
Betula grossa	+	I	III	I	I	.	II	1.9
Carpinus tschonoskii	+	II	III	III	I	II	.	1.9
Quercus acutissima	–	.	.	II	.	II	.	1.7
Clethra barvinervis	+	V	IV	II	II	III	I	1.6
Zelkova serrata	+	.	I	.	.	.	.	1.5
Prunus verecunda	+	I	.	I	I	II	.	1.4
Acanthopanax sciadophylloides	+	II	I	II	III	II	.	1.3
Alunus firma	–	.	I	.	II	I	.	1.3
Carpinus japonica	+	III	III	II	II	II	II	1.2

Acer amoenum	+	III	III	III	II	II	II	1.0
Evodiopanax innovans	–	.	I	.	II	.	.	0.9
Stewartia pseudocamellia	–	.	II	.	.	.	II	0.8
Acer sieboldianum	+	II	III	.	II	I	II	0.7
Styrax japonica	+	II	I	IV	.	II	.	0.7
Pieris japonica	+	I	III	I	II	.	.	0.6
Pterostyrax hispida	–	.	I	.	.	.	.	0.6
Ilex macropoda	+	IV	II	II	II	II	I	0.5
Abies firma	–	I	I	I	III	I	I	0.5
Prunus grayana	+	I	.	I	I	II	.	0.5
Acer mono var. *marmoratum*	+	I	II	II	.	II	.	0.5
Styrax obassia	+	I	I	II	II	II	I	0.4
Acer japonicum	–	.	I	II	I	II	.	0.4
Hamamelis japonica	+	I	I	.	I	II	.	0.4
Lyonia ovalifolia var. *elliptica*	+	.	II	.	.	.	.	0.4
Meliosma myriantha	–	.	II	.	II	.	II	0.4
Fraxinus lanuginose	+	III	III	II	I	.	II	0.4
Carpinus cordata	+	.	II	.	I	.	II	0.3

The study plots in both old-growth and secondary-growth stands were classified into six forest types by cluster analysis based on the basal area of each species. Roman numerals show the constancy class, as follows: I, 1%–20%; II, 21%–40%; III, 41%–60%; IV, 61%–80%; V, 81%–100%. The mean values of relative dominance for each species were the average of relative dominance values among the six forest types calculated based on their total basal area, and indicate the dominance of the species in the local forest vegetation. Plus and minus signs indicate the existence or nonexistence of that species in the OFR (cf. Table 6.1).

Table 4.2. Species composition of each type of understory vegetation on old-growth and secondary-growth forests in and around the Abukuma Mountains, Ibaraki Prefecture

Vegetation type	*Sasamorpha*	*Sasa*	*Pleioblastus*	*Hydrangea*	*Disporum*	*Carex*	Mean value
Number of stands	10	6	6	7	5	4	of relative
Total number of species	104	85	147	141	123	108	dominance (%)
Sasamorpha borealis	V	I	I	.	.	4	17.4
Sasa nipponica	.	V	II	I	.	.	15.7
Pleioblastus chino	I	I	V	III	.	1	11.6
Carex dolichostachya var. *glaberrima*	I	.	I	.	.	4	9.8
Hydrangea hirta	II	III	IV	V	III	3	8.3
Disporum smilacinum	I	II	V	V	V	1	5.7
Athyrium yokoscense	I	IV	III	V	IV	3	3.7
Prenanthes acerifolia	I	I	III	III	.	1	2.1
Pieris japonica	I	.	IV	I	I	.	1.9
Rhododendron semibarbatum	II	II	II	I	I	3	1.9
Stephanandra incisa	II	III	III	II	III	1	1.8
Viburnum dilatatum	I	V	V	V	V	1	1.6
Skimmia japonica	II	.	III	.	.	1	1.5
Rhododendron obtusum var. *kaempferi*	IV	III	IV	III	III	2	1.0
Arundinaria ramosa	.	.	.	I	.	.	0.9
Carpinus laxiflora	I	I	I	III	II	2	0.9
Carex siderosticta	.	II	IV	V	III	.	0.9
Dryopteris erythrosora	.	.	.	.	.	1	0.8
Ilex macropoda	.	II	I	III	IV	3	0.7

Carex morrowii	I	.	.	.	.	4	0.7
Lindera umbellata	I	I	IV	II	III	.	0.7
Viburnum wrightii	I	I	I	I	.	3	0.4
Osmunda japonica	I	I	II	IV	II	.	0.3
Callicarpa japonica	I	IV	V	IV	II	2	0.3
Viburnum phlebotrichum	II	II	V	V	III	1	0.3
Wisteria floribunda	I	II	III	IV	II	2	0.2
Hydrangea petiolais	II	II	.	I	.	2	0.2
Aucuba japonica	I	.	II	I	.	.	0.2
Pertya robusta	.	.	III	III	I	2	0.2
Rhus ambigua	I	II	III	III	IV	3	0.1
Rubus palmatus var. *coptophyllus*	I	I	III	III	III	3	0.1
Oplismenus undulatifolius	I	I	II	III	III	1	0.1
Symplocos coreana	II	II	III	III	III	.	0.1
Acer amoenum	I	II	II	III	III	1	0.1
Aster scaber	.	.	I	III	III	.	0.1
Fraxinus lanuginosa f. serrata	I	.	I	III	I	3	0.1

Roman numerals and the mean values of relative dominance have the same meanings as in Table 4.1. However, relative dominance values were calculated based on the vegetation cover (%). Arabic numerals in the *Carex* type vegetation column are the number of plots in which the species appeared.

type was observed only in old-growth stands, and the *Hydrangea* and *Disporum* types only in the secondary-growth stands.

Both *Sasamorpha* and *Sasa* types of forest-floor vegetation were associated with the *F. crenata* type of forest. The *Sasamorpha* type of vegetation was also associated with both *F. japonica* type and mixed type I forests, and the *Carex* type was associated with the *Carpinus laxiflora* type forest (Suzuki et al. 1997).

In the secondary forests, the *Pleioblastus* type of vegetation may be a substitutional type for the other two dwarf bamboo types after large-scale human disturbances such as clear-cutting and burning. The *Hydrangea* and *Disporum* types may also be regarded as substitutional vegetation for the *Sasa* and *Sasamorpha* types, while this type of forest-floor vegetation is distributed in dry or slightly dry sites on uplands and around ridges. The *Hydrangea* and dwarf bamboo types of forest-floor vegetation were associated with *Carpinus laxiflora* and mixed type II stands, respectively, but the *Pleioblastus* and *Disporum* types were associated with *Q. serrata* type stands (Suzuki et al. 1997).

Old-growth stands in the OFR belonged to the *F. crenata* and *F. japonica* types, and these stands correspond to Sasamorpho–Fagetum crenate Suz.–Tok. 1949 on the phytosociological classification of beech forests in Japan (Hukushima et al. 1995). *Quercus* and *F. crenata* type stands were observed as secondary growth in the OFR. Secondary stands in and around the OFR are the substitutional vegetation of deciduous broad-leaved forest in the lower part of the montane zone, and are regarded as Castaneo–Quercetum serratae Okutomi, Tsuji et Kodaira 1976 (Miyawaki 1986).

4.4 Effects of Land-Use History on Forest Vegetation

4.4.1 Burning, Mowing, and Cattle Grazing

Until 80 years ago, burning in early spring to maintain pastures and meadows had been carried out repeatedly on the Pacific Coast of Japan, which has dry winters with little snow. As a result, the forests had changed to grasslands composed of *Miscanthus sinensis* or *Zoysia japonica* (Ohsako 1937; Numata 1969). In the Yama–Ogawa pasture adjacent to the OFR, the grassland vegetation was composed of both palatable (*Miscanthus sinensis, Imperata cylindrica, Zoysia japonica, Desmodium podocarpum* var. *mandshuricum*) and unpalatable (*Pteridium aquilinum* var. *latiusculum*, *Conyza bonariensis*, *Astilbe thunbergii* var. *congesta*) plants (Tokyo Regional Forest Office 1925a, b).

In spite of repeated and large-scale burning, old-growth natural forests were left in riparian areas and on adjacent hillsides, and small stands of trees were left intentionally as shelters and resting places for cattle (Tokyo Regional Forest Office 1925a, b). Burning also contributed to an increase in the relative importance of *Quercus crispula*, *Q. serrata,* and *Castanea crenata*, which all had thick bark which was resistant to fire, around pastures and meadows. In the southern Abukuma Mountains, the landscape mosaic before the Meiji era (about the 1910s) might be composed of pasture on the gentle slopes, natural forests in the valley bottoms, and small stands of trees in the pastures (Tokyo Regional Forest Office 1925a, b). Cattle grazing also affected the forest-floor vegetation in old-growth stands. Cattle, and especially horses, easily entered the forests and grazed on the dwarf bamboo, which was the main type of forest-floor vegetation in this area. With the decline of the dwarf bamboo communities, regeneration of the broad-leaved trees has been promoted successfully (Nakashizuka and Numata 1982a, b; Maeda 1988).

As horse breeding has declined since World War II, the pasture and meadowland has gradually changed to secondary forests. In abandoned pastures and meadows, *Q. serrata* and *C. crenata* have invaded, and now predominate. Their dominance in the abandoned pastures and meadows is attributed to their species characteristics such as strong light demands, fast growth rate, strong sprouting ability, and rapid sexual reproduction (see Chapters 7 and 12). The parent trees are relatively abundant as the result of historical land-use (Ishizuka 1977). On the floor of the secondary forests, *Sasa nipponica* and *Sasamorpha borealis*, which had declined as a result of burning and grazing, did not appear to recover, and *Pleioblastus chino* and the herbaceous components of the pasture and meadow communities such as Gramineae remained after the reforestations.

4.4.2 Charcoal Making

A coppicing system was adopted in secondary-growth stands in this area at least 500 years ago to make fuelwood and charcoal. The stands were usually clear-cut with a rotation of 20–30 years, and the trees regenerated by sprouting. Fallen branches, dead trees, and understory vegetation in these secondary-growth stands were used as woody fuel for home consumption, and the leaf litter on the forest floor was collected in early winter to make farmyard manure. These secondary-growth stands were indispensable to the lives of the inhabitants, and they grew quite close to the residential areas. Such stands are composed of tree species with a strong sprouting ability, such as *Quercus*, *Castanea*, *Carpinus*, and *Prunus* species, etc. (Kamitani 1993). Most component species of the coppiced forests have a low density or a limited distribution in the valley bottoms and on adjacent hillsides in the old-growth stands. They appear to have expanded their distributions into disturbed areas created by human activities. In contrast, *Fagus crenata* and *F. japonica,* the main components of old-growth stands, decreased their dominance in the coppicing system because of their slower growth rate and later reproductive maturation.

In general, secondary-growth stands tend to have high species diversity (Ito and Miyata 1977; Nagaike 2000). In the Abukuma Mountains, secondary-growth stands where the coppice system has been adopted also show a high species diversity, which includes *Carpinus*, *Acer*, *Purnus*, *Quercus,* and *Castanea* species, but secondary-growth stands with a low species diversity dominated by *Q. serrata* were also observed. This species is useful for fuelwood and charcoal-burning, and during extended coppice management with a short rotation they have been selectively preserved and have multiplied. Another deciduous oak, *Q. acutissima,* is also useful, although this has been planted since the area was used for charcoal-burning.

Dwarf bamboos are the main components of the undergrowth, and have been used as fuel and farmyard manure. They were also used in the building of the charcoal kilns and as fuel in charcoal-burning. As a result, the undergrowth of secondary stands, which had been dominated by dwarf bamboos, changed to a poorer vegetation composed of shrub and herbaceous species which could compete against weeds.

4.5 Human Impacts on Ogawa Forest Reserve

4.5.1 Distribution of Old-Growth and Secondary-Growth Stands

The OFR (98 ha) is a mosaic of species-rich old growth and the surrounding secondary stands, which are 50–80 years old and are dominated by *Quercus serrata*. About 50% of the Forest Reserve is secondary-growth stand at present (Fig. 4.2). According to the topographic map published in 1911 (Land and Survey of Japan), the edges of this land were treeless, suggesting that they were then used as pasture. Therefore,

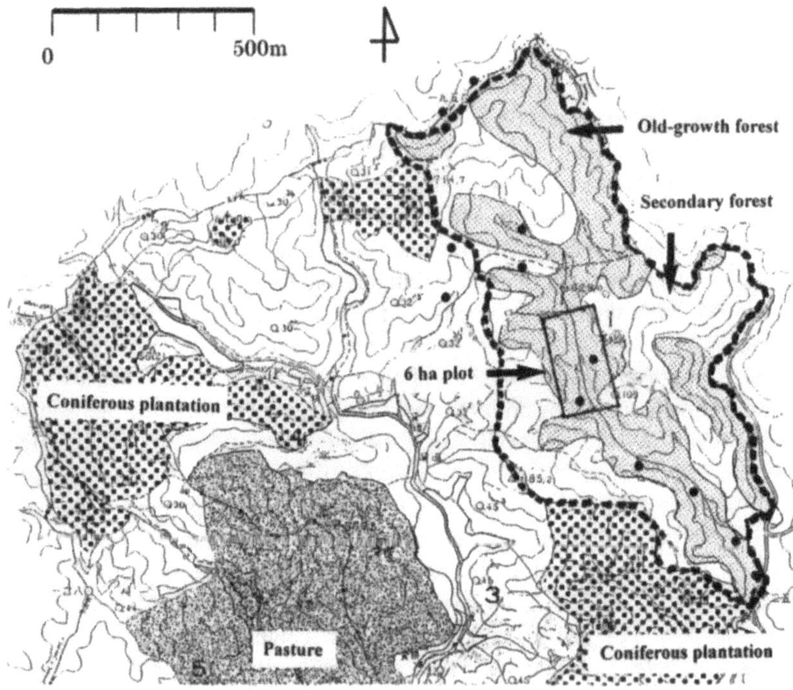

Fig. 4.2. Current land-use map of the area in and around Ogawa Forest Reserve (OFR). The area within the *broken line* is the OFR (98 ha). The *shaded* and *unshaded* areas show old-growth and secondary forests, respectively. *Circles* indicate evidence of old charcoal kilns. The contour interval is 10 m

both in the past and at present, old-growth stands with the original species composition and stand structure were distributed in the valley bottoms and on adjacent hillsides, and secondary-growth stands were mainly distributed on the surrounding ridges. The floristic composition and stand structure of the 6-ha permanent plot established in the old-growth stand in 1986 was characterized as having high species diversity dominated by *Q. serrata*. Some light-demanding and fast-growing tree species, such as *Swida controversa* and *Acer rufinerve,* as well as *Q. serrata*, have appeared in the plot. *Q. serrata,* which is the dominant species in the secondary growth, occupies about 18% of the total ground area of the 6-ha plot, while *Fagus crenata*, which is one of the main components of the old-growth stands, occupies only 8.5% (Masaki et al. 1992; Chapter 6).

4.5.2 Bambusoideae and its Distribution

Because the OFR, which is now protected, was previously surrounded by pasture, animals (initially horses, and then cows after the 1960s) have invaded the reserve and grazed on the undergrowth in the past. As a result, dwarf bamboo (*Sasamorpha borealis* or *Sasa nipponica*), which was the natural vegetation, has now declined dramatically. *Sasa nipponica* was replaced with *Arundinaria ramosa,* which is more resistant to cattle grazing. *Sasamorpha borealis* has also declined, but survives on steep hillsides where cattle cannot reach it. Since the grazing of cattle was stopped in the 1980s, *Sasamorpha borealis* and *Sasa nipponica* have started to spread in the OFR (K. Niiyama, unpublished data).

Because of the decline of the dwarf bamboo, tree seedlings which usually cannot live under the shade of dwarf bamboo began to survive and form seedling banks on the forest floor even under the closed canopy. They were able to regenerate successfully after canopy openings caused by natural disturbances (Masaki et al. 1992; Abe et al. 1995; cf. Nakashizuka and Numata 1982a, b; Suzuki et al. 1989).

4.5.3 Abandoned Charcoal Kilns

Before the area was designated as a Forest Reserve in 1969, no particular conservation management system had been adopted for the OFR. For this reason, the forest had been subjected to partial cutting for charcoal-burning. The existence of large old tree stumps and traces of charcoal kilns are evidence of this land-use history. There are seven abandoned charcoal kilns in the watershed of the OFR, and six kilns have been found in the Forest Reserve. These findings suggest that some part of the OFR had been subjected to coppice-system cutting for charcoal-burning before it was designated as a reserve. Even in the old-growth stands, the dominance of *Quercus serrata* suggests some human impact (see Chapter 6). A simulation trial using Markov models based on canopy–sapling associations predicted the elimination of *Q. serrata*, *Castanea crenata,* and *Q. crispula* from OFR in the steady state assuming single tree-fall-type replacement (see Chapter 6, Masaki et al. 1992).

4.6 Conclusions

In the southern Abukuma Mountains, there may be four types of original forest vegetation: *Fagus crenata* type, which developed on the upland slopes with deep organic soil, *F. japonica* type, which mainly developed on unstable hillsides with rocks and shallow organic soil, *Carpinus laxiflora* type, which developed on and around narrow ridges, and mixed type I, with a high diversity mainly composed of *Fagus*, *Quercus*, *Carpinus*, and *Acer* species, which developed on the hillsides. A few old-growth stands of *Abies firma*, which is a typical natural forest on narrow ridges and steep slopes in the transition zone from warm temperate to cool temperate climates, were also observed. These natural forests had been changed to secondary-growth stands by several kinds of human impact in the past, and now their structure and floristic composition remain only in small areas of old-growth stands.

The floristic composition of secondary-growth stands differs according to their land-use histories. In the abandoned pastures and meadows, *Q. serrata* and *Castanea crenata* invaded rapidly and created stands of low species diversity. On the other hand, the coppice system, with short-rotation clear-cuts, resulted in forests of high species diversity composed of *Carpinus*, *Acer*, *Purnus*, *Quercus*, and *Castanea* species. However, repeated coppicing and the effects of forest fires caused less species diversity, and stands dominated by *Q. serrata.* The effects of land-use history can also be observed in the forest-floor vegetation in the decline of the dwarf bamboo and the remaining grassland flora.

The OFR was composed of old-growth stands in the valley bottoms and the adjacent hillsides, and secondary-growth stands on and around the surrounding ridges. The old-growth stands had also been affected by human activities such as cattle grazing and selective cutting. The influences of these activities could be seen in the floristic composition of the tree cover and understory. However, since the designation of the area as a Forest Reserve, the old-growth stands have been recovering and now closely resemble the original features of the natural forest.

Acknowledgments

I thank K. Osumi, H. Tanaka, K. Hoshizaki, T. Masaki, and T. Nakashizuka for their critical reading of the manuscript. Thanks are also due to H. Nomiya, K. Kitamura, and G. Iwamoto for their assistance in the field work, and to the staff of the Takahagi District Forest Office for providing the data on land-use history and forest management plans.

References

Abe S, Masaki T, Nakashizuka T (1995) Factors influencing sapling composition in canopy gaps in a temperate deciduous forest. Vegetatio 120:21–32

Hanawa-machi (ed) (1986) History of Hanawa Town (in Japanese). Hanawa Town

Hukushima T, Takasuna H, Mataui T, Nishio T, Kyan Y, Tsunetomi Y (1995) New phytosociological classification of beech forests in Japan (in Japanese with English summary). Jpn J Ecol 45:79–98
Ibaraki Prefectural Forest Experiment Station (1980) Map of actual vegetation of Ibaraki Prefecture. Ibaraki Prefecture
Ishizuka K (ed) (1977) Distribution of plant communities and environment (in Japanese). Asakura-shoten, Tokyo
Ito S, Miyata I (1977) Species diversity of plant community (in Japanese). In: Ito S, Ishizuka K (eds) Species composition and structure of plant community. Asakura-shoten, Tokyo, pp 76–111
Japan Fuel Wood Association (1960) Japanese history of charcoal (in Japanese)
Kamitani T (1993) Ecological studies on regeneration of beech (*Fagus crenata* Blume) coppice forests in a heavy snowfall region (in Japanese with English summary). Mem Fac Agric Niigata Univ 30:1–108
Kashimura T (1968) Natural forest communities in Abukuma Mountains. Ecol View 17:75–85
Kawano T (1981) Inquiry of Hitachi-fudoki (in Japanese). Tsukuba-shorin, Tsuchiura
Maeda T (1988) Studies on natural regeneration of beech (*Fagus crenata* Blume) (in Japanese with English summary). Spec Bull Agric Utsunomiya Univ 46:1–79
Masaki T, Suzuki W, Niiyama K, Iida S, Tanaka H, Nakashizuka T (1992) Community structure of a species-rich temperate forest, Ogawa Forest Reserve, central Japan. Vegetatio 98:97–111
Miyawaki A (ed) (1986) Vegetation of Japan, Kanto (in Japanese). Shinbundo, Tokyo
Nagaike T (2000) The effect of human disturbance on forested landscape structure and plant species diversity in a *Fagus crenata* region (in Japanese with English summary). Bull Yamanashi For For Pro Res Inst 21:29–85
Nakashizuka T, Numata M (1982a) Regeneration process of climax beech forest. I. Structure of a forest with an undergrowth of *Sasa*. Jpn J Ecol 32:57–67
Nakashizuka T, Numata M (1982b) Regeneration process of climax beech forest. II. Structure of a forest under the influences of grazing. Jpn J Ecol 32:473–482
Numata M (1969) Progressive and retrogressive gradients of grassland vegetation measured by degree of succession. Ecological judgement of grassland condition and trend. IV. Vegetatio 19:96–127
Ohsako M (1937) Studies of meadows and pastures in Japan (in Japanese). Japan Forest Technical Association, Tokyo
Suzuki T (1952) Forest vegetation of eastern Asia (in Japanese). Kokin Shoin, Tokyo
Suzuki W, Nakashizuka T, Maeda T (1989) Structure and regeneration of remaining natural beech (*Fagus crenata*) forests in the southern part of Ibakaki Prefecture (in Japanese). Trans Jpn For Soc 100:365–367
Suzuki W, Nomiya H, Nakashizuka T (1997) Community structure and succession of secondary forests regarding to human impacts, central Japan. In: IAVS'97 Symposium Conference abstracts, pp 92–93, IAVS'97 Symposium, České Budějoice
Taga Livestock Industry Cooperation (1982) History of horse breeding in Taga County (in Japanese)
Tokyo Regional Forest Office (1925a) Second forest management plan for Takahagi District National Forests (in Japanese)
Tokyo Regional Forest Office (1925b) Management plan for pastures and meadows of the Takahagi District Forest Office (in Japanese)

5 Ground Design of the Research Site

HIROSHI TANAKA and TOHRU NAKASHIZUKA

5.1 Introduction

Following the methodology of studying a tree community described in Chapter 1, we designed this study (1) to cover the entire life cycle of the trees, (2) to integrate the population studies of component species into a community-level target for the forest, and (3) to be able to continue our long-term observations. We set up a permanent core plot in Ogawa Forest Reserve (OFR), and carried out additional investigations in and around that plot.

Our main purpose was to understand the mechanisms which maintained the tree species diversity in the forest, or the coexistence mechanisms of the tree community, and to elucidate the dynamics of this species-rich, temperate, deciduous forest community. Since the response to a particular environment or disturbance varies between the developmental stages of a given tree species, as well as between species (Harcombe 1987), it is important to know which life stage(s) are the most important in allowing the coexistence of species within a forest community (Chapter 1; Nakashizuka 1996). Therefore, this plot was designed to give an estimate of the demographic parameters of the trees at each stage of their entire life cycle (flowering, fruiting, seed production and dispersal, seedling emergence, and the growth and survival of seedlings, saplings, and mature trees). We tried to cover the environmental heterogeneity (Chapter 3) and past human activities and disturbance (Chapter 4) that are associated with the climatic conditions of this area (Chapter 2). We also selected methods that could be continued in the long term in order to maintain the same degree of accuracy.

5.2 Research Design for Tree Demography Incorporating Canopy-Gap Dynamics in Ogawa Forest Reserve

Ogawa Forest Reserve is a remnant (98 ha), species-rich temperate deciduous forest in central Japan. In the central part of the Reserve, a 6-ha (200 m × 300 m) permanent

Ecological Studies, Vol. 158
Nakashizuka, Matsumoto (eds.) Diversity and Interaction in a Temperate Forest Community: Ogawa Forest Reserve of Japan

plot was established in April 1987. Different sampling methods were adopted to estimate the demographic parameters at different developmental stages (Fig. 5.1). We divided the tree life cycle into five stages: seed, first-year seedling, seedling (< 30 cm high), sapling (> 30 cm high, < 5 cm diameter at breast height (DBH), i.e., at 1.3 m), and adult (> 5 cm DBH). All the adult trees in the 6-ha plot were identified, tagged, and mapped. We measured their DBH every 2 years until 1993 and every 4 years thereafter, as well as the DBH of all newly recruited stems. Saplings were identified and tagged in 651 quadrats (2 m × 2 m) located at all the intersections of the 10 m × 10 m grids throughout the 6-ha plot. The heights and DBH (if measurable) of the saplings were also recorded every 2 years.

To investigate seed and seedling demography, we established 263 combinations of a seed trap (0.5 m^2) and an adjacent 1 m × 1 m quadrat in the central 1.2 ha of the 6-ha plot (Fig. 5.2). These were arranged regularly in a matrix at a distance of 7.1 m from each other. Seeds which fell into the seed traps were collected every 2–4 weeks (every 2 weeks in the seeding season, and every 4 weeks over the rest of the year), identified, counted, and classified according to their appearance and content: sound, empty, immature, or suffered predation by insects or other animals). First-year seedlings in the quadrats were identified and marked, and their appearance (with or without cotyledons, number of leaves, eaten or not eaten, unusual leaf color, etc.) was recorded every 2–4 weeks (every 2 weeks from April to August, every 4 weeks after August). Older seedlings in the quadrats were identified and marked, and their height was measured every year. Environmental factors which might influence seed germi-

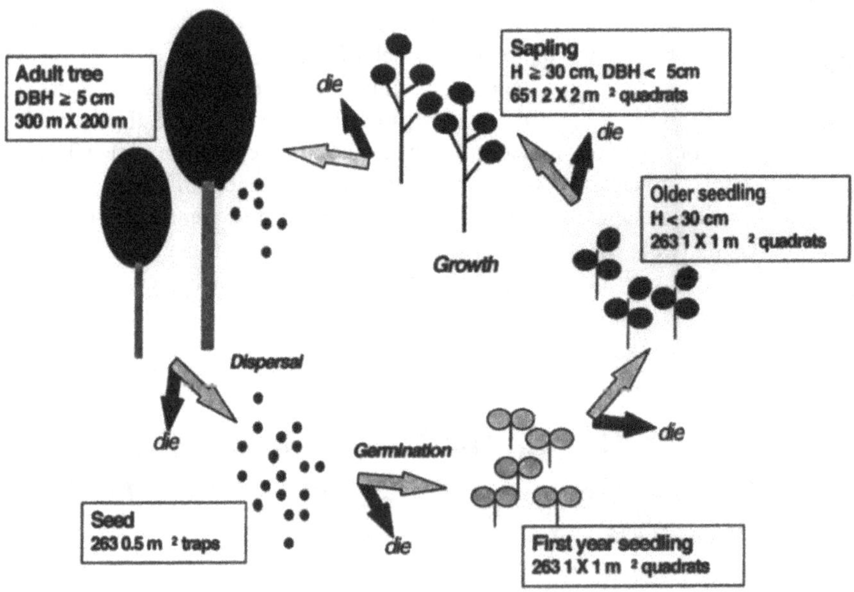

Fig. 5.1. The life cycle of a tree and the methods applied to estimate the demographic parameters at each stage. *H*, height; *DBH*, diameter at breast height (i.e., 1.3 m)

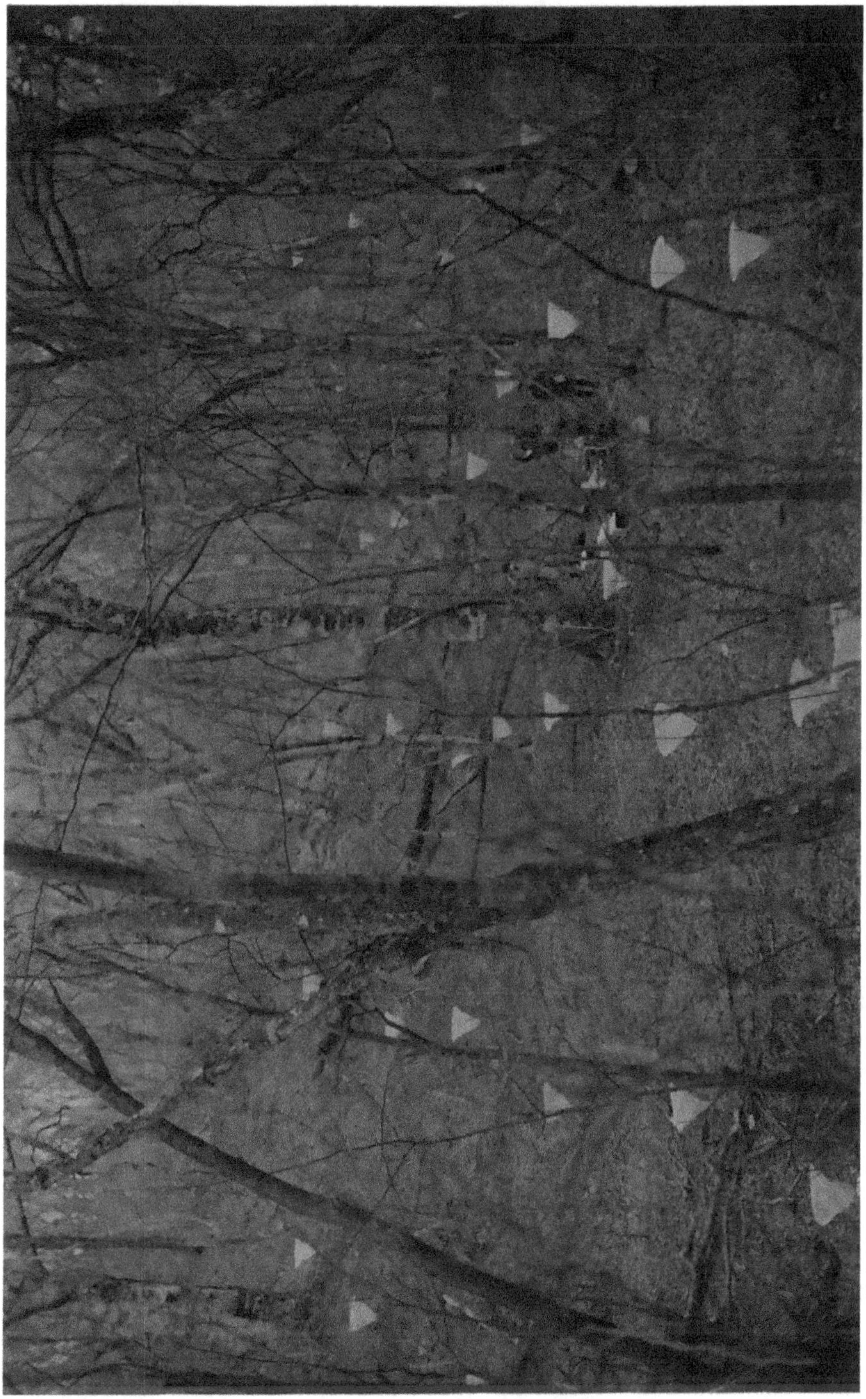

Fig. 5.2. Seed traps in a regular arrangement in the central 1.2-ha area of Ogawa Forest Reserve. A 1 m×1 m seedling quadrat was set at the side of each seed trap. (Photograph by Hiroshi Tanaka)

Fig. 5.3. Seed traps set in a newly created gap in the canopy. A 1 m × 1 m seedling quadrat was set at the side of each seed trap. (Photograph by Kaoru Niiyama)

nation or seedling survival and growth (mineral soil exposure, soil moisture, light condition estimated by hemispherical photographs, distance from the nearest conspecific adult tree, etc.) were measured for every 1-m^2 quadrat. Masaki et al. (1994), Shibata and Nakashizuka (1995), and Tanaka et al. (1998) have described these field procedures in detail.

Canopy gaps created by single or multiple tree falls are the major disturbances in this forest (Nakashizuka et al. 1992; Tanaka and Nakashizuka 1997). The distribution of canopy gaps in the 6-ha plot has also been monitored since 1987 at a spatial resolution of every 5 m × 5 m subquadrat. From 1987 to 1991, canopy gaps (defined as an area without a canopy higher than 10 m above the ground) were monitored every 2 years. Since 1991, the "vegetation profile technique" (Karr 1971; Hubbell and Foster 1986) has been applied, and the canopy height in every subquadrat has been classified and recorded in five height classes: 0–2 m, 2–5 m, 5–10 m, 10–15 m, and >15 m (Nakashizuka et al. 1995).

In four newly created gaps, seed and seedling investigations have been conducted with a similar system to that used in the 1.2-ha subplot. In total, 45 extra sets of seed traps (0.5 m^2) and 1 m × 1 m seedling quadrats were established in those new gaps (Fig. 5.3). The seedfall and the emergence and survival of seedlings there have been studied in the same way as in the 1.2-ha subplot. For saplings in other gaps, more than 100 quadrats (5 m × 5 m) were established in gaps in and around the 6-ha plot, which varied in size, age, and topographic location. Saplings in these gaps were also measured in 1990, 1992, 1994, and 1996 following the same methods as those used in the 2 m × 2 m sapling quadrats (Abe et al. 1995).

The results of the population dynamics study of constituent species are reported elsewhere in this book. Matrix model analyses will be applied to estimate the population growth rate and the relative importance of each developmental stage. Metapopulation analyses will be applied to estimate the relative importance of gaps for the maintenance of the population (Chapter 13; Abe et al. 1998). Finally, individually based models can be constructed based on the demographic data obtained in the research system used in the OFR (Chapter 14).

5.3 Other Ecological Studies in OFR

Multiple interactions between trees and other organisms work as mechanisms to maintain or promote species coexistence, and to sustain ecosystem functions such as nutrient and carbon dynamics. In OFR, several aspects of this interaction between trees and other organisms have been studied. One is the role of ectomycorrhizal fungi in determining tree distribution and regeneration (Chapter 21). The nutrient status of the forest floor formed by interacting ectomycorrhiza and white rot fungi was also studied (Chapter 8). There are reports of understory herbaceous plants whose matter production is strictly limited through the interception of radiation by canopy trees (Kawarasaki and Hori 1999, 2001). The behavioral ecology of the two common forest mice species (K. Shioya and Y. Koizumi unpublished data, 2001), and the seasonal and annual dynamics of the bird community (H. Tojo unpublished data,

2001) have also been studied in the 6-ha plot. The contributions of forest mice and birds as seed predators and dispersers were also studied based on the demographic data obtained in the 1.2-ha subplot as well as additional field experiments (Masaki et al. 1994; Iida 1996; Tanaka 1995; Iida and Nakashizuka 1997; Shibata et al. 1998).

The spatial heterogeneity of (abiological) environmental factors was investigated using a unit spatial scale of a 10 m × 10 m grid in the 6-ha plot. The spatial patterns of the soil's physical and chemical characteristics were investigated in relation to the microtopography and plant distribution (Chapters 3 and 8). A study of the nitrogen dynamics of the forest is now ongoing (K. Hirai unpublished data, 2001), and the spatial pattern of the denitrification processes has been studied in the 6-ha plot (Ishizuka et al. 2000).

The genetic structure and diversity of adult trees of the major component species such as *Fagus japonica*, *F. crenata*, *Carpinus laxifolia,* and *C. tschonoskii* were clarified based on mapping data in the 6-ha plot by using allozyme analyses (Kitamura et al. 1992a, b). For *F. crenata*, more intensive analyses were conducted for the seedling, sapling, and adult populations around the 6-ha plot, and the genetic structures and diversity of different subpopulations were compared (Chapter 19; Kitamura et al. 1997a, b; Kawano and Kitamura 1997). The reproductive characteristics of sparsely distributed tree species such as *Magnolia obovata* and *Aesculus turbinata* were investigated in the 100-ha area of the OFR, including the 6-ha plot (Chapter 20; Isagi et al. 2000).

There are many mechanisms involved in maintaining tree species and genetic diversity, and they work at various stages of the life cycles of the trees (Chapter 1; Nakashizuka 1996). For each tree population, the most important mechanism and the life stage at which it works may be different. Studies which focus on specific aspects of these mechanisms should be integrated in order to understand the entire pattern of the maintenance mechanisms of forest tree diversity in a community.

References

Abe S, Masaki T, Nakashizuka T (1995) Factors influencing sapling composition in canopy gaps of a temperate deciduous forest. Vegetatio 120:21–32

Abe S, Nakashizuka T, Tanaka H (1998) Effects of canopy gaps on the demography of the subcanopy tree *Stylax obassia.* J Veg Sci 9:787–796

Harcombe PA (1987) Tree life table: simple birth, growth, and death data encapsulate life histories and ecological roles. BioScience 37:557–568

Hubbell SP, Foster RB (1986) Canopy gaps and the dynamics of a neotropical forest. In: Crawley MJ (ed) Plant ecology. Blackwell Scientific, Oxford, pp 77–96

Iida S (1996) Quantitative analysis of acorn transportation by rodents using a magnetic locator. Vegetatio 124:39–43

Iida S, Nakashizuka T (1997) Spatial and temporal dispersal of *Kalopanax pictus* seeds in a temperate deciduous forest, central Japan. Plant Ecol 135:243–248

Isagi Y, Kanazashi T, Suzuki W, Tanaka H, Abe T (2000) Microsatellite analysis of the regeneration process of *Magnolia obovata.* Heredity 84:143–151

Ishizuka S, Sakata T, Tanikawa T, Ishizuka K (2000) N_2O emission and spatial distribution in a Japanese deciduous forest (in Japanese with English summary). J Jpn For Soc 82:62–71

Karr JR (1971) Structure of avian communities in selected Panama and Illinois habitats. Ecol Monog 41:207–233

Kawano S, Kitamura K (1997) Demographic genetics of the Japanese beech, *Fagus crenata*, at Ogawa Forest Reserve, Ibaraki, Central Honshu, Japan. III. Population dynamics and genetic substructuring within a metapopulation. Plant Species Biol 12:157–177

Kawarasaki S, Hori Y (1999) Effect of flower number on the pollinator attractiveness and the threshold plant size for flowering in *Pertya triloba* (Asteraceae). Plant Species Biol 14:69–74

Kawarasaki S, Hori Y (2001) Flowering phenology of understory herbaceous species in a cool temperate deciduous forest in Ogawa Forest Reserve, central Japan. J Plant Res 114:19–23

Kitamura K, Okuizumi H, Seki T, Niiyama K, Shiraishi S (1992a) Isozyme analysis of the mating system in natural populations of *Fagus crenata* and *F. japonica* (in Japanese with English summary). Jpn J Ecol 42:61–69

Kitamura K, Okuizumi H, Suzuki W, Shiraishi S (1992b) Hardy–Weinberg equilibrium with allozyme loci among natural population of *Carpinus laxiflora* in Ogawa Forest Reserve, central Japan. J Jpn For Soc 74:346–349

Kitamura K, Shimada K, Nakashima K, Kawano S (1997a) Demographic genetics of the Japanese Beech, *Fagus crenata*, at Ogawa Forest Reserve, Ibaraki, Central Honshu, Japan. I. Spatial genetic substructuring in local populations. Plant Species Biol 12:107–135

Kitamura K, Shimada K, Nakashima K, Kawano S (1997b) Demographic genetics of the Japanese beech, *Fagus crenata*, at Ogawa Forest Reserve, Ibaraki, Central Honshu, Japan. II. Genetic substructuring among size classes in local populations. Plant Species Biol 12:137–155

Masaki T, Kominami Y, Nakashizuka T (1994) Spatial and seasonal patterns of seed dissemination of *Cornus controversa* in a temperate forest. Ecology 75:1903–1910

Nakashizuka T (1996) Factors maintaining forest tree diversity. In: Turner IM, Diong CH, Lim SSL, Ng PKL (eds) Biodiversity and the dynamics of ecosystems. DIWPA Series Volume 1:51–62

Nakashizuka T, Iida S, Tanaka H, Shibata M, Abe S, Masaki T, Niiyama K (1992) Community dynamics of Ogawa Forest Reserve, a species-rich deciduous forest, central Japan. Vegetatio 103:105–112

Nakashizuka T, Katsuki T, Tanaka H (1995) Forest canopy structure analyzed by using aerial photographs. Ecol Res 10:13–18

Shibata M, Nakashizuka T (1995) Seed and seedling demography of four co-occurring *Carpinus* species in a temperate deciduous forest. Ecology 76:1099–1108

Shibata M, Tanaka H, Nakashizuka T (1998) Causes and consequences of mast seed production of four co-occurring *Carpinus* species in Japan. Ecology 79:54–64

Tanaka H (1995) Seed demography of three co-occurring *Acer* species in a Japanese temperate deciduous forest. J Veg Sci 6:887–896

Tanaka H, Nakashizuka T (1997) Fifteen years of canopy dynamics analyzed by aerial photographs in a temperate deciduous forest, Japan. Ecology 78:612–620

Tanaka H, Shibata M, Nakashizuka T (1998) A mechanistic approach for evaluating the role of wind dispersal in tree population dynamics. J Sust For 6:155–174

Part 3

Tree Community Structure and Dynamics

6 Structure and Dynamics

TAKASHI MASAKI

6.1 Overview of the Structure and Trends in the Dynamics of Ogawa Forest Reserve

This chapter analyzes the structure and dynamics of the tree community of Ogawa Forest Reserve (OFR).The main focus is on the relationships between species richness and size structure in local stands, and an explanation is proposed of how the high species richness of OFR is maintained (or how it might be lost). The study used the 1987 and 1997 census data of trees with a diameter at breast height (DBH) ≥ 5 cm (see Chapter 5). The 1987 data were used for the static analyses, and the data from both years were used for the analyses of community dynamics.

The 6-ha plot in the OFR (see Chapter 5) included 55 tree species in 1987. The basal area (BA) and stem density of the plot were 32.3 m^2 ha^{-1} and 822.7 ha^{-1}, respectively. During the 10-year period from 1987 to 1997, these parameters increased to 33.5 m^2 ha^{-1} and 836.8 ha^{-1}, respectively. Two species, *Clerodendrum trichotomum* and *Euonymus oxyphyllus,* initially had a DBH < 5 cm, but during the 10-year period they grew large enough to be included. However, one species (*Acer crataegifolium*) was eliminated during the 10 years, resulting in 56 species within the plot in 1997. The community turnover time was estimated to range from 58 to 240 years (see Chapter 7 and Nakashizuka et al. 1992). The graph of the size distribution of the trees within this forest is L-shaped and the maximum DBH was > 120 cm (Fig. 6.1), giving the forest an old-growth appearance.

Table 6.1 lists 46 of the 56 tree species in the plot according to their life-history type (as defined below). The other ten species were shrub species by nature (maximum DBH < 10 cm), but they had happened to exceed the minimum census size (i.e., DBH 5 cm). The most dominant species in 1987 (i.e., the species with the largest BA) was *Quercus serrata* (8.6 m^2 ha^{-1}), followed by *Fagus japonica* (6.7 m^2 ha^{-1}) and *F. crenata* (2.8 m^2 ha^{-1}) (Table 6.1). There were 15 subdominant species (BA > 0.2 m^2 ha^{-1}), which belonged to Fagaceae (*Quercus crispula, Castanea crenata*), Betulaceae (*Carpinus laxiflora, C. tschonoskii, C. cordata, C. japonica, Betula grossa, Ostrya japonica*), Aceraceae (*Acer mono* var. *marmoratum* f. *dissectum, A. rufinerve,*

Ecological Studies, Vol. 158
Nakashizuka, Matsumoto (eds.) Diversity and Interaction in a Temperate Forest Community: Ogawa Forest Reserve of Japan

Table 6.1. Bias of species distribution among young, developing, and old stand types

Life history guild[a]	Species	Habitat guild[b]	Y[c]	D[c]	O[c]
TC	*Fagus japonica*	A		++	
	Acer sieboldianum	Bc		++	
	Carpinus cordata	A		++	++
	Acer mono var. *marmoratum* f. *dissectum*	Ca			++
	Fagus crenata	Ca			++
	Sorbus alnifolia	Bb	+		
	Acer amoenum	Ca	+		+
	Carpinus laxiflora	Ba			+
IC	*Quercus serrata*	Ba	++		
	Swida controversa	Cb	++		
	Carpinus tschonoskii	Cb	++		
	Castanea crenata	Ba	++		
	Quercus crispula	Ba	++		
	Carpinus japonica	Ca		++	
	Sorbus japonica	Bc			++
	Prunus verecunda	Ba	+		+
	Ostrya japonica	Cb	+		+
	Betula grossa	Bb		+	
TU	*Styrax obassia*	Ba	++		
	Acer tenuifolium	A	++	++	
	Pieris japonica	Bb		++	
	Acer nikoense	Cc		++	++
	Fraxinus lanuginosa f. *serrata*	Bb			++
	Styrax japonica	Ba	+		
	Meliosma myriantha	A	+	+	
	Clethra barbinervis	Bb		+	
	Hamamelis japonica	A		+	
	Benthamidia japonica	Bb		+	+
	Ilex macropoda	Bc		+	+
	Acanthopanax sciadophylloides	Bc		+	+

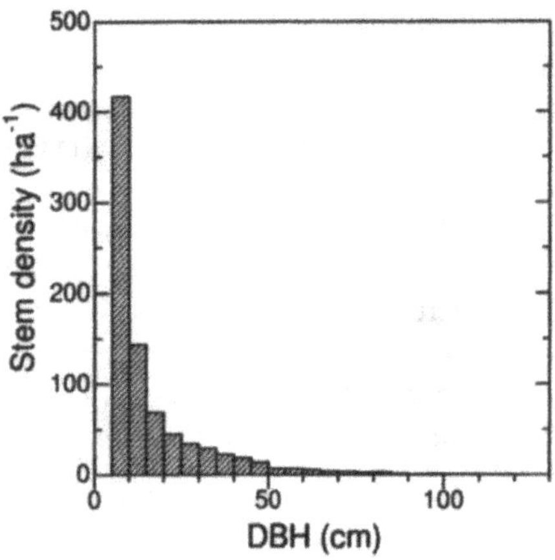

Fig. 6.1. Size (diameter at breast height; DBH) distribution of all species in the Ogawa Forest Reserve (OFR). Stem densities were calculated at 5-cm intervals above 5 cm DBH

Table 6.1. *Continued*

Life history guild[a]	Species	Habitat guild[b]	Y[c]	D[c]	O[c]
IU	*Acer rufinerve*	Bb		++	++
Occasional species	*Zelkova serrata*		+		
	Prunus buergeriana		+		
	Prunus grayana		+		
	Acer distylum		+		
	Magnolia obovata		+		
	Euonymus sieboldianus		+		
	Acer carpinifolium		+		
	Acer cissifolium			+	
	Rhus trichocarpa			+	
	Morus australis			+	
	Alnus hirsuta var. *sibirica*			+	
	Malus tschonoskii			+	
	Lyonia ovalifolia var. *elliptica*			+	+
	Kalopanax pictus			+	+
	Phellodendron amurense				+

[a]TC, shade-tolerant canopy species; IC, shade-intolerant canopy species; TU, shade-tolerant understory species; IU, shade-intolerant understory species. Categorization of species into these four guilds was done for common species (with stem densities ≥ 5 ha^{-1}). If a species is not common, it was categorized as an occasional species. These classifications were done for species with maximum DBH ≥ 10 cm.

[b]Classification types are from Masaki et al. (1992). A, lower slope; Ba, upper slope (the result of a larger-scale disturbance); Bb, upper slope (the results of patchy disturbances); Bc, upper slope (local "climax" patches); Ca, topography-generalist (local "climax" patches); Cb, topography-generalist (the results of patchy disturbances); Cc, topography-generalist (relations to disturbances are unclear). The habitat guilds to which occasional species belong cannot be determined owing to small sample sizes.

[c]Distribution biases of stems are shown as ++ (significant in χ^2-test) or + (not significant) when the observed density within a stand type exceeded the expected value. Y, young stand; D, developing stand; O, old stand.

A. amoenum), and other families (*Swida controversa*, *Stylax obassia*, *Prunus verecunda*, *Sorbus japonica*) (Table 6.1). Thirty-one species had a stem density ≥ 5 ha^{-1} (Table 6.1). In this chapter, all these species are defined as "common species". Another 15 species are defined as "occasional species" (Table 6.1). Regeneration of the most dominant species, *Quercus serrata*, and 10 of the 15 subdominant species was found to be unsuccessful, as shown below. This implies that a compositional shift is now occurring in this forest despite its old-growth appearance. This is often found to be the case in other temperate forests also (e.g., Christensen and Peet 1984; Masaki et al. 1999; Parker et al. 1985; Peterken and Jones 1987).

OFR shows a higher species richness than other cool-temperate forests in Japan. For example, an old-growth temperate forest in Kanumazawa, in northern Japan, whose temperature and methodology is almost the same as those of OFR, has fewer than 20 tree species in a 1-ha plot compared with about 35 in the OGR (Fig. 6.2).

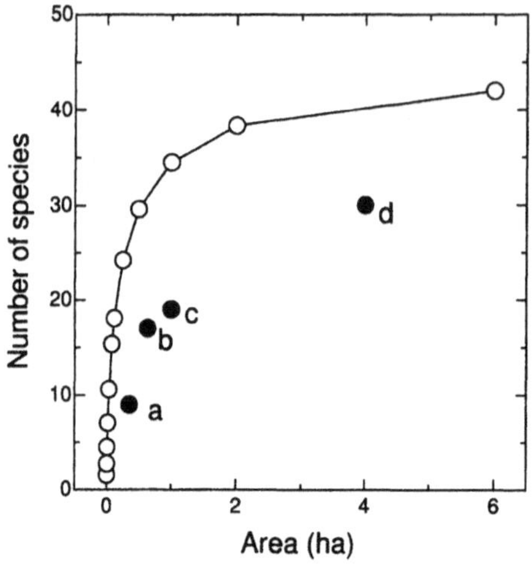

Fig. 6.2. Change in the number of tree species (DBH ≥ 5 cm except for shrub species) as a function of the sample area within the 6-ha plot in OFR (*open circles* and *solid line*). This is a modification of Fig. 4 in Masaki et al. (1999). *Solid circles* represent the numbers of tree species in other temperate forests composed of deciduous trees in various parts in Japan: (*a*) Kayano-Daira in central Japan (DBH ≥ 4 cm, including shrub species) (Nakashizuka and Numata 1982a); (*b*) Mt. Moriyoshi in northern Japan (DBH ≥ 5 cm, including shrub species) (Nakashizuka and Numata 1982b); (*c*) Kanumazawa in northern Japan (DBH ≥ 5 cm, except for shrub species) (Masaki et al. 1999); (*d*) Mt. Daisen in southwestern Japan (DBH ≥ 4 cm, including shrub species) (Hara et al. 1995). Note that in this figure the number of species in OFR is underestimated compared with the other forests because it does not include shrub species

Other temperate forests in southwestern, central and northern areas of Japan also have fewer species than OFR (Fig. 6.2). The species diversity is also higher in OFR than in Kanumazawa: Shannon's diversity index (H') in an area of one ha is 3.97 in OFR and 3.30 in Kanumazawa (Masaki et al. 1999).

6.2 Life-History Guild

As a first step toward explaining the community organization of OFR, common species were categorized based on two-life history traits: life form and regeneration mode. For life form, each species was classified as either an understory (or subcanopy) species or a canopy species depending on the maximum DBH of that species in the plot. The species with a smaller maximum DBH are understory (or subcanopy) species, and those with larger DBH values are canopy species.

For regeneration mode, each species was classified as either a shade-tolerant species or a shade-intolerant species (cf. Swain and Whitmore 1988). Evaluation is based on the size distribution index (SDI), which is the third moment of the DBH distribution around the midpoint of the DBH range.

$$\mathrm{SDI} = \frac{1}{n}\sum_{i=1}^{n}(x_i - 0.5)^3 \quad (1)$$

$$x_i = (d_i - 5)/(D - 5) \quad (2)$$

where d_i and x_i are the DBH and standardized DBH of the ith stem, respectively, and D is the species-specific maximum DBH. When smaller trees are abundant, the SDI has a small value, and when smaller trees are scarce, the SDI becomes larger. Nakashizuka et al. (1992) referred to shade-tolerant species in OFR as L-shaped species, and shade-intolerant species as Bell-shaped species, based on their size distribution pattern. Based on a plot of the SDI of each of the 31 common species against its maximum DBH, the species were categorized into four groups: shade-tolerant canopy (TC) species, shade-intolerant canopy (IC) species, shade-tolerant understory (TU) species, and shade-intolerant understory (IU) species (Fig. 6.3). The threshold maximum DBH was set at 40 cm based on field observations. The thresholds of the SDI were set at –0.05 and –0.01 for canopy species and understory species, respectively, bearing in mind the negative correlation between maximum DBH and SDI (Masaki et al. 1999) (Fig. 6.3).

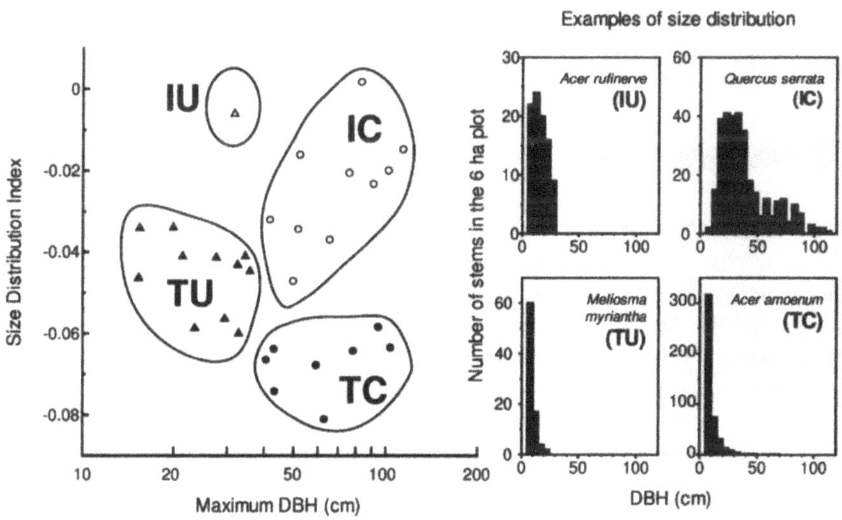

Fig. 6.3. Classification of species using species-specific maximum DBH and the size distribution index (SDI), which is the third moment of the DBH distribution around the midpoint of the DBH range. In this figure, the species are classified into four groups: shade-tolerant canopy (*TC*) species (*closed circles*), shade-intolerant canopy (*IC*) species (*open circles*), shade-tolerant understory (*TU*) species (*closed triangles*), and shade-intolerant understory (*IU*) species (*open triangles*). The size distribution of an example of a species from each group is also shown. In the graphs, the horizontal axis shows the DBH (cm) and the vertical axis shows the number of stems in the plot (at 5-cm intervals)

Nakashizuka et al. (1992) and Masaki et al. (1999), using census data from 1987 to 1991 and from 1987 to 1993, respectively, showed that the above classification of the regeneration mode based on the SDI was valid. TC and TU species have recruitment rates that are almost equivalent to, or significantly larger than, mortality rates. This implies that the population size of TC and TU species is almost maintained or is increasing under present conditions. In other words, the regeneration of these species is successful or continuous under the current disturbance regime without large (roughly ≥ 400 m²) gaps (Abe et al. 1995). On the other hand, for most of the IC and IU species, the mortality rate is higher than the recruitment rate. This means that the population size of these species is generally decreasing, i.e., their regeneration is unsuccessful. For the establishment of recruits, these species are considered to require some episodic events, such as the formation of large gaps (i.e., 1,500 m², see below), which have been rare during the last 50 years at least (Abe et al. 1995). Thus, the classification of species based on their size distributions correctly reflects their current regeneration modes.

The species in each group (based on the 1987 census) are listed in Table 6.1. Two *Fagus* species and four subdominant species are TC species, while the most dominant species, *Quercus serrata*, and nine subdominant species are IC species. Each of the TU and IU species includes one subdominant species. Thus, at the time of the first census (1987), OFR was dominated by shade-intolerant species rather than by shade-tolerant species.

Markovian simulation based on tree-to-tree replacement patterns also suggested that most IC species would be eliminated, and that TC species would persist in the projected equilibrium state of this forest (Masaki et al. 1992).

6.3 Habitat Guild

Masaki et al. (1992) found seven habitat guilds of species in the tree community of OFR using cluster analysis based on interspecific spatial correlations. The habitat guilds mainly reflect topography: they include a lower-slope guild (A), three upper-slope guilds (Ba, Bb, Bc), and three topography-generalist guilds (Ca, Cb, Cc) (as defined below). These guilds were closely related to the life-history guilds. They also roughly correspond to the forest types classified by Suzuki (Table 4.1, Chapter 4), including the secondary forests around OFR. Guilds A, B, and C consist of the stands of *F. japonica* type, mixed type I (for a definition see Chapter 4), and *F. crenata* type, respectively, in Table 4.1 (Chapter 4).

Guilds A, Bc, and Ca are mainly composed of TC and TU species, implying that these guilds were local "climax" patches that had recovered from past disturbances (Table 6.1). On the other hand, Ba, Bb, and Cb are mainly composed of typical IC and IU species; these guilds are probably the result of recent disturbances (Table 6.1). In Ba and Bb, the constituent species occurr in different clump sizes; many species in Ba have large (i.e., ≥ 1,500 m²) clump sizes and those in Bb have smaller (< 1,500 m²) clump sizes. Masaki et al. (1992) concluded that guild Ba is the result of larger-scale disturbances, such as forest fires, that occurred in the area surrounding the plot 70–80 years ago (see Chapter 4). Guilds Bb and Cb are probably the results of more moderate disturbances such as gaps created by several falling trees.

Thus, the community structure of OFR is described as being composed of different habitat guilds with different topographic preferences. Within a habitat, a mosaic of patches of various sizes and ages has been formed, and each patch is mainly composed of species of a life-history guild which can be closely related to its disturbance history.

6.4 Structural Condition and Its Dynamics

As discussed above and elsewhere (Chapters 4 and 7), OFR has been affected by large-scale disturbances in the past. We can identify the effects of such events by analyzing the local stand structure. A young stand can be characterized by the small size of leading trees (i.e., the largest trees within a stand), reflecting a recent initiation of stand development, and a less positively skewed size distribution (i.e., larger SDI), reflecting simultaneous regeneration. Without further disturbances, such a stand will change to one with larger leading trees and with more positively skewed size distribution (i.e., lower SDI; more abundant small trees) (see Parker et al. 1985). Based on this idea, the 6-ha plot was divided into 600 cells of 10 m × 10 m, and maximum DBH and SDI were calculated for each cell in order to identify its structural conditions. Even though the cells were covered by a dense canopy from adjacent trees, some of the cells did not contain any trees. For these cells, some of the parameters used in this study could not be calculated. To avoid this problem, four 5 m × 10 m areas adjacent to each focal cell, one on each side, were included in the calculations.

A negative correlation was found between SDIs and maximum DBH ($r = -0.64$, $P < 0.05$). These cells were classified into seven structural-condition groups depending on these two parameters. First, the developmental stage of the cells was defined according to their maximum DBH: cells with maximum DBH < 50 cm, ≥ 50 cm and < 80 cm, and ≥ 80 cm were classified as "young", "developing", and "old", respectively. The threshold values (50 cm and 80 cm) were based on the mean and median maximum DBH (60.7 cm and 58.9 cm, respectively) and on field observations, in an effort to make the number of cells in each stage as equivalent as possible. Second, the cells were also classified according to their SDI: the small trees in each cell were classified as "abundant" (SDI < –0.06), "intermediate" (SDI ≥ –0.06 and < –0.04), or "scarce" (SDI ≥ –0.04). The threshold values (–0.04 and –0.06) were based on the mean and median values of the SDI (–0.050 and –0.052, respectively), while also considering the need for equivalence in the number of cells among groups.

Thus, each cell was categorized into one of nine groups: YA (young and abundant), YI (young and intermediate), YS (young and scarce), DA (developing and abundant), DI (developing and intermediate), DS (developing and scarce), OA (old and abundant), OI (old and intermediate), and OS (old and scarce). Because of the negative correlation between the two parameters (developmental stage and the scarcity of small trees), there were only a few cells in the YA and OS groups. Therefore, YA and YI were pooled and re-defined as YA. Similarly, the OI cells were included

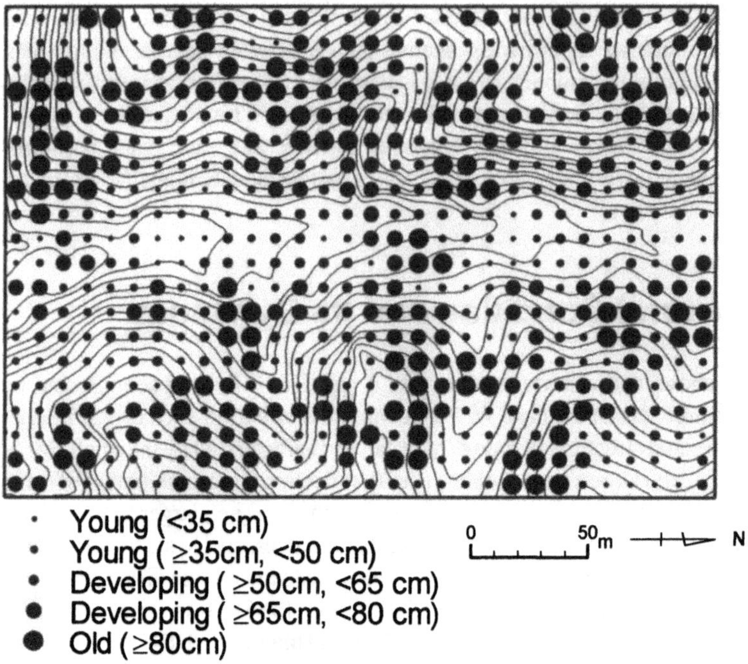

Fig. 6.4. Distribution of stands at various stages along the developmental sere (from young to old stands) within the 6-ha plot (300 m × 200 m). The size of each *dot* is proportional to the maximum DBH within each 10 m × 10 m cell. The *contour lines* are at 2-m intervals

with the OS cells. Thus, the cells were finally classified into seven structural condition types (YA, YS, DA, DI, DS, OA, and OS).

Many young stands aggregate close to the edge of the plot, reflecting the spatial pattern of the large-scale disturbances discussed earlier (Fig. 6.4), which were probably the result of past human activity. This is also evident in aerial photographs taken before 1981 (Tanaka and Nakashizuka 1997) (see Chapter 7). However, some young stands were distributed along a small channel near the center of the plot and in low-lying areas (Fig. 6.4), partly reflecting the topographic constraints (i.e., there are no large trees in the channel).

The transition probabilities among the seven stand types were calculated from the 1987 and 1997 census data (Fig. 6.5). The values in the figure show the probability that a cell of a particular stand type would change into a different stand type in this 10-year period. Higher probabilities are indicated by thicker lines. The major fluxes show that structural development begins at YS and tends to undergo a change to YA, DI, DA, and OA in that order. Stands DS and OS can be understood as a kind of "side trip". Assuming that these transition probabilities will persist, future stand dynamics can easily be projected using a simple matrix model (Fig. 6.6). This analysis showed that visible changes would cease in about 300 years, and the forest structure would reach equilibrium in about 500 years from now. Although OFR is currently dominated by developing and young stands, when it reaches the projected

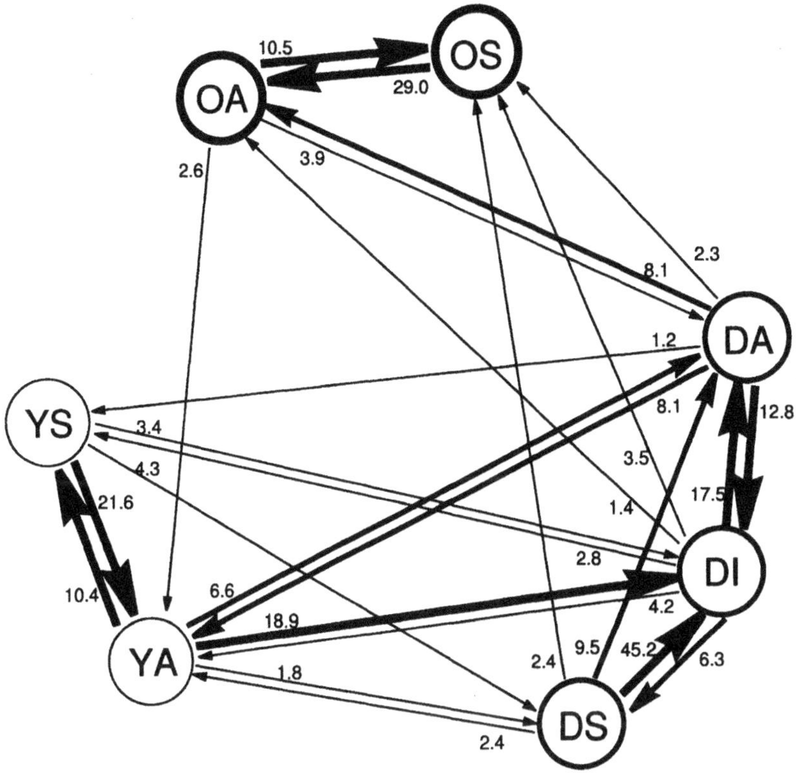

Fig. 6.5. Transitions between different pairs of structural stand types observed over a 10-year period, shown by *arrows* (directions) and figures (probabilities (%)). Older stands are marked by *thicker circles,* and higher probabilities are shown by *thicker arrows*

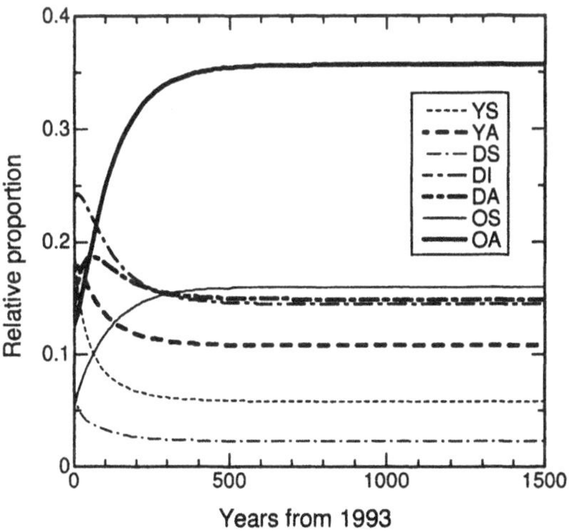

Fig. 6.6. Simulation of future changes in the relative proportions of seven structural stand types in OFR. *Solid lines,* old stands; *dotted* and *dashed lines,* developing stands; *dashed lines,* young stands. The thickness of the lines is in reverse proportion to the SDI

equilibrium, it will be dominated by old stands and the proportion of young, DS, and DI stands will decrease.

Thus, the total structural condition of OFR is not currently in a stable state. The disproportionately dominant young stands are now decreasing. Tanaka and Nakashizuka (1997) presented similar results after analyzing canopy height dynamics during a 15-year period from 1976 to 1991 (see also Chapter 7). During those years, the number of sites with lower canopies (≤ 15 m) continued to decrease, and the number of sites with higher canopies (> 15 m) continued to increase.

6.5 Structural Condition and Species Distribution

The mean number of species within each stand type was compared with that of the whole OFR as a function of the sample areas (log m^2), and their differences were tested by ANCOVA (Fig. 6.7). Among the seven stand types, YS, YA, and DA showed significantly higher species accumulation rates than did the whole OFR. For YA and DA, this difference might be due to higher densities of stems because the species accumulation rate on a stem basis was almost the same as that of the whole OFR. However, this was not the case for YS because its stem density was significantly lower than that for the whole OFR. Young stands (YA + YS) showed significantly higher rates of species accumulation than did the whole OFR, while developing (DA + DI + DS) and old (OA + OS) stands did not.

The distribution bias of each species among young, developing, and old stands was tested by χ^2-test (Table 6.1). Many of the TC and TU species showed distributions that were significantly biased toward developing or old stand types (i.e., the

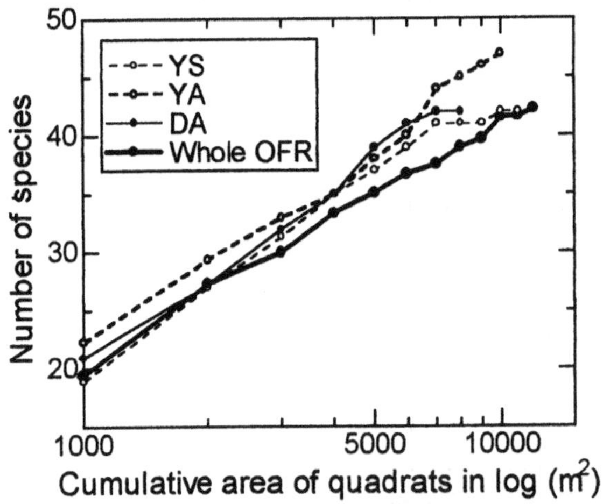

Fig. 6.7. Mean number of species in the whole OFR and in the DA, YA, and YS stands as a function of sample areas (log m^2). DA, YA, and YS were significantly different in species richness from the total reserve (ANCOVA, $P < 0.05$). The other four stands (DS, DI, OS, OA) did not show significant differences and are not shown in this figure

stem densities of these species were higher than expected in these stand types). Although more than half of the TU species showed no significant trends in distribution pattern, that pattern appeared to be biased toward developing or old stands. On the other hand, many of the IC species showed a significant bias toward young stands. IU species (*Acer rufinerve*), contrary to expectation, showed a significant bias toward developing and old stands.

In the context of the habitat guilds, species which constitute a "climax" patch (i.e., guilds A, Bc, Ca; 13 species) all showed distributions biased toward developing or old stands (and eight of these were significant). Species depending on moderate disturbances i.e., guilds Bb (seven species) and Cb (three species), showed biases toward developing stands (three were significant) and young stands (two were significant), respectively. Most of the "large-scale disturbance specialists," i.e., guild Ba (7 species), showed a distribution biased toward young stands (and four were significant).

Although it is impossible to test the distribution of occasional species owing to small sample sizes, most of them occurred in young and developing stands with higher densities than expected. Therefore, the higher rate of species accumulation in young stands should be due to IC species ("large-scale disturbance specialists") and occasional species.

6.6 Why Is OFR High in Species Richness?

OFR currently includes a large number of tree species. This is partly because of habitat differentiation among the species guilds, which is in contrast to the lesser importance of habitat differentiation among species in a tropical forest with a much higher species richness (Hubbell and Foster 1986). Other analyses of OFR reported in this chapter suggest that species richness is due to the existence of abundant shade-intolerant species, which are thought to have become established as young stands after recent large-scale disturbances. In addition, young stands contribute greatly to the local species richness by fostering occasional species.

However, the Markovian projection of species composition suggested that dominant and subdominant IC species will have been eliminated in the future community. The analysis of structural conditions also supported this result: the projection of the future stand structure of OFR suggested that these young stands become less important. Thus, the current species richness in OFR depends largely on the presence of "large-scale disturbance specialists." This mechanism contrasts with that in tropical forests, which are mostly composed of generalist species with respect to the light environment (Welden et al. 1991).

A patchy disturbance regime often causes a more equilibrated species composition (e.g., Lusk and Smith 1998; Stewart and Rose 1990). OFR currently appears to be shifting toward a state dominated by shade-tolerant species. In such a state, it will be difficult for shade-intolerant species to maintain their populations because the light is at a low level and limited to a narrow range when the gap size is small (Busing and White 1997). Yet, when large disturbances occur, partitioning of the habitat along a wide gap–shade gradient might occur among tree species (Brokaw 1987), allowing the establishment of IC species and occasional species within the community. However, if the species richness in this forest is maintained, it will be in a nonequilibrium manner (e.g., Chesson and Case 1986; Hubbell 1979) through species coexistence by disturbance-mediated environmental fluctuations. The cycle of these fluctuations will be long enough to allow the existence of shade-tolerant species, and short enough to allow the existence of shade-intolerant and occasional species. This is consistent with the "intermediate-disturbance hypothesis" (Connell 1978).

Uchiyama (1990) reported that *Quercus* species, the species that are most representative of IC species, have been increasing in this district during the last 1,000 years. This corresponds to anthropogenic history (see Chapter 4). Therefore, it is likely that past human activity, together with natural forces, are responsible for the "intermediate-disturbance" pattern in and around OFR (cf. Nakashizuka and Iida 1995). Monitoring future changes in composition and disturbance events, including man-made disturbances, will provide useful data for testing the above hypothesis about the mechanisms affecting community organization in OFR.

References

Abe S, Masaki T, Nakashizuka T (1995) Factors influencing sapling composition in canopy gaps of a temperate deciduous forest. Vegetatio 120:21–32

Brokaw NVL (1987) Gap-phase regeneration of three pioneer tree species in a tropical forest. J Ecol 75:9–19

Busing RT, White PS (1997) Species diversity and small-scale disturbance in an old-growth temperate forest: a consideration of gap partitioning concepts. Oikos 78:562–568

Christensen NL, Peet RK (1984) Convergence during secondary forest succession. J Ecol 72:25–36

Chesson PL, Case TJ (1986) Overview: nonequilibrium community theories: chance, variability, history, and coexistence. In: Diamond J, Case TJ (eds) Community ecology. Harper & Row, New York, pp 229–239

Connell JH (1978) Diversity in tropical rain forests and coral reefs. Science 199:1302–1310

Hara T, Nishimura N, Yamamoto S (1995) Tree competition and species coexistence in a cool-temperate old-growth forest in southwestern Japan. J Veg Sci 6:565–574

Hubbell SP (1979) Tree dispersion, abundance, and diversity in a tropical dry forest. Science 203:1299–1309

Hubbell SP, Foster RB (1986) Biology, chance, and history and the structure of tropical rain forest tree communities. In: Diamond J, Case TJ (eds) Community ecology. Harper & Row, New York, pp 314–329

Lusk CH, Smith B (1998) Life history differences and tree species coexistence in an old-growth New Zealand rain forest. Ecology 79:795–806

Masaki T, Suzuki W, Niiyama K, Iida S, Tanaka H, Nakashizuka T (1992) Community structure of a species-rich temperate forest, Ogawa Forest Reserve, central Japan. Vegetatio 98:97–111

Masaki T, Tanaka H, Tanouchi H, Sakai T, Nakashizuka T (1999) Comparative study of structure, dynamics and disturbance regime of five temperate broad-leaved forests in Japan. J Veg Sci 10:805–814
Nakashizuka T, Iida S (1995) Composition, dynamics and disturbance regime of temperate deciduous forests in Monsoon Asia. Vegetatio 121:23–30
Nakashizuka T, Numata M (1982a) Regeneration process of climax beech forests. I. Structure of a beech forest with the undergrowth of *Sasa*. Jpn J Ecol 32:57–67
Nakashizuka T, Numata M (1982b) Regeneration process of climax beech forests. II. Structure of a forest under the influences of grazing. Jpn J Ecol 32:473–482
Nakashizuka T, Iida S, Tanaka H, Shibata M, Abe S, Masaki T, Niiyama K (1992) Community dynamics of Ogawa Forest Reserve, a species-rich deciduous forest, central Japan. Vegetatio 103:105–112
Parker GR, Leopold DJ, Eichenberger JK (1985) Tree dynamics in an old-growth, deciduous forest. For Ecol Manag 11:31–57
Peterken GF, Jones EW (1987) Forty years of change in Lady Park Wood: the old-growth stands. J Ecol 75:477–512
Stewart GH, Rose AB (1990) The significance of life history strategies in the developmental history of mixed beech (*Nothofagus*) forests, New Zealand. Vegetatio 87:101–114
Swain MD, Whitmore TC (1988) On the definition of ecological species groups in tropical rain forests. Vegetatio 75:81–86
Tanaka H, Nakashizuka T (1997) Fifteen years of canopy dynamics analyzed by aerial photographs in a temperate deciduous forest, Japan. Ecology 78:612–620
Uchiyama T (1990) Palynological studies of alluvial sediments in the mid-temperate zone of Japan. II. Northeastern area of the Tohoku district. Jpn J Palynol 36:17–32
Welden CW, Hewett SW, Hubbell SP, Foster RB (1991) Sapling survival, growth, and recruitment: relationship to canopy height in a neotropical forest. Ecology 72:35–50

7 Disturbance Regimes

Tohru Nakashizuka

7.1 Disturbance Regimes and Forest Ecosystems

The effects of disturbances on forest ecosystems include many aspects with different scales, from the ecophysiology of individual organs to ecosystem functions or further landscape-level processes (Pickett and White 1985). Most plant species are adapted to some kind of disturbance. Pioneer or gap-phase species require safe sites that are created by disturbances (see Part 4), while even a shade-tolerant species may respond to the change in light conditions after a gap has been created (see Part 5). Changes in tree composition trends are predicted to be greatly dependent on the present and assumed future disturbance regimes (see Chapter 6). The investigation of disturbance regimes, therefore, gives information necessary to understand the structure and functions of populations and communities.

In Ogawa Forest Reserve (OFR), the dominant causes of disturbance are tree-fall gap creation and fire. Disturbance regimes resulting from these two causes are different in quality, distribution, intensity (spatial scale), and frequency (temporal variation). In this chapter, I describe each of these disturbance regimes, referring to their influence on forest structure and dynamics.

7.2 Tree-Fall Gaps

A tree-fall gap is a disturbance commonly seen in many mature forests, from the tropical to the boreal zone (Brokaw 1985; Runkle 1982, 1992; Kanzaki 1984; Liu and Hytteborn 1991; Yamamoto 1989, 1992). Many reports on gap disturbance regimes in temperate forests have been published, but most of them are based on a one-time census and little information exists on dynamic features of gap disturbances. In OFR, canopy gap creation in the forest has been investigated by several methods, including repeated measurements.

Ecological Studies, Vol. 158
Nakashizuka, Matsumoto (eds.) Diversity and Interaction in a Temperate Forest Community: Ogawa Forest Reserve of Japan

7.2.1 Census Method for Canopy Gap Investigation

Gap formation has been studied mainly by inventorying individual canopy gaps ("census method"). In this method, canopy gaps are recorded when they are observed on a census line or in a census area. Although very large areas or long lines are required for the reliable estimation of the gap formation regime, this method is rather convenient for counting large numbers of gaps with the least field labor (Runkle 1992). However, this method has some problems: (1) What constitutes a gap may be defined differently for different forest types or by different researchers; (2) Small or old gaps may be overlooked; and (3) it is difficult to obtain comparable data from repeated censuses (Nakashizuka et al. 1995).

This method was not applied to the investigation of canopy gap disturbance regimes in the OFR, but it was used once to investigate species composition of seedlings and saplings and their distribution patterns with respect to age, size, and topographic location of the gap (Abe et al. 1995). In that study, almost all the gaps in and around a 6-ha plot were surveyed and their ages estimated. The ages of the gaps were determined by analyses of the annual rings of the saplings or subcanopy trees in and around the gaps (Chapter 12, Abe et al. 1995).

7.2.2 Canopy Profile Method

Canopy structure can also be investigated by drawing a canopy profile ("profile method") using data such as those obtained from a digital elevation model (DEM) of the canopy surface. This method was developed by Karr (1971), who called it the "vegetation profile diagram method," and it has been applied in tropical forests (Hubbell and Foster 1986; Brokaw and Grear 1991) and as well as in the OFR (Nakashizuka et al. 1995; Tanaka and Nakashizuka 1998).

Data for DEMs in the OFR can be obtained either from ground observations or from aerial photographs. In the first case, the canopy height is measured at the center of every 5 × 5 m subquadrat in a permanent 6-ha plot, as was done in the forest community on Barro Colorado Island (Hubbell and Foster 1986). The canopy height up to 15 m was measured with a pole, and categorized into 5 height classes (0–2, 2–5, 5–10, 10–15, and over 15 m, up to the maximum canopy height in this forest, about 35 m). Measurements have been performed in the spring (March to May, before leaf flush) every second year since 1991.

In the second case, the digital elevation of the canopy surface is obtained from aerial photographs taken in summer time. In a deciduous forest such as OFR, a DEM of the ground surface can also be obtained from aerial photographs taken in winter, when the trees do not have leaves. The difference between the elevation of the canopy surface and that of the ground surface at any given point gives the canopy height at that point (Fig. 7.1). The accuracy of this method was checked by comparing the DEM obtained by using ground observations in the permanent 6-ha plot (Nakashizuka et al. 1995) with that obtained for the same area by using aerial photographs. Another advantage of using aerial photographs is that aerial photographs taken in the

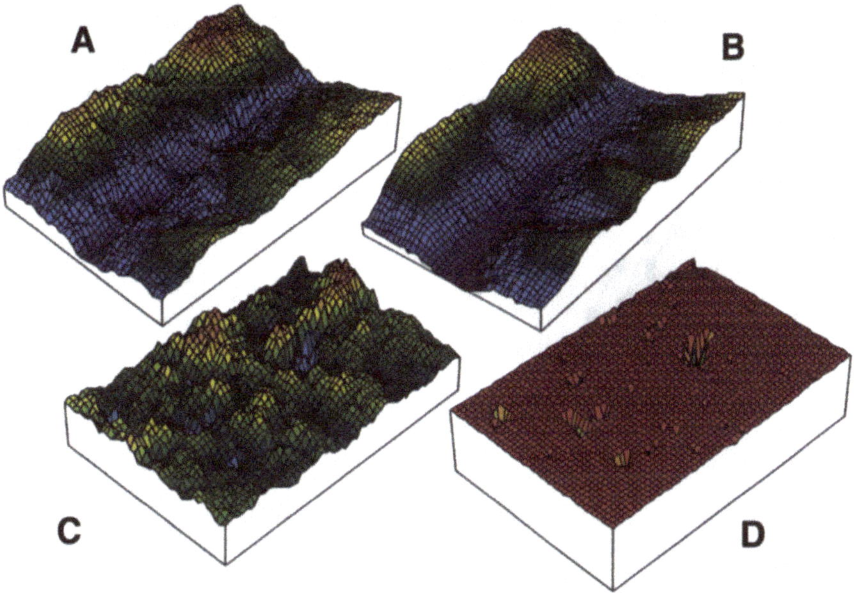

Fig. 7.1. The method used to obtain data for the digital elevation model (DEM) of canopy height in the 6-ha plot in OFR. The DEMs for canopy (**A**) and ground surface (**B**) were obtained from aerial photographs taken in summer and winter, respectively. The difference in elevation between the two DEMs gives the DEM of canopy height (**C**). If we define a canopy gap an area with a canopy no higher than a given height (here, 15 m), we can obtain the distribution of the canopy gaps (**D**). Modified after Nakashizuka et al. (1995)

past can also be used. Because OFR is a National Forest of Japan, aerial photographs have been taken every five years. The photographic resolution, however, was not high enough before 1970 for this kind of analysis. Therefore, DEMs were obtained for the canopy for 1976, 1981, 1985, and 1991. (Tanaka and Nakashizuka 1998)

In the profile method, gaps can be defined as subquadrats in which the canopy is no higher than a given height. Compared with the census method, the profile method recognizes gaps much more accurately and is more reliable. Furthermore, repeated measurements can be made by different observers.

7.2.3 Gap Size Distribution

Gap size distribution suggests the intensity and the scale of tree-fall disturbance. A gap was defined as an area with a canopy no higher than 10 m, which is similar to definitions adopted for other temperate forests, and the figures can be compared with data obtained by the census method (Nakashizuka et al. 1995).

Gap size distribution in OFR (Fig. 7.2) is quite similar to that of other temperate deciduous forests (Runkle 1982; Nakashizuka 1987; Yamamoto 1989; Hiura 1995) and temperate evergreen forests (Yamamoto 1992). The relative area occupied by

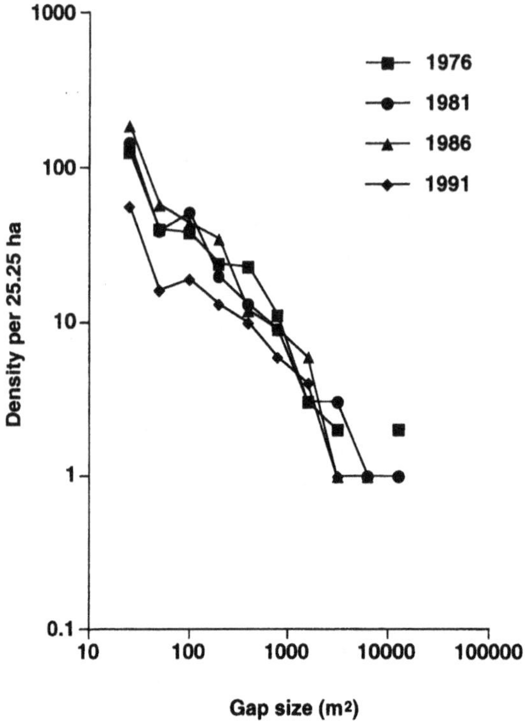

Fig. 7.2. Gap size distributions in a 25.25 ha area in OFR. The gaps were detected by using aerial photographs taken in the past. Gap size is plotted on a logarithmic scale. Modified after Tanaka and Nakashizuka (1998)

gaps (canopy height < 15 m) was 6.4 % (by ground measurements in the 6-ha plot) or 4.8 % (by aerial photographs in the area of the 6-ha plot) in 1991. The maximum size of the gaps was about 1 ha, rather larger than has been found in other forests, partly because of past fire disturbance, which we describe in detail below. Another reason for the large gaps is the presence of the stream. A few large gaps were distributed along the stream, creating relatively large areas (up to about 800 m^2) without any trees. In general, there were many gaps in the small size classes, and the number decreased as the gap size increased.

At any given time, a gap may not be the size it was originally, when it was first formed, because it might have been formed by several episodes of tree falls. The maximum size of gaps newly created during 5-year intervals was smaller (usually less than 500 m^2), reflecting the actual size of gaps at the time of formation (Fig. 7.3).

Size class distributions of gaps defined by different canopy height limits closely follow a power function model, suggesting the fractal nature of the canopy structure (Fig. 7.4). This trend is similar to that shown by data from a tropical forest on Barro Colorado Island, where the profile method was applied to a 50-ha area (Hubbell and Foster 1986).

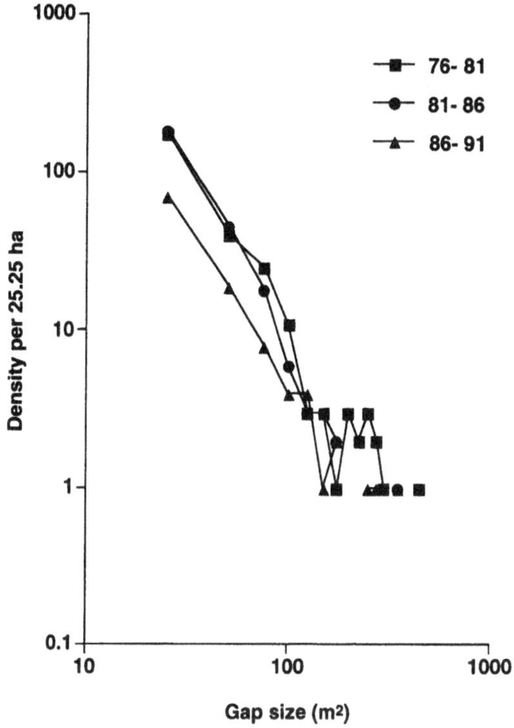

Fig. 7.3. Size distributions of newly created canopy gaps. The new gaps created during each 5-year interval were detected by identifying differences in canopy height at the beginning and end of each interval. Modified after Tanaka and Nakashizuka (1998)

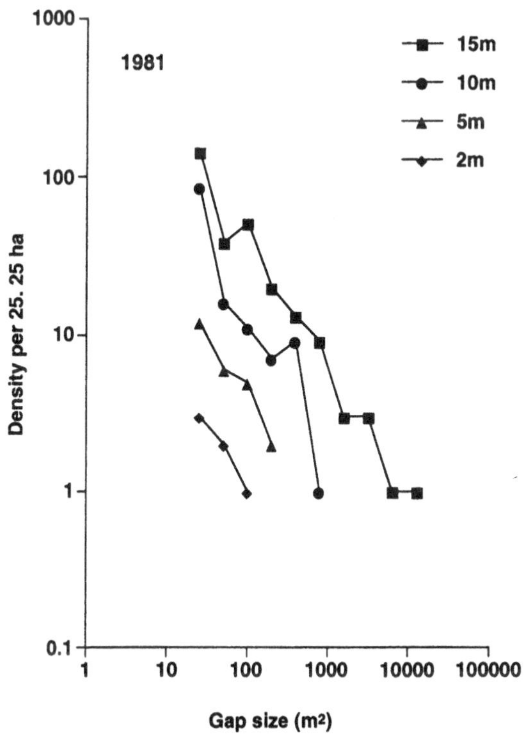

Fig. 7.4. Size distribution of gaps defined for different canopy heights. Canopy gaps defined as the area with canopy trees no higher than 15, 10, 5, and 2 m are shown. Modified after Tanaka and Nakashizuka (1998)

7.2.4 Causes of Gaps

A survey of 55 gaps in and near the 6-ha plot found that the gaps had been caused by broken trunks (38%), uprooted trees (34%), fallen branches (13%), and standing dead trees (16%) (Abe et al. 1995), suggesting that the main cause of tree falls in the OFR is strong winds, particularly those associated with autumn typhoons and very low pressure weather systems in late winter (Abe et al. 1995). A similar pattern of causes has been reported for other temperate deciduous forests in Japan (Nakashizuka 1987; Yamamoto 1989).

7.2.5 Rates of Gap Formation and Closure

The rates of gap formation and closure observed by the profile method with ground-based observations were 0.3%–0.9% and 0.9%–1.5% yr^{-1}, respectively, from 1991 to 1997, and those obtained with aerial photographs were 0.4%–1.1% and 0.3%–1.1% yr^{-1}, respectively, from 1976 to 1996 (Fig. 7.5). Based on these figures, the turnover time of the canopy, which is equivalent to the mean disturbance recurrence time, is 90–330 years. These values are comparable with those estimated for other temperate

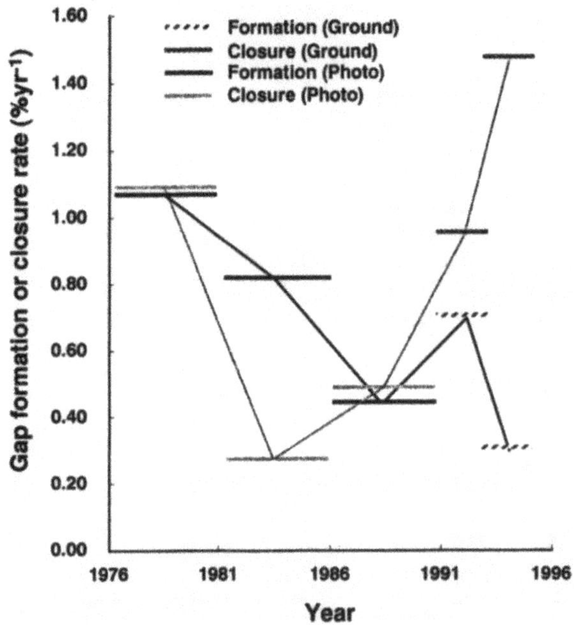

Fig. 7.5. Changes in the rates of formation and closure of canopy gaps. Canopy gaps (area with canopy no higher than 15 m) were studied by using aerial photographs taken between 1976 and 1991, and by ground observations made between 1991 and 1994. Gap formation and closure were in balance in 1976, although more recently, the closure rate has exceeded the formation rate

Table 7.1. Canopy turnover rate, population turnover of canopy trees (DBH > 10 cm), and biomass turnover rate (in basal area) in deciduous forests in Japan (after Nakashizuka and Iida 1995)

		Tree density		Basal area	
	Gap formation rate (% yr^{-1})	Recruitment (% yr^{-1})	Mortality (% yr^{-1})	Gain (% yr^{-1})	Loss (% yr^{-1})
Moriyoshi, Akita[a]	*0.32–0.62*[e]			*1.35*[e]	*0.97*[e]
Kayanodaira, Nagano[b]		0.31	0.42	0.84	0.62
Ohdaigahara, Nara[c]	0.31 *0.41–0.82*[e]	0.10	0.74	0.61	0.68
Ogawa, Ibaraki[d]	0.42	1.19	1.20	1.12	0.88

[a]Nakashizuka (1984).
[b]Watanabe (1993).
[c]Nakashizuka (1991).
[d]Nakashizuka et al. (1992).
[e]Estimated from the forest structure.

Table 7.2. Transitions between canopy height categories from year *t* to year *t* + 5

		Canopy Height Class in year *t* + 5 (m)				
		0–2	2–5	5–10	10–15	15–
Canopy Height Class in year *t* (m)	0–2	0	0.319	0.389	0.148	0.144
	2–5	0.030	0.197	0.338	0.254	0.181
	5–10	0.003	0.027	0.194	0.529	0.247
	10–15	0.001	0.003	0.034	0.470	0.492
	15–	0.001	0.001	0.006	0.039	0.953

forests, and with the biomass turnover time estimated by gain and loss of basal area in the OFR (Table 7.1).

The probability of transitions between canopy height classes during the observation period can be summarized by a probability matrix (Table 7.2), which shows that high canopies tended to be stable and low canopies changed rapidly. This trend indicates that the gaps were quickly filled, thus moving into higher canopy height classes. The eigenvector of the matrix gives the equilibrium height distribution of the canopy height in the OFR, assuming that the transition probabilities during the observation period did not change. The distributions obtained were slightly different between matrices based on ground observations and those based on aerial photographs. Matrices based on ground observations included a lower frequency of low canopy height classes than those based on aerial photographs. The gap-area equilibrium percentages (canopy height < 10 m) were 9.5% and 5.6%, for the ground and aerial photograph methods, respectively. These values are not largely different from the present observation.

However, the longer-term data obtained from aerial photographs show that there were large temporal variations in gap formation and closure. In the late 1970s to early 1980s, the gap formation rate was higher than the closure rate, although since then, until recently, gap formation has been very slow. This variation in the gap formation rate is also supported by gap census data from the 6-ha plot (age distributions determined by tree ring analyses), showing that many gaps were made before the early 1980s (Fig. 7.6). In the late 1970s and early 1980s, there are many records of strong winds over 25 m sec^{-1} in instantaneous velocity, while the frequency of such winds became quite low in the late 1980s and early 1990s (Abe et al. 1995).

Gap formation also varied spatially. Gaps tended to be created more frequently on the periphery of already existing gaps, and they frequently closed quickly at gap margins. The former suggests that the canopy trees adjacent to gaps are more prone to damage by strong winds, as estimated by BCI data (Hubbell and Foster 1986). The latter tendency indicates that gap closure occurs by rapid branch extension by the canopy trees adjacent to the gaps.

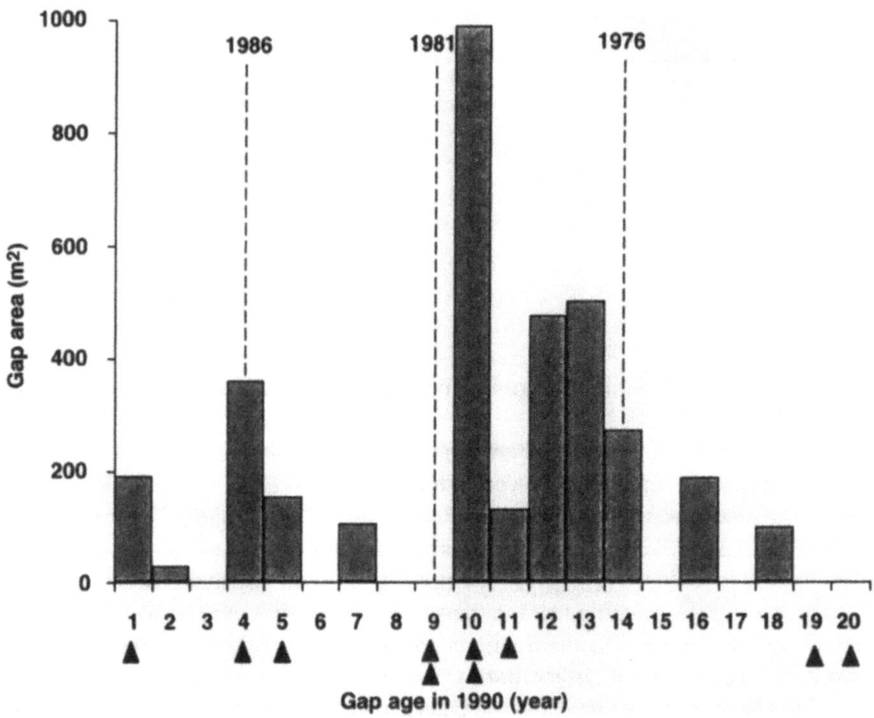

Fig. 7.6. The distribution of gap ages determined by annual ring analyses of trees and saplings growing in and around the gaps. Triangles indicate years during which strong winds (instantaneous velocity more than 20 m sec^{-1}) were recorded. Modified after Abe et al. (1995)

Thus, the disturbance regime attributed to tree fall is characterized by gaps up to about 500 m^2 in area caused by tree falls either of a single tree or of a few canopy trees, mostly from their trunks being broken by typhoons or strong winter winds, and the mean recurrence time is about 100–300 years.

7.3 Fire Disturbance

Fire has seldom been studied as a major cause of disturbance in the temperate deciduous forests of Japan, although fire is mentioned in a few very old reviews (e.g., Honda 1907). Although fires are now strictly controlled by the forest management policy, there is evidence that understanding fire disturbance is important for understanding the structure and dynamics of the OFR.

7.3.1 Forest Structure and Predicted Dynamics

There are some young stands dominated by *Quercus serrata* distributed mostly on the ridges in OFR. Analyses of the annual rings of trees in these stands showed that these trees mostly regenerated between 1900 and 1920 (Peters, unpublished data). The stands have a uniform structure, with all trees having a similar stem diameter (Masaki et al. 1992). Some species relatively abundant in the OFR, such as *Quercus* spp., *Carpinus* spp., and *Castanea crenata*, are known as dominants in the secondary forests in this region, indicating that these stands were established after large-scale disturbances (Chapter 4). They have large populations of adult (reproductive-age) trees, although there are a very few regenerated saplings in OFR. Predictions made by using Markovian transition matrices suggest that the dominance of *Quercus serrata* will diminish, assuming canopy replacement as indicated by the present forest structure (Masaki et al. 1992).

The dominant species, *Q. serrata*, has the ability to grow many sprouts from stumps of logged trees (Sakai and Sakai 1998), although there are few coppices created this way in OFR. This fact suggests that the established *Quercus* population in the OFR was not regenerated by resprouting after logging, but by seedlings germinated after some disturbance, probably fire.

Pollen analyses conducted in Kameyachi Bog, which is about 6 km southwest of OFR, showed that tree compositions similar to the present forest have existed continuously for the last 3000 years. Furthermore, ash or charcoal was found in almost every layer of the peat in the bog (Ikeda, personal communication). Charcoal fragments have also been found in the surface soil layers of the OFR (see Chapters 3 and 8). The land surrounding the OFR has been utilized as pasture for cattle and horses for a long time, and people continue to use fire to maintain the grasslands (see Chapter 4). Some of the large trees in OFR have fire scars visible in their annual rings. These facts all indicate that the forests in this region have been influenced by fire disturbances over a long time period.

7.3.2 Fire Disturbance Regime in OFR

The young stands on the ridges of the OFR sometimes are several to several tens of hectares in size (Fig. 4.2 in Chapter 4), indicating the spatial scale of fire disturbance. However, estimating the recurrence time is difficult, and most recent fires were caused by humans (Chapter 4). The largest example of *Q. serrata,* located near the valley bottom, is about 110 cm DBH (diameter at breast height), and is estimated to be 200–300 years old (the exact age was impossible to determine because of its hollow center). This suggests that the recurrence time is greater than 200–300 years in at least some parts of the forest. Probably, fire was more frequent on the ridges (recurrence time, several decades) than in the valley bottom (several hundred years), as discussed in Chapter 4 of this volume. Some fires may have burned only along the ground, as suggested for *Quercus* forests in North America (Abrams 1992). Most fires do not seem to have killed the more fire-resistant species.

7.4 Geographical Configuration of the Disturbance Regimes

The disturbance regimes of the OFR can be discussed in the geographical context of the deciduous forests in Japan and eastern Asia. Temperate deciduous forests have been classified by several researchers in different ways, but mainly they fall into three types in eastern Asia (Table 7.3). Here, I follow the definitions suggested by Kira (1991) and in this Chapter.

The three forest types are different both with respect to climate and species composition (Ching 1991; Yim 1977; Nozaki and Okutomi 1990). The Warm Temperate Deciduous Forest (WDF) is characterized by the dominance of *Quercus* and *Carpinus* species in a climate with a dry winter. This forest type is distributed in southern continental China, the Korean Peninsula, and the eastern (Pacific Ocean) side of the main range in Japan. The Cool Temperate Mixed Broadleaf/Coniferous Forest (CMF) is characterized by the dominance of *Quercus*, *Acer,* and some coniferous species, and is distributed in areas with relatively low temperatures and dry winters such as northern continental China and Hokkaido. The Cool Temperate Deciduous Forest (CDF) is characterized by the overwhelming dominance of *Fagus*, and is distributed only in areas with a humid climate throughout the year and heavy snowfall. The main region in eastern Asia with this forest type is the western (Sea of Japan) side of the main range in Japan.

The disturbance regimes of the three forest types are also different. All three forest types have disturbances cause by tree falls, particularly in mature stands, while WDF and CMF, which are found in regions having a dry winter climate, show fire disturbances as well. Actually, fire records from 1945 to 1950 (Inoue 1950) show a surprising coincidence with the distribution of WDF on Honshu Island (Fig. 7.7). In this climatic area, fires usually occur at the end of the dry winter (February–April). On the other hand, large-scale blowdowns of trees have been reported only for CMF

Table 7.3. Temperate deciduous forests in monsoonal Asia (after Nakashizuka and Iida 1995)

China	Korea	Japan		Whole area
Hou (1983)	Yim (1977)	Yoshioka (1973)	Nozaki and Okutomi (1990)	Kira (1991)
Mixed coniferous and deciduous broad-leaved forest	Northern deciduous broad-leaved forest	Boreal mixed coniferous and deciduous broad-leaved forest	Upper-temperate forest	Cool-temperate mixed deciduous broadleaf/conifer forest
*	*	Temperate deciduous broad-leaved forest	*Fagus crenata* forest	Cool-temperate deciduous broadleaf forest
Deciduous broad-leaved forest	Central deciduous broad-leaved forest		Intermediate-temperate forest	Warm-temperate deciduous broadleaf forest
	Southern deciduous broad-leaved forest			

* Corresponding forest type is not found in this region.

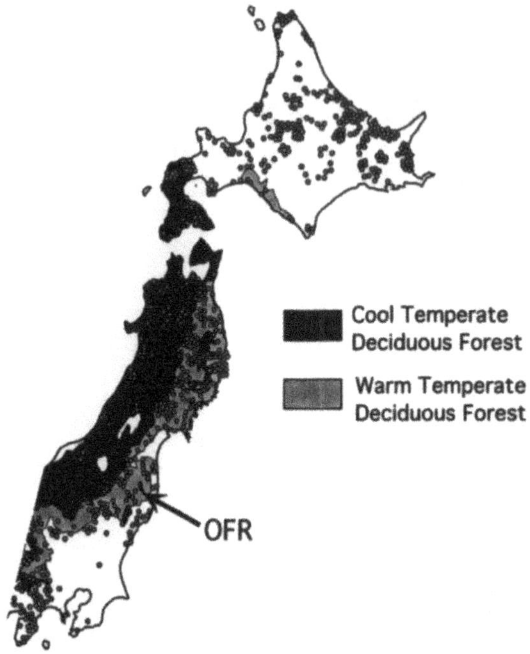

Fig. 7.7. Geographical distribution of forest fires that burned more than 10 ha between 1945 and 1950. The distributions of Cool-Temperate and Warm-Temperate Deciduous Forests are also shown. Modified after Nakashizuka and Iida (1995)

Table 7.4. Summary of deciduous temperate forest types in eastern Asia (modified after Nakashizuka and Iida 1995)

	Warm-temperate	Cool-temperate	
	Deciduous broadleaf forest	Deciduous broadleaf forest	Mixed broadleaf/ conifer forest
Dominants	*Quercus* *Carpinus*	*Fagus*	*Quercus* *Acer* Conifers
Species richness	high	low	high
Disturbances	Tree-fall gap Fire	Tree-fall gap	Tree-fall gap Fire Big blowdown

(Tamate et al. 1977; Watanabe et al. 1990). The species composition and disturbance regimes of the three forest types are summarized in Table 7.4. The forest type in OFR is WDF, being dominated by *Quercus* and *Carpinus* with fewer *Fagus* than a typical CDF and having dry winters with little snow.

The associations between climate and disturbance regimes for the forest types are analogous to those found in North America (Runkle 1990). In North America, forests dominated by *Fagus* are distributed only in areas with a humid climate, and the predominant cause of disturbance is tree fall. In drier areas, the dominance of *Fagus* decreases, and *Quercus* becomes more dominant. Fire plays an important role in these areas. Some conifers mix with hardwood species in the Northern Hardwood Forest, and all three kinds of disturbances, tree fall, fire, and blowdown, are important there.

In summary, the structure and species composition of forests are closely related to the disturbance regimes as well as to the climate. Disturbances play an important role in the processes of population, community, and ecosystem as described in subsequent chapters of this volume.

References

Abe S, Masaki T, Nakashizuka T (1995) Factors influencing sapling composition in canopy gaps of a temperate deciduous forest. Vegetatio 120:21–32

Abrams MD (1992) Fire and the development of oak forests. Bioscience 42:346–353

Brokaw NVL (1985) Treefalls, regrowth, and community structure in tropical forests. In: Pickett STA, White PS (eds) The ecology of natural disturbance and patch dynamics. Academic, New York, pp 53–69

Brokaw NVL, Grear JS (1991) Forest structure before and after Hurricane Hugo at three elevations in the Luquillo Mountains, Puerto Rico. Biotropica 23:386–392

Ching KK (1991) Temperate deciduous forests in East Asia. In: Rohrig E, Ulrich B (eds) Temperate deciduous forests. Elsevier, Amsterdam, pp 539–555

Hiura T (1995) Gap formation and species diversity in Japanese beech forests: a test of the intermediate disturbance hypothesis on a geographic scale. Oecologia 104:265–271

Honda S (1907) Zonation of forest plants in Japan (in Japanese). Miura Shoten, Tokyo

Hou H (1983) Vegetation of China with reference to its geographical distribution. Ann Mo Bot Gard 70:509–548

Hubbell SP, Foster RB (1986) Canopy gaps and the dynamics of a neotropical forest. In: Crawley M J (ed) Plant ecology. Blackwell, Oxford, pp 77–96

Inoue K (1950) Statistic analyses on forest fires in Japan (in Japanese). Forest Agency, Japan

Kanzaki M (1984) Regeneration in subalpine coniferous forests. I. Mosaic structure and regeneration process in a *Tsuga diversifolia* forest. Bot Mag Tokyo 97:297–311

Karr JR (1971) Structure of avian communities in selected Panama and Illinois habitats. Ecol Monogr 41:207–232

Kira T (1991) Forest ecosystems of east and southeast Asia in a global perspective. Ecol Res 6:185–200

Liu Q, Hytteborn H (1991) Gap structure and regeneration in a primeval *Picea abies* forest. J Veg Sci 2:391–402

Masaki T, Suzuki W, Niiyama K, Iida S, Tanaka H, Nakashizuka T (1992) Community structure of a species-rich temperate forest, Ogawa Forest Reserve, central Japan. Vegetatio 98:97–111

Nakashizuka T (1984) Regeneration process of climax beech (*Fagus crenata* Blume) forests IV. Gap formation. Jpn J Ecol 34:75–85

Nakashizuka T (1987) Regeneration dynamics of beech forests in Japan. Vegetatio 69:169–175

Nakashizuka T (1991) Population dynamics of coniferous and broad-leaved trees in a Japanese temperate mixed forest. J Veg Sci 2:413–418

Nakashizuka T, Iida S (1995) Composition, dynamics and disturbance regime of temperate deciduous forests in Monsoon Asia. Vegetatio 121:23–30

Nakashizuka T, Iida S, Tanaka H, Shibata M, Abe S, Masaki T, Niiyama K (1992) Community dynamics of Ogawa Forest Reserve, a species-rich deciduous forest, central Japan. Vegetatio 103:105–112

Nakashizuka T, Katsuki T, Tanaka H (1995) Forest canopy structure analyzed by using aerial photographs. Ecol Res 10:13–18

Nozaki R, Okutomi, K (1990) Geographical distribution and zonal interpretation of intermediate-temperate forests in eastern Japan (in Japanese with English summary). Jpn J Ecol 40:57–69

Pickett STA, White PS (1985) The ecology of natural disturbance and patch dynamics. Academic, New York

Runkle JR (1982) Patterns of disturbance in some old-growth mesic forests of eastern North America. Ecology 63:1533–1546

Runkle JR (1990) Gap dynamics in an Ohio *Acer-Fagus* forest and speculations on the geography of disturbance. Can J For Sci 20:632–641

Runkle JR (1992) Guidelines and sample protocol for sampling forest gaps. General Technical Report PNW-GTR-283. Department of Agriculture, Forest Service, Pacific Northwest Research Station. Portland, OR, USA

Sakai A, Sakai S (1998) A test for the resource remobilization hypothesis: tree sprouting using carbohydrates from above-ground parts. Ann Bot 82:213–216

Tamate S, Kashiyama T, Sasanuma T, Takahashi K, Matsuoka H (1977) On the distribution maps of forest wind damage by typhoon No. 15, 1954 in Hokkaido (in Japanese with English summary). Bull For Exp Sta 289:43–67

Tanaka H, Nakashizuka T (1998) Fifteen years of canopy dynamics analyzed by aerial photographs in a temperate deciduous forest, Japan. Ecology 78:612–620

Watanabe R (1993) Forest structure of Kayanodaira beech forest of the Institute of Natural Education, Shinshu University. 2. Growth of forest trees within a period (1982–1992) (in Japanese with English summary). Bull Inst Nat Educ Shiga Height 30:33–41

Watanabe S, Shibata S, Kawahara S, Shibano S, Kurahashi A, Satoo Y, Anazawa C, Takada N, Takahashi Y (1990) A memoir on the actual situation of the forest wind-damaged by typhoon No. 15 in 1981 in the Tokyo University Forest in Hokkaido (in Japanese). Miscellaneous information, Tokyo University Forests 27:79–221

Yamamoto S (1989) Gap dynamics in climax *Fagus crenata* forests. Bot Mag Tokyo 102:93–114

Yamamoto S (1992) Gap characteristics and gap regeneration in primary evergreen broad-leaved forests of western Japan. Bot Mag Tokyo 105:29–45

Yim YJ (1977) Distribution of forest vegetation and climate in the Korean Peninsula. IV. Zonal distribution of forest vegetation in relation to thermal climate. Jpn J Ecol 27:269–278

Yoshioka K (1973) Plant geography (in Japanese). Kyoritsu Shuppan, Tokyo

8 Effect of Soil Conditions on the Distribution and Growth of Trees

TAKASHI MASAKI, YOJIRO MATSUURA, and MASAMICHI TAKAHASHI

8.1 Introduction

Soil conditions and/or landforms have critical effects on the growth (Ricklefs 1996; Tilman 1988; Tokuchi et al. 1999) and distribution of plants (Nagamatsu and Miura 1997; Whittaker 1967). They affect plant biomass and the species composition of the plant community by controlling vegetation development. For forests, many studies have suggested that there is a close relationship between soil conditions and the vegetation pattern. As an example, Tilman (1988) suggested that the nitrogen mineralization rate affected the species composition of forests. Soil moisture, which is correlated with topography or landforms, also affects the species composition of forests (Bray and Curtis 1957; Whittaker 1967) and plant biomass (Tokuchi et al. 1999). However, in most of these previous studies, the study plots were of relatively small size and discontinuously sampled within a landscape or watershed. In the organization of the local community, spatial heterogeneity of soils or landforms should play an important role.

As described in Chapter 3, the Ogawa Forest Reserve (OFR) includes eight soil types (Bl_D, Bl_E, lBl_D(d), B_B, B_D(d), B_D, B_E, and G), based on *Classification of Forest Soil in Japan* (Forest Soil Division 1976), within a small area (i.e., 6 ha). These soil types correspond to particular landforms and show a very heterogeneous pattern of distribution. Masaki et al. (1992) found species assemblies, within which species were spatially correlated with each other, that appeared to reflect particular landforms (also see Chapter 6). In this chapter, we investigate the effects of soil type on forest structure, tree growth, and tree species distribution in more detail and in a more direct way than we did in past studies at the OFR by combining data on soil and landform distributions with tree census data.

8.2 Field Methods and Analyses

While the system of soil taxonomy proposed by the (U.S.) Soil Survey Staff (1992) well reflects the conditions of the soil surface and is useful for investigating the

Ecological Studies, Vol. 158
Nakashizuka, Matsumoto (eds.) Diversity and Interaction in a Temperate Forest Community: Ogawa Forest Reserve of Japan

interrelationship between the distribution of herbaceous species or seedlings and soil conditions, the soil classification system proposed by the (Japanese) Forest Soil Division (1976) well explains the distribution pattern and growth of trees of larger size than seedlings or saplings (Chapter 3).

Soil types were identified by comparison with reference soil profiles and related landforms (Chapter 3). The distribution of soil types within the 6-ha plot were confirmed by soil auguring at grid points (grid size, 10 m × 10 m, for a total of 651 (31 × 21) points). The 6-ha plot was subdivided into 2400 cells (each 5 m × 5m; only one corner of each cell was at a grid point). Each cell was characterized by the soil type identified at the nearest grid point. In the soil survey at grid points, however, one of the soil types, B_E, was not sampled. Therefore, the other seven types (Bl_D, Bl_E, lBl_D(d), B_B, B_D(d), B_D, and G) were used to designate the cells. The characteristics of each soil type are described in detail in Chapter 3.

Tree census data from 1987 was used to analyze tree distribution patterns, and data from 1997 was also used to analyze the diameter at breast height (DBH) of each stem. The methods used for the tree census are described in Chapter 5.

The basal area ($\Sigma(DBH^2 \pi / 4)$) and population size were calculated for each soil type and for each of the common species (a total of 31 species; see Chapter 6), and the differences between these values compared with the expected values were tested. We tested the statistical discrepancy between a randomly distributed artificial community and the actual one as follows. We constructed the artificial community with a random distribution of trees using the census data in 1987 and calculated the basal area and the population size for each soil type and for each species; this process was repeated 2000 times. On the basis of these calculations, the mean value and 95% confidence limits of the basal area and population size of each species on each soil type were estimated. Thus, expected ranges of the values based on the null hypothesis were obtained. The observed values were compared with these expected values to test whether the bias among soil types was significant.

The growth rate over a 10-year period was also calculated as the mean annual increment in DBH for each surviving tree by the equation $(DBH_{97}–DBH_{87}) / 10$, where DBH_{97} and DBH_{87} are the DBHs of a tree in 1997 and 1987, respectively. In order to eliminate the effects of light and size on the growth rate, two categories of trees were defined: canopy tree, that is, a tree whose crown reached the canopy layer and that was not overtopped by adjacent trees; and overtopped tree, that is, a tree whose crown did not protrude above the canopy layer and that was overtopped by adjacent trees. Thus, a canopy tree is a tree of larger size that receives abundant light ($n = 908$), and an overtopped tree is a tree of smaller size with low access to the light resource ($n = 2813$). Trees of smaller size that received abundant light (i.e., juveniles under the canopy gaps) were not included in this study because their number was too small ($n = 215$) to obtain significant results. In addition, trees that had experienced a change in the light environment during the 10-year period (e.g., gap formation above an overtopped tree) ($n = 486$) were also excluded from the analyses because it was difficult to distinguish the effects of light from the effects of the soil type on the growth rate of these trees. Thus, about 84% of the total number of trees of common species were used for the analysis of DBH growth. Variations in the growth rate among soil types were tested by ANOVA for each of the two categories (canopy trees and overtopped trees).

8.3 Distribution of Trees

The distribution of basal area and population size including all species (i.e., the whole community) reflected the soil distribution pattern in the OFR (Fig. 8.1a); both values were highest for soil type B_D, a moderately humid soil that was dominant in the OFR (Chapter 3). On the other hand, both values were lowest for soil type lBl_D(d), a moderately moist soil (but somewhat drier than soil type B_D), and the least common soil type in the OFR.

However, there were some significant biases in the distribution of basal area and population size. Observed values for the xeric soil types B_B and B_D(d) exceeded the expected range (Fig. 8.1a). On the other hand, observed values for the mesic soil types B_D and G were lower then expected. This result appears to be somewhat irregular considering that B_B soils were relatively deficient in available resources such as N and soil moisture compared with the other soil types (Chapter 3). This subject will be discussed later with the analyses of growth rates.

Masaki et al. (1992) recognized three assemblies of tree species: species distributed on the lower part of a slope (group A); those distributed on the upper part of a slope (group B); and those displaying little preference for a particular topographic position (group C). Species of groups B and C were further subdivided into three subgroups (a, b, c), probably reflecting the regeneration pattern (Masaki et al. 1992, also see Chapter 6).

According to the relationship between landform and soil (Chapter 3), the distribution of group A is expected to correspond to soil types Bl_D, Bl_E, B_D, and G, and that of group B is expected to correspond to soil types lBl_D(d), B_B, and B_D(d). In the following discussion, we examine the distribution of each group of tree species.

The results of the statistical tests are summarized in Table 8.1. The species of group A showed a significant preference for soil types B_D or G. They showed a negative preference for B_B and B_D(d) soils. Furthermore, most showed a negative preference for mesic sites with Black soil (Bl_D and Bl_E). Representative distribution patterns of basal area and population size are shown in Figs. 8.1b and c. *Carpinus cordata* showed a positive preference for B_D and less preference for the xeric sites B_B and B_D(d). *Acer tenuifolium* showed a more remarkable pattern; most of its population was distributed on soil type G, found at the site with the most abundant soil moisture, and none of it was distributed at the xeric sites with soil types B_B and B_D(d). Thus, species of group A mostly showed the expected distribution patterns in relation to soil types.

The species of group B also showed the expected distribution patterns, with a significant preference for soil types lBl_D(d), B_B, and B_D(d), and they showed less preference for the mesic sites of soil types Bl_D, Bl_E, B_D, and G. Representative distribution patterns are shown in Figs. 8.1d and e. The population size distribution of *Quercus serrata* had a peak at soil type B_D(d), while that of *Clethra barbinervis*

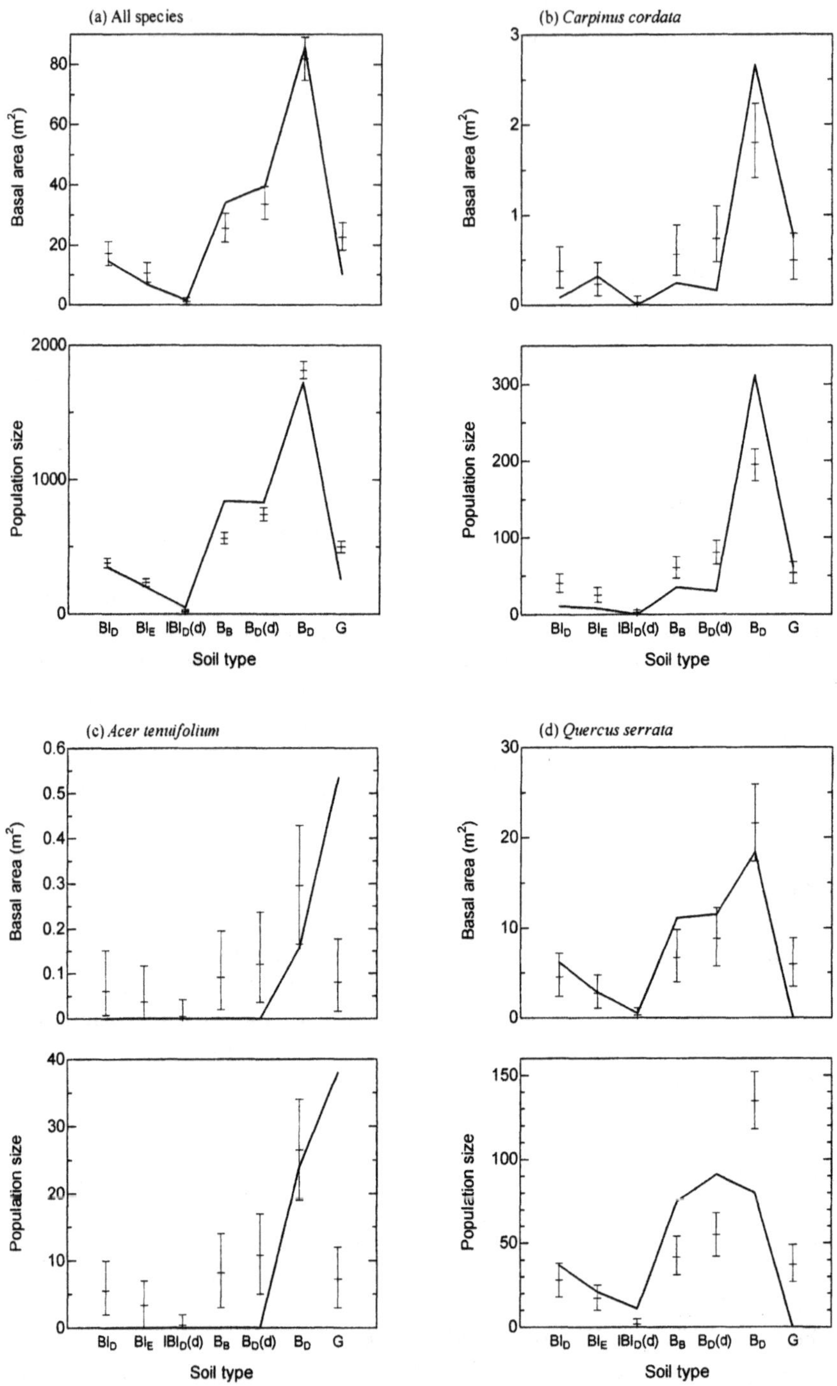

Fig. 8.1. The distribution of basal area (*upper panels*) and population size (*lower panels*) with respect to seven soil types: **a** all species inclusive; **b**, **c** species of group A; **d**, **e** species of group B; and **f**, **g** species of group C. *Solid lines* represent observed values and *vertical bars* represent the expected range of values. Identification of groups was by cluster analysis using spatial correlation among species (cf. Masaki et al. 1992). Soils are classified according to Forest Soil Division (1976)

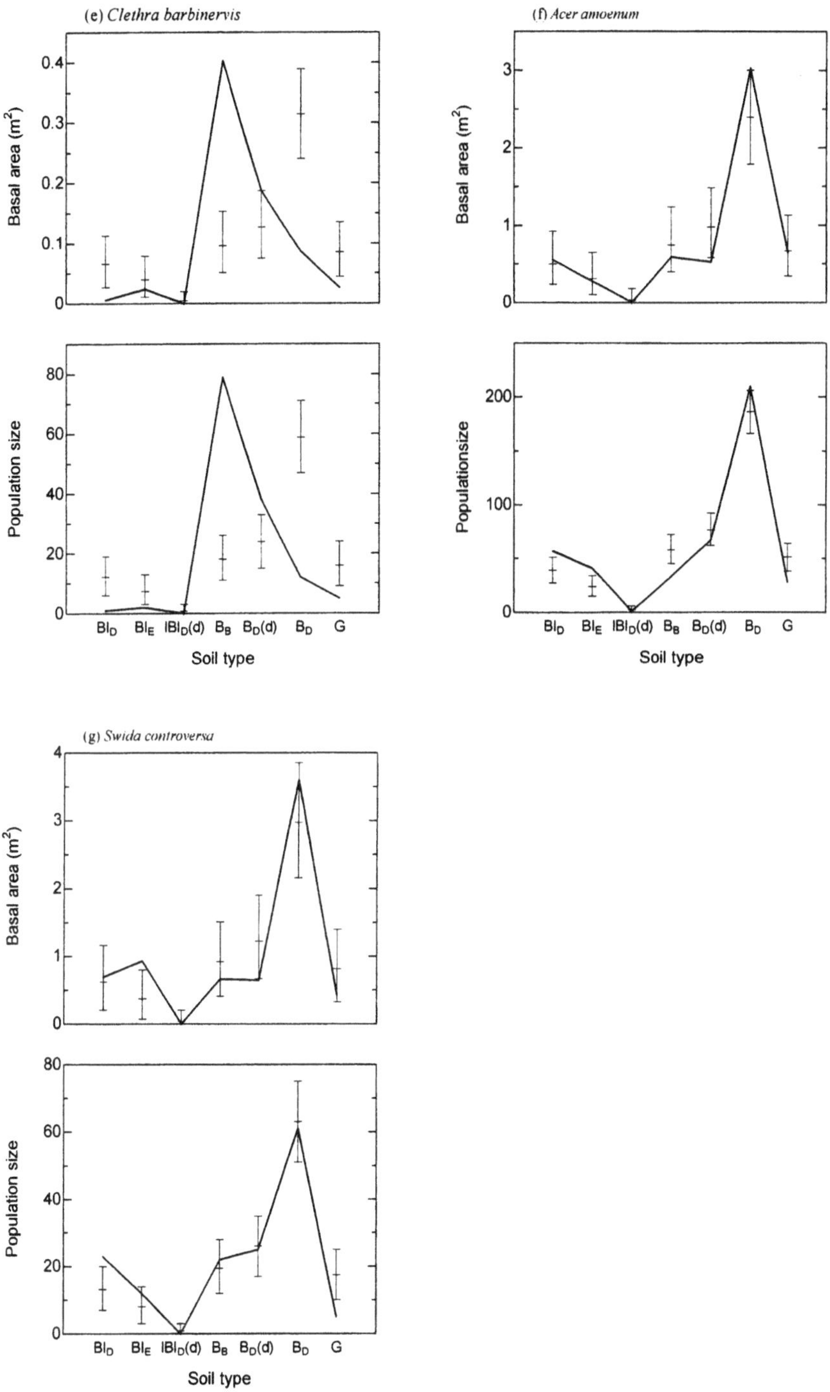

Fig. 8.1. *Continued*

Table 8.1. Biases of basal area and population size of 31 common tree species among seven soil types. The degrees of bias are shown by ++, +, –, and – –[a]

Group[b]	Species	Soil type						
		Bl_D	Bl_E	$lBl_D(d)$	B_B	$B_D(d)$	B_D	G
A	*Fagus japonica*	– –	– –				++	
	Carpinus cordata	– –	–		– –	– –	++	
	Acer tenuifolium	– –			– –	– –	–	++
	Meliosma myriantha		– –		– –		+	
	Hamamelis japonica							++
B-a	*Carpinus laxiflora*			+	++	++	–	– –
	Quercus serrata			+	++	+	–	– –
	Castanea crenata				++		–	– –
	Quercus crispula				+		–	– –
	Prunus verecunda				+	+	– –	– –
	Styrax obassia	+			++		– –	– –
	Styrax japonica						– –	+
B-b	*Sorbus alnifolia*				+		–	
	Betula grossa				+		–	
	Pieris japonica				++	+	– –	
	Fraxinus lanuginosa f. *serrata*	– –			++	+	– –	–
	Clethra barbinervis	– –	–		++	++	– –	– –
	Benthamidia japonica				+			– –
	Acer rufinerve	– –			+	++		– –
B-c	*Acer sieboldianum*							
	Sorbus japonica				+			
	Ilex macropoda			+	+	++	– –	
	Acanthopanax sciadophylloides	+		+	–	– –		
C-a	*Acer mono* var. *marmoratum* f. *dissectum*		–		– –	–	++	–
	Fagus crenata		–					– –
	Acer amoenum	+	+		–	–	++	–
	Carpinus japonica				+			
C-b	*Swida controversa*	+	+			–		–
	Carpinus tschonoskii					–		–
	Ostrya japonica				– –		+	
C-c	*Acer nikoense*				– –	– –	++	– –

[a] "+" (or "–") indicates that an observed value (either basal area or population size) was higher (or lower) than the expected range. "++" (or "– –") indicates that both values were higher (or lower) than expected.

[b] Groups were identified by cluster analysis using spatial correlation among species. Detailed information on the method is found in Masaki et al. (1992).

showed a more remarkable pattern with a distinct peak at the most xeric soil type B_B, both in basal area and population size.

A difference in distribution among subgroups B-a, B-b, and B-c was found in the bias to soil types B_D and G. All of the species in subgroup B-a were significantly and negatively biased both toward soil types B_D and G (but *Styrax japonica* at G was exceptional). Thus, the species of subgroup B-a strongly avoided the lower slopes and the valley bottom. This might reflect the regeneration pattern of B-a species, which are considered to have become established after a forest fire (Masaki et al. 1992; also see Chapter 6). It is probable that forest fires have little effect on the moist sites where soil types B_D and G are found. On the other hand, sites with soil types Bl_D and Bl_E might be more vulnerable to forest fire than sites with B_D and G soils because the former soils were distributed at the valley-head (Chapter 3) close to the crestslope and upper sideslope.

Some of the species of group C (site generalists) showed significant biases for soil types Bl_D, Bl_E, and B_D (Table 8.1). However, these biases were not conspicuous (Figs. 8.1f, g); the differences between the observed values for basal area and population size and the top of the 95% confidence interval were not large. Soil type B_D was most dominant in the OFR. In a previous study, the species in group C did not show any apparent bias to specific sites (cf. Figure 2c in Masaki et al. 1992), but this may have been because soil type B_D was dominant in the plot studied.

8.4 Growth of Trees

The growth rate (i.e., the mean annual increment in DBH) calculated for all species showed significant variation among soil types, both for canopy trees and for overtopped trees (Fig. 8.2a). The growth of trees was higher on soil types Bl_D and Bl_E for both categories. This might reflect the more abundant available nitrogen in Black soil compared with Brown forest soil (Chapter 3). On the other hand, canopy trees also showed higher growth rates on B_B, despite this soil type having the least amounts of available nitrogen and soil moisture (Chapter 3).

The positively biased distribution and higher growth rate on B_B were unexpected patterns considering the available resources in the soil. There are two probable explanations. First, the DBH growth rate of each tree was not strongly affected by soil type. Rather, soil conditions may affect the growth rate with respect to tree height (e.g., Tange et al. 1989). Second, the differences in growth rate among soil types may have been due merely to differences in species composition; as discussed below, intraspecific variations in growth rate among soil types were not very conspicuous. To test the second hypothesis, the growth rates of canopy trees of two dominant species, *Fagus japonica* (which preferred mesic sites) and *Quercus serrata* (xeric sites) were compared for each soil type. This comparison showed that the growth rate of *Q. serrata* exceeded that of *F. japonica* (by approximately twofold) on any soil type (significant at B_B, B_D(d), and B_D), suggesting that the second explanation is probably correct.

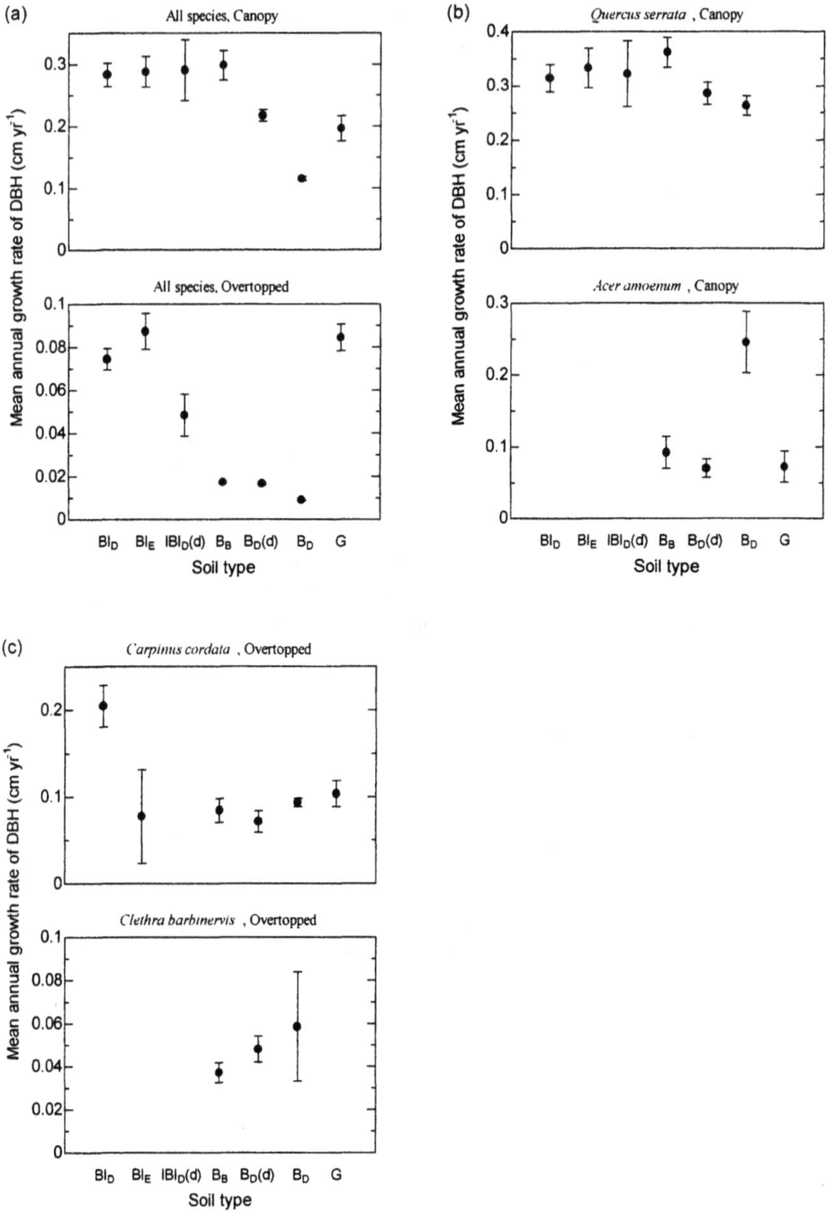

Fig. 8.2. The mean annual growth rate of DBH (*solid circles*) with standard error (*vertical bars*): **a** canopy trees (*upper panels*) and overtopped trees (*lower panels*) of all species; **b** canopy trees of *Quercus serrata* (*upper panel*) and *Acer amoenum* (*lower panel*); and **c** overtopped trees of *Carpinus cordata* (*upper panel*) and *Clethra barbinervis* (*lower panel*). See text for definitions of "canopy" and "overtopped" trees. Soils are classified according to Forest Soil Division (1976)

Table 8.2. Effect of soil type on the growth rate of DBH tested by ANOVA

Group	Species	Canopy tree[a]		Overtopped tree[a]	
		Significance level[b]	Preferred site[c]	Significance level	Preferred site
A	*Fagus japonica*	ns		< 0.05	B_B
	Carpinus cordata	.		< 0.05	Bl_D
	Acer tenuifolium	.		ns	
	Meliosma myriantha	.		ns	
	Hamamelis japonica	.		ns	
B-a	*Carpinus laxiflora*	ns		ns	
	Quercus serrata	< 0.05	B_B	ns	
	Castanea crenata	ns		.	
	Quercus crispula	ns		ns	
	Prunus verecunda	ns		ns	
	Styrax obassia	.		< 0.1	Bl_E
	Styrax japonica	.		ns	
B-b	*Sorbus alnifolia*	ns		ns	
	Betula grossa	ns		ns	
	Pieris japonica	.		ns	
	Fraxinus lanuginosa f. *serrata*	.		< 0.01	B_B
	Clethra barbinervis	.		< 0.01	B_D
	Benthamidia japonica	.		ns	
	Acer rufinerve	ns		< 0.01	B_B, B_D(d)
B-c	*Acer sieboldianum*	.		ns	
	Sorbus japonica	< 0.05	B_D	ns	
	Ilex macropoda	.		ns	
	Acanthopanax sciadophylloides	ns		ns	
C-a	*Acer mono* var. *marmoratum* f. *dissectum*	ns		ns	
	Fagus crenata	ns		ns	
	Acer amoenum	< 0.1	B_D	ns	
	Carpinus japonica	.		ns	
C-b	*Swida controversa*	ns		ns	
	Carpinus tschonoskii	ns		ns	
	Ostrya japonica	ns		< 0.01	Bl_E
C-c	*Acer nikoense*	.		ns	

ns, not significant; ., sample size too small for the statistical test.

[a]A canopy tree is a tree whose crown reaches the canopy layer and which is not be overtopped by adjacent trees. An overtopped tree is a tree whose crown does not protrude above the canopy layer and which is overtopped by adjacent trees.

[b]If the effect of soil type on the growth rate was significant or marginally significant, P is shown as < 0.1, < 0.05, or < 0.01.

[c]If the effect of soil type was significant, the soil type on which the highest growth rate was observed is shown in this column.

The results of the ANOVA for each species are summarized in Table 8.2. Of 31 species, 10 species showed significant or marginally ($P < 0.1$) significant variation in growth rate among soil types for either category of tree. *Quercus serrata* showed the highest growth rate at B_B sites (Fig. 8.2b), where both basal area and population size of this species were higher than expected. *Acer amoenum* showed the highest growth rate on soil type B_D (Fig. 8.2b), where the highest basal area and population size of this species were observed. The higher growth rates of these species at particular sites should have a positive effect on the occupancy of such sites. *Fraxinus lanuginosa* f. *serrata* and *Acer rufinerve* also showed similar patterns.

Carpinus cordata and *Clethra barbinervis* showed their highest growth rates on Bl_D and B_D soils, respectively (Fig. 8.2c). However, each species was distributed with less dominance than expected on each soil type (Table 8.1, Figs. 8.1b, e). For these species, it appeared that higher growth rates at particular sites might have negatively affected occupancy at such sites. On the other hand, no particular trends in growth variation among soil types were found in four other species (i.e., *Fagus japonica*, *Styrax obassia*, *Sorbus japonica,* and *Ostrya japonica*).

8.5 Conclusion

The analyses in this chapter suggest that the spatial patterns of species assemblies well reflected the soil types in the OFR. Therefore, variation in response to soil types among tree species should partly facilitate species coexistence within the tree community of the OFR. In other words, heterogeneity in soil conditions results in heterogeneity in the community structure and contributes to a certain degree to the maintenance of species diversity in OFR (Chapter 1).

On the other hand, the variation in the distribution pattern of each species cannot be fully explained by growth variations among juvenile or adult trees. Some other factors must control the process of habitat differentiation among tree species. One likely explanation is that the growth response to soil type might be more critical during the seedling/sapling stage than during the juvenile/adult stage (Shibata and Nakashizuka 1995). The growth pattern in relation to soil conditions at early stages of life history might eventually affect the distribution pattern of future adult populations. It is also probable that heterogeneity in the availability of some mineral, such as exchangeable aluminum, might affect which trees become established. Further analyses of soil conditions, including factors not considered in this study, will provide more valuable information on the role of soil in the organization of the OFR tree community.

References

Bray JR, Curtis JT (1957) An ordination of the upland forest communities of southern Wisconsin. Ecol Monogr 27:325–349

Forest Soil Division (1976) Classification of forest soil in Japan 1975 (in Japanese with English summary). Bull Gov For Exp Sta 280:1–28
Masaki T, Suzuki W, Niiyama K, Iida S, Tanaka H, Nakashizuka T (1992) Community structure of a species-rich temperate forest, Ogawa Forest Reserve, central Japan. Vegetatio 98:97–111
Nagamatsu D, Miura O (1997) Soil disturbance regime in relation to micro-scale landforms and its effects on vegetation structure in a hilly area in Japan. Plant Ecol 133:191–200
Ricklefs RE (1996) Ecology, 3rd ed. WH Freeman, New York
Shibata M, Nakashizuka T (1995) Seed and seedling demography of four co-occurring *Carpinus* species in a temperate deciduous forest. Ecology 76:1099–1108
Soil Survey Staff (1992) Keys to soil taxonomy, 5th edition. SMSS technical monograph No. 19. Pocahontas Press, Blacksburg, Virginia
Tange T, Matsumoto Y, Mashimo Y, Sakura T (1989) Courses of height growth of *Cryptomeria japonica* trees planted on a slope (In Japanese with English summary). Bull Tokyo Univ For 81:39–51
Tilman D (1988) Plant strategies and the dynamics and structure of plant communities. Princeton University Press, Princeton, New Jersey
Tokuchi N, Takeda H, Yoshida K, Iwatsubo G (1999) Topographical variations in a plant–soil system along a slope on Mt. Ryuou, Japan. Ecol Res 14:361–369
Whittaker RH (1967) Gradient analysis of vegetation. Biol Rev 49:207–264

Part 4

Tree Demography

9 Reproductive Traits of Trees in OFR

MITSUE SHIBATA and HIROSHI TANAKA

9.1 Introduction: Reproductive Schedules of Tree Species

When plants start to reproduce (when they reach a certain size or the age of onset of flowering and fruiting), how often and how many of the plants produce offspring during their lifetime (frequency of reproduction) and how many offspring are produced in a year (magnitude of reproduction) are important factors of plant demography, because these factors determine the lifetime fecundity of the plants. Concerning the critical size of reproduction of plants, many studies have been conducted for herbaceous perennials and monocarpic species. There are intraspecific variations in the size of reproductive individuals, and it has been pointed out that not only the size but also the growth rate of individual plants relate to the timing of their reproduction (Reekie et al. 1997; Wesselingh et al. 1997). Several models of the critical size of reproduction that take into account growth rate (Kohyama 1982; Nakashizuka et al. 1997) show that both size and growth rate affect the resource. Differences in reproductive traits are important factors contributing to the ability of tree species to coexist in a forest (Kohyama 1993; Thomas 1996). However, the long life span, large individual size, and diverse environmental conditions of tree habitats make it difficult to analyze and model the reproductive traits of tree species in a forest community. It is important to construct a simple model of the critical conditions of reproduction of tree species based on empirical data to understand tree life histories and coexistence mechanisms in a forest community.

Annual fluctuations of flowering and seeding, that is, the frequency and magnitude of reproduction, are also an important part of the reproductive schedule of tree species, which are usually polycarpic. It is known that seed-crop size has a large effect on local seed-dispersal patterns (see Chapter 10). A theoretical model suggests that the optimal reproductive strategy of a perennial plant is to produce seeds constantly after the plant reaches a given critical size (Iwasa and Cohen 1989). However, the intermittent production of large seed crops by a population of plants, called "masting" (Kelly 1994), has been reported for many tree species (Feret et al. 1982; Allen and Platt 1990; Sork et al. 1993; Koenig et al. 1994; Ashton et al. 1988; Herrera

Ecological Studies, Vol. 158
Nakashizuka, Matsumoto (eds.) Diversity and Interaction in a Temperate Forest Community: Ogawa Forest Reserve of Japan

et al. 1998; Nakashinden 1995; Sakai et al. 1999). The studies in Ogawa Forest Reserve (OFR) have also revealed that large numbers of species, such as species of Fagaceae (see Chapter 22), the genera *Acer* (Tanaka 1995) and *Carpinus* (Shibata et al. 1998), *Swida controversa* (Masaki et al. 1998), and *Kalopanax pictus* (Iida and Nakashizuka 1998) show annual fluctuations in seed production. Several hypotheses regarding the ecological and evolutionary significance of masting have been proposed, including resource matching, pollination efficiency (wind pollination), animal pollination, animal dispersal, predator satiation, seed dispersal, and environmental prediction (Kelly 1994). However, masting behavior and its significance have not yet been well documented, because few studies have used sufficient quantitative data on seed production. To understand the annual fluctuation of seed production as an adaptive reproductive schedule, it is necessary to know the long-term variation in annual seed production (how many species produce seeds constantly or variably) for various species and to test the different masting hypotheses.

In this chapter, we first examine how the reproductive schedule relates to tree life history, and then focus on the following two objectives: (1) the construction of a simple model indicating the critical conditions of reproduction; and (2) analysis of the annual fluctuation of seed production in representative species in OFR and investigation of the possible ecological advantages of that fluctuation.

9.2 Monitoring of Flowering and Fruiting

Reproduction and individual growth of major component species were intensively monitored in a 6-ha plot in the OFR. During the flowering seasons from 1988 to 1995, we observed sample trees of 15 species with binoculars to check whether they were flowering. We also checked whether 12 of these species were fruiting during the fruiting season. Sample trees were selected for each species to include a wide range of tree sizes of at least 5 cm DBH (diameter at breast height). The trees that flowered (fruited) at least once during the observation period were classified as flowering (fruiting) trees, considering their intermittent flowering habit. We measured DBH of the sample trees in 1987 and 1993 and calculated the average relative growth rate (RGR) of DBH [ratio of log (DBH) in 1993 to log (DBH) in 1987, divided by six] of individuals for the six-year period of the study.

We estimated the population-level production of staminate aments (male flowers) and seed fall of the major component species in a 1-ha subplot located at the center of the 6-ha plot (See Fig. 5.2). We placed 121 seed traps in a regular matrix (10 m × 10 m) in the 1-ha subplot and collected the contents of the seed traps and brought them back to the laboratory every 4 weeks from April to August and every 2 weeks from September to December. In the laboratory, we counted all the seeds of 16 species from the contents of the seed traps and cut the seeds open to classify them into five basic categories: (1) sound (attained mature seed size and with a sound embryo); (2) immature (failed to attain mature seed size); (3) empty (attained mature seed size but with either no embryo or an undeveloped embryo); (4) suffered predation by insects (with a gnawing mark or hole); and (5) broken (mainly caused by vertebrate predation). For the 11 species of Fagaceae and Betulaceae, which are wind pollinated except for *Castanea crenata,* we sorted and weighed the staminate aments. The dry mass of staminate aments was used as an index of the amount of pollen. Detailed descriptions of these methods are available in Masaki et al. (1994) and Shibata et al. (1998).

9.3 Critical Size of Reproduction

9.3.1 Linear Discriminant Function of Reproduction

The reproductive allocation depends on the amount of resource production (Klinkhamer et al. 1992; Nakashizuka et al. 1997). It is also suggested that the growth rate, which is influenced by environmental conditions such as light, nutrients, moisture, and temperature, affects reproduction at a given size (Kohyama 1982). Our model of the critical size of reproduction is a linear discriminant function with two parameters:

$$R = \mathrm{a}D + \mathrm{b}$$

Where D and R are log (DBH) and log (RGR of DBH), respectively, and a and b are constants. We used D and R as indicators of plant production. In general, there is a trend for RGR of DBH to decrease as the DBH of the individual tree increases. We expected this discriminant function to have a negative slope, implying that reproduction is positively dependent both on individual size and growth rate. Individual trees begin to flower and/or fruit at a certain DBH, which would vary among the individuals with different growth rates. An individual with a higher RGR of DBH will begin to flower (and/or fruit) at a smaller size.

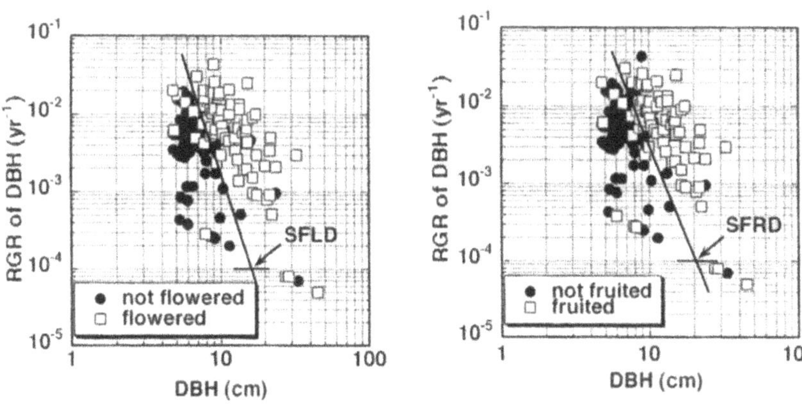

Fig. 9.1. The relationship between individual tree size (DBH) and growth rate (RGR of DBH) for flowering or fruiting of *Carpinus cordata*. We named the point of the discriminant function intersected by the minimum value of RGR the "stable flowering DBH" (*SFLD*) or the "stable fruiting DBH" (*SFRD*)

We tested whether this model could explain the critical conditions of tree reproduction with the data obtained from observations of flowering and fruiting and successive measurements of DBH for each tree. As a result, we were able to divide the individuals of each species into two groups, flowered (fruited) and nonflowered (nonfruited) trees, by the linear discriminant function with two parameters, DBH of individual trees and RGR of DBH of individual trees, and the function had a negative slope, as we predicted (Fig. 9.1). Correct discrimination of the flowering trees was achieved for between 71% and 100% (average discriminant score was 88%) of all sample trees of each species. Discriminant scores for the classification of fruiting trees were from 68% to 100% (average, 87%). These results suggest that this simple linear discriminant function model can predict the reproductive conditions of each individual tree.

9.3.2 Relationship Between the Size at Reproduction and the Maximum Size

We defined the point of the discriminant function line intersected by the hypothesized minimum value of RGR (RGR = 0.0001 yr^{-1}) as the "stable flowering DBH" (SFLD) or "stable fruiting DBH" (SFRD). The SFLD (SFRD) can be regarded as the minimum size at which a tree flowers (fruits) irrespective of its growth rate (Fig. 9.1). SFLD (SFRD) is a demographic parameter incorporating both size and growth that represents the flowering (fruiting) schedule. From direct observation of flowering and fruiting of individual trees in the 6-ha plot, we identified the smallest flowering (fruiting) trees in a mature forest for each species and defined the "minimum flowering DBH" (MFLD), or "minimum fruiting DBH" (MFRD).

We found that the SFLD, SFRD, MFLD, and MFRD of each species were approximately proportional to the maximum size of the species (Fig. 9.2). In these

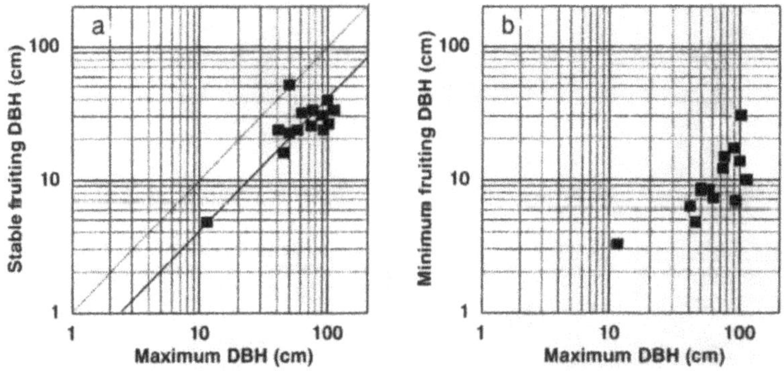

Fig. 9.2. The relationship between maximum DBH of the main component tree species and its stable fruiting DBH (SFRD) (**a**), and minimum fruiting DBH (MFRD) (**b**). The *broken line* indicates an SFRD (MFRD) the same as the maximum DBH. The *solid line* indicates that an SFRD and MFRD 40% and 10% of the maximum DBH, respectively

analyses, the DBH of the largest individuals in this forest was assumed to be the maximum size of the respective species. This relationship means that the smaller tree species (shrub and subcanopy species) can potentially begin to reproduce earlier than the larger tree species (canopy species). And we found that all trees that exceeded 30%–40% of the maximum size for their species reproduced irrespective of their growth rate (Fig. 9.2).

This finding that the maximum DBH is a good indicator of the reproductive schedule has implications for the trade-off between life history traits of coexisting species. The forest architecture hypothesis (Kohyama 1993) suggests that the trade-off between maximum tree size and the recruitment rate (of saplings) allows tree species to coexist in a forest community that has vertical stratification (size structure). A positive relationship between SFLD or SFRD and maximum DBH supports this hypothesis, because species that began to reproduce at a smaller size would have a higher recruitment rate.

9.4 Annual Fluctuations in Reproduction in a Forest Community

Over 9 years of observation, many species in this temperate deciduous forest were found to have rather large annual fluctuations in seed production (Fig. 9.3). Only 4 years, i.e., 1988, 1990, 1993, and 1995, were good seed years for many species. The coefficient of variation (CV, presented as a ratio) of annual sound seed production from 1987 to 1995, which was used as an index of masting (Silvertown 1980; Kelly 1994), was more than 1.5 for 10 out of 16 species (Fig. 11.2 in Chapter 11). *Fagus crenata* (CV, 2.7) had the largest annual fluctuation in seed production. Four co-occurring *Carpinus* species (*C. laxiflora*, *C. tschonoskii*, *C. japonica*, and *C. cordata*) also showed relatively large fluctuations in seed production (CV of sound seed production, 2.2, 1.7, 2.2, and 2.4, respectively). Synchronized annual seed production among the species in this genus was also observed (Fig. 9.3, Shibata et al. 1998). Annual seed production of *Acer mono*, *A. amoenum*, and *Swida controversa* fluctuated moderately (CV slightly less than 1.5). In contrast, *Quercus serrata* and *Q. crispula* (CV, 0.6 and 0.8, respectively) produced seeds in nearly constant numbers. Fluctuating and synchronized annual seed production was found to be more common than constant seed production among the tree species in this forest community. This trend for seed production in a community to fluctuate annually is common in other forest types as well, for example, a Mediterranean montane forest (Herrera 1998) and a tropical rain forest in Southeast Asia (Sakai et al. 1999).

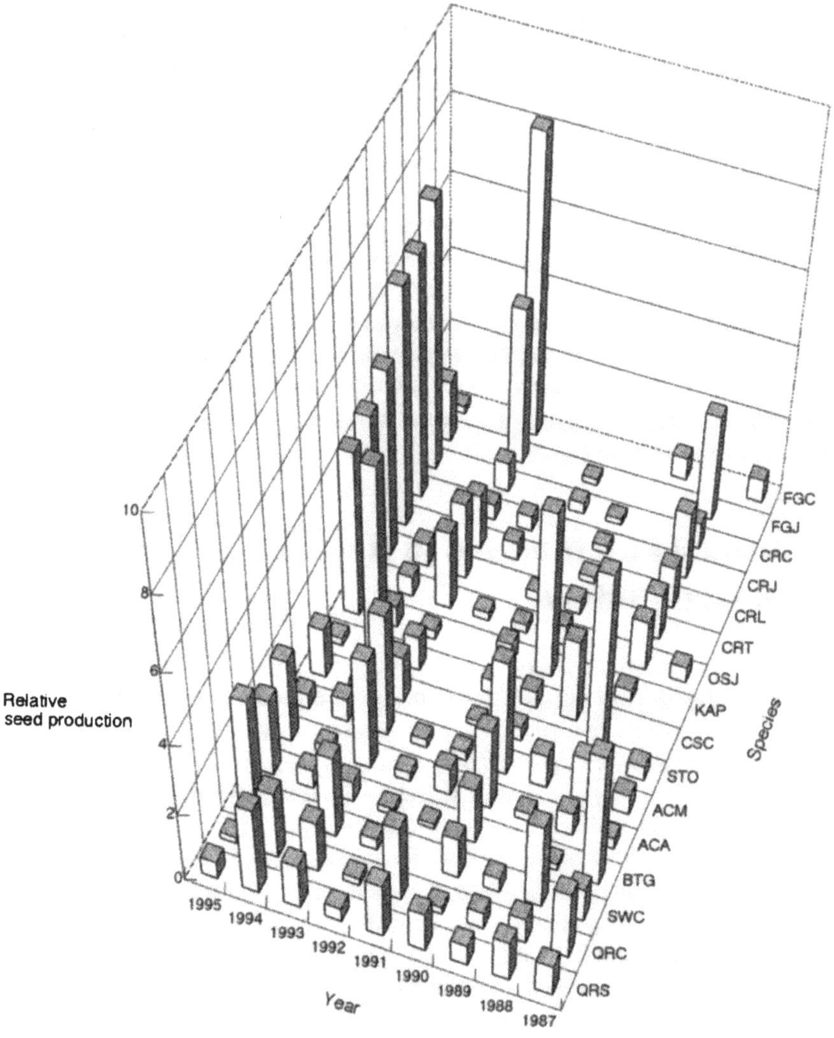

Fig. 9.3. Annual fluctuation of sound seed production of the main component tree species from 1987 to 1995. Relative values of sound seed production are shown. The average amount of seed production from 1987 to 1995 is set as 1. *FGC, Fagus crenata*; *FGJ, Fagus japonica*; *CRC, Carpinus cordata*; *CRJ, Carpinus japonica*; *CRL, Carpinus laxiflora*; *CRT, Carpinus tschonoskii*; *OSJ, Ostrya japonica*; *KAP, Kalopanax pictus*; *CSC, Castanea crenata*; *STO, Styrax obassia*; *ACM, Acer mono*; *ACA, Acer amoenum*; *BTG, Betula grossa*; *SWC, Swida controversa*; *QRC, Quercus crispula*; *QRS, Quercus serrata*

9.5 Ecological Advantages of Mast Seed Production: A Case Study of Four Co-occurring *Carpinus* Species

Several hypotheses, e.g., pollination efficiency (wind pollination), predator satiation, animal pollination, animal dispersal, and environment prediction, have been proposed to explain the selective advantages of masting (see Kelly 1994 for a review). Norton and Kelly (1988) and Kelly (1994) categorized most of these hypotheses as "economy of scale" hypotheses, because they suggest that a larger reproductive effort is more efficient, and thus an occasional large effort rather than a more regular, smaller one is favored. Until recently, pollination efficiency and predator satiation were best supported among these hypotheses (see Kelly 1994; Nilsson 1985; Nilsson and Wästljung 1987; Allison 1990).

The pollination efficiency (wind pollination) hypothesis states that masting evolved owing to a disproportionate increase in fertilization and seed set during mast years (Smith et al. 1990). This hypothesis was proposed because many of the species that had been reported to mast are wind-pollinated species (Smith et al. 1990). According to the predator satiation hypothesis, masting is an antipredator adaptation that allows for the survival of seeds by alternately starving and satiating seed predators (Janzen 1971; Silvertown 1980). This hypothesis has also drawn attention because it is one of the major ecological studies of interaction between plants and animals (see Chapter 22).

Carpinus species have wind-pollinated flowers and comparatively small, wind-dispersed seeds that suffer from predispersal predation by insects. It is highly probable that synchronized annual fluctuations within and among *Carpinus* species can be explained by the pollination efficiency and predator satiation hypotheses (Shibata et al. 1998). To test these two hypotheses, we investigated flowering, seed production, and seed fall in each population for 9 years (1987–1995). We also monitored the seasonal development of pistillate aments in 1995.

9.5.1 Pollination Efficiency

Smith et al. (1990) tested the potential advantages of mast flowering for pollination efficiency in wind-pollinated species using a theoretical model. They pointed out two necessary conditions for their model: (1) fluctuations in male and female reproductive efforts must positively correlate across years (be synchronous); and (2) the cost of sexual reproduction through female function must be the same regardless of fertilization success. We tested whether the four *Carpinus* species satisfy these conditions.

We observed the demography of female flowers (pistillate aments) and ovaries from the flowering season to the seed-fall season. In the OFR, *Carpinus* species flower in early May. We found that all staminate aments fell after the flowering season, but many pistillate aments grew into fruiting branches and did not fall until the end of August (Shibata et al. 1998). At the level of the individual female flower,

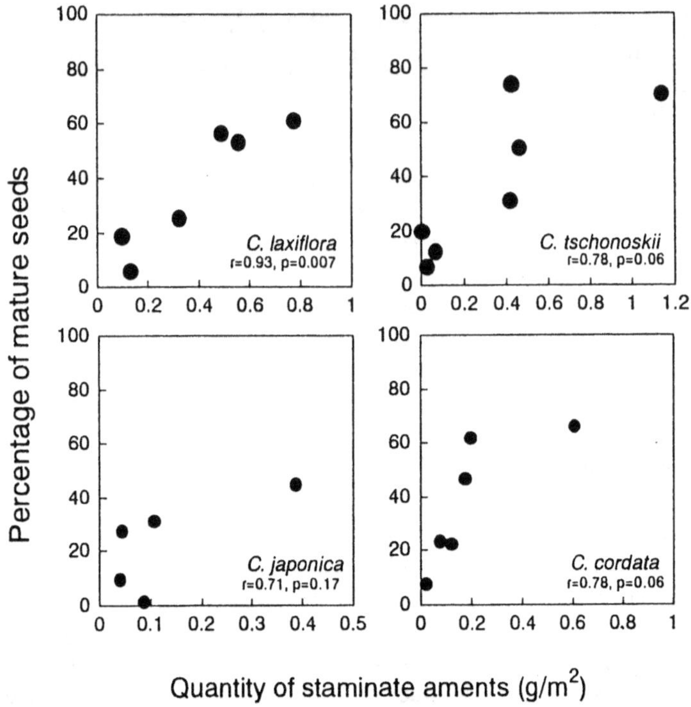

Fig. 9.4. The relationship between the quantity of staminate aments and the percentage of mature seed production. After Shibata et al. (1998)

few ovaries or seeds in pistillate aments had fallen by the end of August, and most ovaries grew to normal, mature-seed size, irrespective of whether the seeds were sound or empty (Shibata et al. 1998). Thus, fruiting branches were not shed when fertilization failed, and even empty seeds attained normal seed size, suggesting a high cost of seed production regardless of fertilization.

The percentage of mature seeds (sound, showing predation by insects, broken seeds) was positively correlated with staminate ament production in *C. laxiflora* (Fig. 9.4). In the case of *C. tschonoskii* and *C. cordata*, the value of the correlation coefficient (r) was large, but not significant. In *C. japonica*, r was lower than for the other species. The larger percentages of mature seeds were accompanied by lower percentages of empty seeds.

There were positive correlations between staminate ament (pollen) production and both total seed and mature seed production (Fig. 9.5). Because most ovaries in each pistillate ament grow into mature seeds of normal size, the total number of seeds is equivalent to the number of female flowers (ovaries). Thus male and female flower production in the *Carpinus* species we studied were positively correlated.

These *Carpinus* species satisfied the two necessary conditions of the model developed by Smith et al. (1990). The annual fluctuation in seed production of the *Carpinus* species was well explained by the pollination efficiency hypothesis (Smith

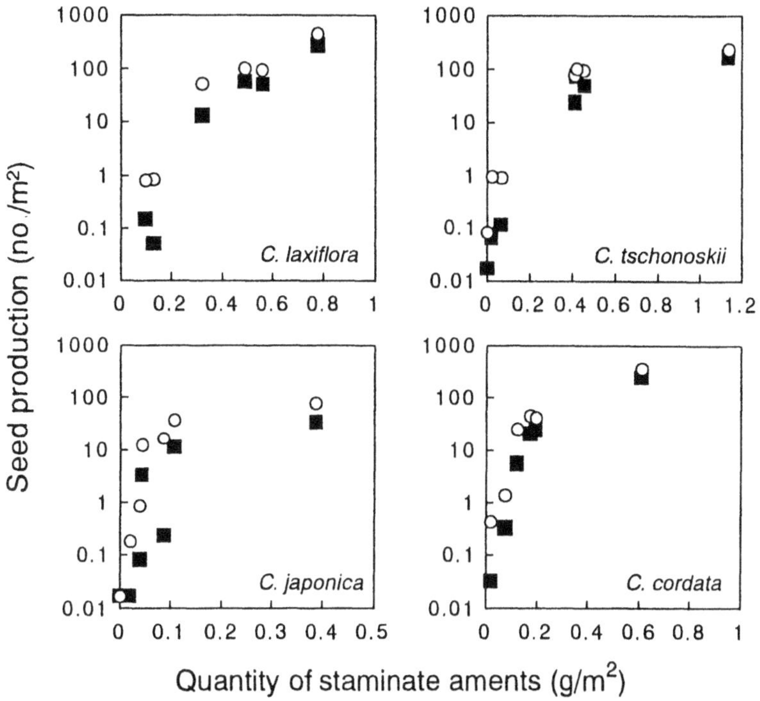

Fig. 9.5. The relationship between the quantity of staminate aments and seed production. *Open circles*, total seed production, including sound seeds, seeds showing predation by insects, broken seeds, immature, and empty seeds; *solid squares*, mature seed production, including sound seeds, seeds showing predation by insects, and broken seeds (indicating predation by vertebrates). After Shibata et al. (1998)

et al. 1990). Wind pollination is common among trees in temperate forests. Do other wind-pollinated species in temperate forests satisfy the conditions of the pollination efficiency model? *Fagus crenata*, *F. japonica,* and *Ostrya japonica,* which grow their empty seeds to normal seed size, probably do, but *Quercus* species, which abort and shed unfertilized seeds, likely do not (as discussed by Sork et al. 1993).

9.5.2 Predator Satiation

Silvertown (1980) presented three necessary conditions to satisfy the predator satiation hypothesis: (1) the production of enough seeds to satiate predators, ensuring that some seeds escape; (2) a minimum interval between mast years, which allows the predator population to have decreased by the next large production year; and (3) synchronous seed production, both among individuals of the same population and among populations of sympatric species sharing the same seed predators. We tested whether the four *Carpinus* species satisfied these three conditions.

Most of the seeds showing predation by insects had a small hole in the seed coat

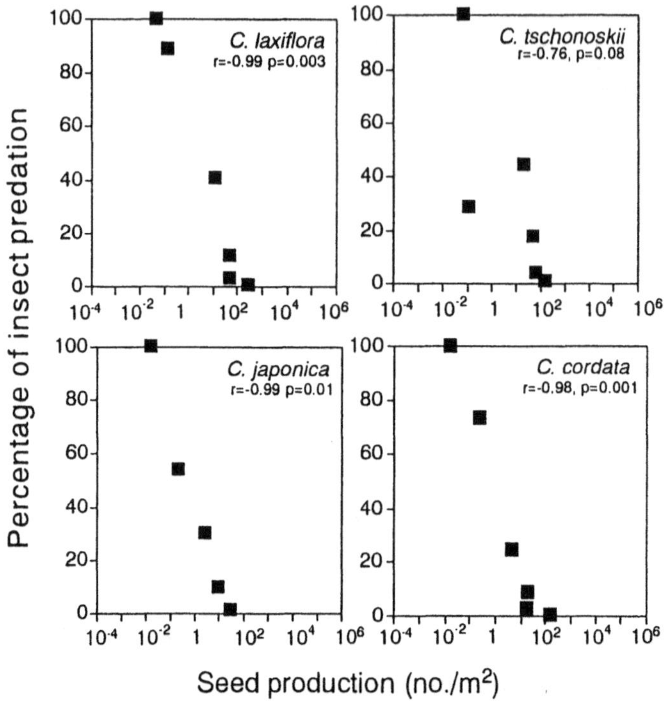

Fig. 9.6. The relationship between annual seed production and the percentage of insect predation for each species. Seed production in this figure is the sum of sound seeds and seeds showing predation by insects. After Shibata et al. (1998)

(0.3–0.5 mm diameter). Although the predators could not be identified by species, the evidence suggested curculionid weevils (Family Curculionidae) as a highly possible predator (K. Maeto, personal communication). Curculionid larvae usually invade and consume the seed contents in summer, when seed coats are still soft.

To seek evidence for predator satiation, we correlated seed production (the sum of sound seeds and seeds showing predation by insects) with the percentage of insect predation, which was calculated as the ratio of seeds showing predation by insects to the sum of sound seeds and seeds showing predation.

The ratio of current year seed production to that of the previous year can be regarded as a relative measure of predator starvation owing to year-to-year fluctuation in seed production. If seed predation depends on the density of predators, which is controlled by the seed production of the previous year, a negative correlation is expected between this ratio and the probability of seeds showing predation (Mattson 1978). To test for a potential starving effect of fluctuating annual seed production, we correlated the ratio of the current year seed production to that of the previous year with the percentage of insect predation.

As a result, predispersal seed predator satiation at the population level was found to be statistically significant in three of four species, i.e., in *C. cordata*, *C. japonica*,

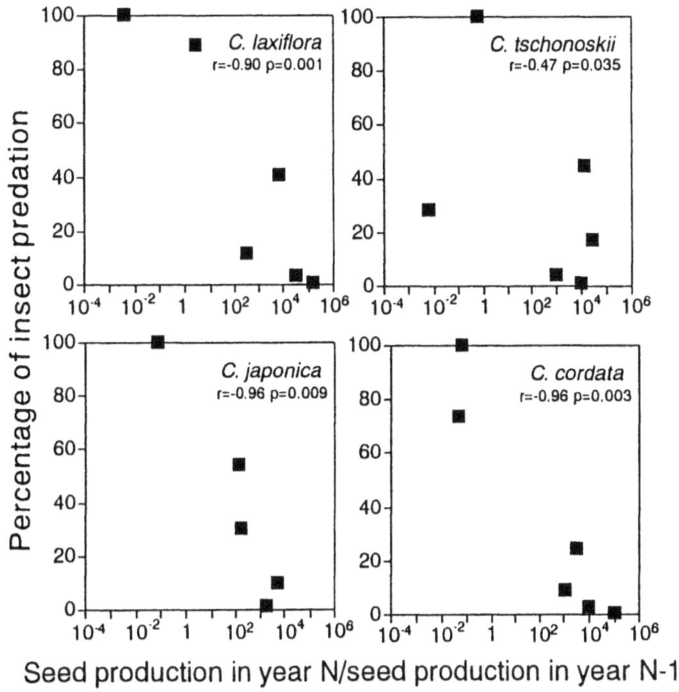

Fig. 9.7. The relationship between the ratio of current year seed production and that of the previous year and the percentage of insect predation in the current year for each species. After Shibata et al. (1998)

and *C. laxiflora* (Fig. 9.6 and Fig. 9.7). Predators are satiated in large seed crop years, but suffer reduced populations in poor seed years. This results in high seed survival in the following good seed years. Although *C. tschonoskii* did not show significant correlations, the correlation coefficients were negative. We think that satiation and starvation of species-specific predators at the predispersal seed stage might operate for this species too, and will be detected when we have further long-term data.

The four *Carpinus* species showed synchronous seed production, both among individuals in the same population and among populations. When the seeds of the four species were pooled, the percentage of insect predation had significantly negative correlations with the sum of sound seeds and seeds showing predation by insects (Fig. 9.8a), and had a marginally significant correlation with the ratio of the current year seed production to the previous year seed production (Fig. 9.8b). If common curculionid species are the predators of these co-occurring *Carpinus* species, then these species will mutually benefit from predator satiation by synchronizing their seed production.

In the OFR, *Q. serrata*, *Q. crispula*, and *Castanea crenata* share a common seed predator (Chapter 22; Ueda 1996). *Fagus crenata* and *F. japonica* also probably share common seed predators, because seed predation marks made by insects are

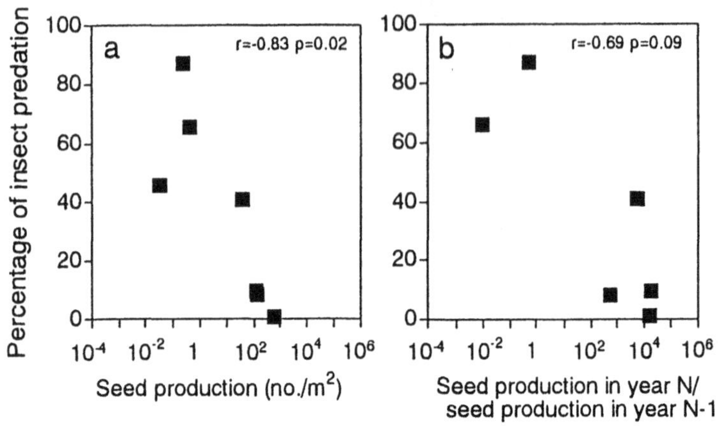

Fig. 9.8. The relationship between t insect predation (%) and annual seed production (a) and the ratio of current year seed production to that of the previous year (b) for the four *Carpinus* species combined. After Shibata et al. (1998)

very similar for both species. Moreover, these *Fagus* species exhibited synchronized mast seed production (Fig. 9.3). It is possible that these co-occurring, closely related species also support the predator satiation hypothesis at the genus level in a forest community.

9.6 Conclusions

In this chapter, we examined two factors affecting the reproductive schedule, i.e., individual size at onset of flowering (fruiting), and annual fluctuation in seed production among reproducing individuals. From a simple model of critical conditions for reproduction, we concluded that tree reproduction is dependent on resource production (DBH growth rate). The schedule of reproduction varied (from relatively constant to large annual fluctuations in seed production patterns) among tree species in a natural temperate forest. Such variation reflects the differences in reproductive cost and resource allocation patterns among different species. Detailed physiological and matter economic studies of the pattern of resource allocation to reproduction will be important for understanding these two factors affecting the reproductive schedule.

We found that the four *Carpinus* species in the forest had synchronized annual fluctuations in seed production, not only at the population level but also at the level of the genus. The pollination efficiency hypothesis explained the regenerative advantage of synchronized annual fluctuations in seed production only at the population level. On the other hand, the predator satiation hypothesis explains the advantage of reproductive synchronicity at both the population and genus levels. Both of these selective advantages can operate independently to promote masting behavior in these species.

We have observed that several species belonging to different taxa have similar synchronized patterns of annual seed production in this forest (Fig. 9.3). A question to be investigated in the future is whether these hypotheses also explain the annual seed production patterns of the other species in the forest, and whether reproduction is synchronized among the broader species groups at the community level.

References

Allen RB, Platt KH (1990) Annual seedfall variation in *Nothofagus solandri* (Fagaceae), Canterbury, New Zealand. Oikos 57:199–206

Allison TD (1990) Pollen production and plant density affect pollination and seed production in *Taxus canadensis*. Ecology 71:516–522

Ashton PS, Givnish TJ, Appanah S (1988) Staggered flowering in the Dipterocarpaceae: new insights into floral induction and the evolution of mast fruiting in the aseasonal tropics. Am Nat 132:44–66

Feret PP, Kreh RE, Merkle SA, Oderwald RG (1982) Flower abundance, premature acorn abscission, and acorn production in *Quercus alba* L. Bot Gaz 143:216–218

Herrera CM (1998) Long-term dynamics of Mediterranean frugivorous birds and fleshy fruits: a 12-year study. Ecol Monogr 68:511–538

Herrera CM, Jordano P, Guitian J, Traveset A (1998) Annual variability in seed production by woody plants and the masting concept: reassessment of principles and relationship to pollination and seed dispersal. Am Nat 152:576–594

Iida S, Nakashizuka T (1998) Spatial and temporal dispersal of *Kalopanax pictus* seeds in a temperate deciduous forest, central Japan. Plant Ecol 135:243–248

Iwasa Y, Cohen D (1989) Optimal growth schedule of a perennial plant. Am Nat 133:480–505

Janzen DH (1971) Seed predation by animals. Annu Rev Ecol Syst 2:465–492

Kelly D (1994) The evolutionary ecology of mast seeding. Trends Ecol Evol 9:465–470

Klinkhamer PGL, Meelis E, de Jong TJ, Weiner J (1992) On the analysis of size-dependent reproductive output in plants. Funct Ecol 6:308–316

Koenig WD, Mumme RL, Carmen WJ, Stanback MT (1994) Acorn production by oaks in central coastal California: variation within and among years. Ecology 75:99–109

Kohyama T (1982) Studies on the *Abies* population of Mt. Shimagare II. Reproductive and life history traits. Bot Mag Tokyo 95:167–181

Kohyama T (1993) Size-structures tree populations in gap-dynamic forest—the forest architecture hypothesis for the stable coexistence of species. J Ecol 81:131–143

Masaki T, Kominami Y, Nakashiuka T (1994) Spatial and seasonal patterns of seed dissemination of *Cornus controversa* in a temperate forest. Ecology 75:1903–1910

Masaki T, Tanaka H, Shibata M, Nakashizuka T (1998) The seed bank dynamics of *Cornus controversa* and their role in regeneration. Seed Sci Res 8:53–63

Mattson WJ (1978) The role of insects in the dynamics of cone production of red pine. Oecologia 33:327–349

Nakashinden I (1995) Fruit years and form of cone production by the Japanese stone pine (*Pinus pumila* Regel) estimated by the cone scars method (in Japanese with English summary). Jpn J Ecol 45:113–120

Nakashizuka T, Takahashi Y, Kawaguchi H (1997) Production-dependent reproductive allocation of a tall tree species *Quercus serrata*. J Plant Res 110:7–13

Nilsson SG (1985) Ecological and evolutionary interactions between reproduction of beech *Fagus silvatica* and seed eating animals. Oikos 44:157–164

Nilsson SG, Wästljung U (1987) Seed predation and cross-pollination in mast-seeding beech (*Fagus sylvatica*) patches. Ecology 68:260–265

Norton DA, Kelly D (1988) Mast seeding over 33 years by *Dacrydium cupressinum* Lamb. (rimu) (Podocarpaceae) in New Zealand: the importance of economies of scale. Funct Ecol 2:399–408

Reekie E, Parmiter D, Zebian K, Reekie J (1997) Trade-offs between reproduction and growth influence time of reproduction in *Oenothera biennis*. Can J Bot 75:1897–1902

Sakai S, Momose K, Yumoto T, Nagamitsu T, Nagamasu H, Hamid AA, Nakashizuka T, Inoue T (1999) Plant reproductive phenology over four years including an episode of general flowering in a lowland dipterocarp forest, Sarawak, Malaysia. Am J Bot 86:1414–1436

Shibata M, Tanaka H, Nakashizuka T (1998) Causes and consequences of mast seed production of four co-occurring *Carpinus* species in Japan. Ecology 79:54–64

Silvertown JW (1980) The evolutionary ecology of mast seeding in trees. Biol J Linn Soc 14:235–250

Smith CC, Hamrick JL, Kramer CL (1990) The advantage of mast years for wind pollination. Am Nat 136:154–166

Sork VL, Bramble J, Sexton O (1993) Ecology of mast-fruiting in three species of North American deciduous oaks. Ecology 74:528–541

Tanaka H (1995) Seed demography of three co-occurring *Acer* species in a Japanese temperate deciduous forest. J Veg Sci 6:887–896

Thomas SC (1996) Relative size at onset of maturity in rain forest trees: a comparative analysis of 37 Malaysian species. Oikos 76:145–154

Ueda A (1996) Insect infestation on acorns of Japanese chestnut, *Castanea crenata* sieb. et zucc., in natural forest (in Japanese). Trans Jpn For Soc 107:233–236

Wesselingh RA, Klinkhamer PGL, de Jong TJ, Boorman LA (1997) Threshold size for flowering in different habitats: effects of size-dependent growth and survival. Ecology 78:2118–2132

10 Seed Dispersal

Hiroshi Tanaka and Yohsuke Kominami

10.1 Introduction

Seed and pollen dispersal are known to be the only ways plants "move." Plants can disperse their offspring via these two processes. Pollen dispersal may be most important for interbreeding with other distant individuals. The transport of seeds for long distances to expand the species range is often mentioned as an advantage conferred by seed dispersal methods. However, with respect to selective advantages, local dispersal at the spatial scale of a single plant community must be explained as a critical and sufficient evolutionary cause of seed dispersal (Howe and Smallwood 1982). Alternative advantages of local seed dispersal that have been emphasized include escaping high seed and seedling mortality around the parent plants, increasing the probability of colonizing favorable sites (disturbances) or finding a physically suitable establishment site or "safe site," and nonrandom directional dispersal to a safe site (Howe and Smallwood 1982; Willson 1992; Nakashizuka et al. 1995).

Several studies have integrated demography and dispersal processes to test the validity of these proposed advantages to seed dispersal (Augspurger 1984; Howe et al. 1985; Alvarez-Buylla and Martinez-Ramos 1990; McPeek and Holt 1992; Schupp 1993; Houle 1994). These studies adopted germination or seedling survival as evaluating functions for fitness, because tree species usually have long life spans and it is difficult to get good estimates of demographic parameters throughout their life cycle (Schupp 1993). Postdispersal seed and seedling stages are the stages that show the most drastic demographic changes in the tree life cycle (Harcombe 1987), and the effect of dispersal becomes more difficult to analyze for later stages. Thus, investigating postdispersal seed and seedling survival rates might be a tractable and efficient way to evaluate dispersal success (Schupp 1993).

In Ogawa Forest Reserve (OFR), the process from seed dispersal to seedling emergence and survival has been monitored intensively for the major component tree species in the community. The spatiotemporal pattern of seed dispersal was clarified by using a set of seed traps regularly arranged in an area large enough to cover the seed shadows (the spatial distribution of dispersed seeds around their source; Janzen 1971; Willson 1992) of the tree species. Frequent observations of subsequent

Ecological Studies, Vol. 158
Nakashizuka, Matsumoto (eds.) Diversity and Interaction in a Temperate Forest Community: Ogawa Forest Reserve of Japan

seedling emergence and survival in the same area allowed the fates of the dispersed and nondispersed seeds to be estimated. Additional controlled experiments were conducted both in the field and laboratory to further explain the mechanisms of the processes occurring in the field. In this chapter, we will show the diverse patterns of seed dispersal in this forest community and discuss the validity of the above-mentioned hypotheses regarding the selective advantages of seed dispersal for several tree species, incorporating the demographic traits of those species, especially in the earlier stages of regeneration (Masaki et al. 1994; Nakashizuka et al. 1995; Iida 1996; Tanaka et al. 1998).

10.2 Seed Dispersal Modes and Community Organization

Among the 105 woody species (including shrub species) that constituted the forest community of OFR, 55 are frugivore-dispersed (endozoochorous), 36 are wind-dispersed (anemochorous), 8 are dispersed by food-hoarding animals (dyszoochorous), 2 are dispersed by the plant itself (autochorous), and 4 are nondispersed (gravity-dispersed) species (barochorous) (Kominami unpublished data; see Chapter 11). None of the dyszoochorous species have special organs for dispersal (wings or fleshy fruits), but are potentially dispersed by animals that show caching behavior (food-hoarding mammals or birds). The proportion of endozoochorous species in Ogawa Forest Reserve (52%) was the highest among reported values for temperate deciduous forests (9.5%–53.8%, Jordano 1992). Among tree species with diameter at breast height (DBH) more than 5 cm, the trend was almost the same: among 53 tree species, 21 were endozoochorous species, 23 were anemochorous, 7 were dyszoochorous, 1 was autochorous, and 1 was barochorous. However, the three most dominant tree species in this forest (*Quercus serrata*, *Fagus japonica*, and *F. crenata*), which together accounted for 56% of the total basal area, were dyszoochorous. Actually, dyszoochorous species constituted 66%, anemochorous species 22%, and endozoochorous species only 11% of the total basal area of trees > 5 cm DBH. The trend for wind-dispersed or scatter-hoarded dispersed (by food-hoarding animals) tree species to be dominant among canopy trees and for endozoochorous species to be dominant among smaller-sized trees is common in temperate deciduous forests (Howe and Smallwood 1982).

The seed size of the tree species varied in a wide range from 0.5 mg (*Betula grossa*) to 2 g (*Quercus crispula*), that is, by more than three orders of magnitude. Both anemochorous species and endozoochorous species had smaller (lighter) seeds than dyszoochorous species in this community (K. Niiyama et al. unpublished data). It is claimed that species in shaded and late-successional habitats tend to have larger seeds than those of early-successional habitats (Salisbury 1942; Baker 1972; Foster and Jahnson 1985). However, the relationship between seed size and adult tree shade tolerance as reflected by the population size structure (Masaki et al. 1992) was unclear in this forest (Nakashizuka et al. unpublished data). As will be discussed later, seed size is related only to survival at seed and seedling stages. The probability that at least one seed will travel a given distance from the parent tree is proportional to the number of seeds released; thus seed-crop size has a major effect on the size of the seed shadow in general (Willson 1993). The number of dispersed seeds per tree varied greatly among species, among individuals, and among years in this forest (see Chapter 9). Seed shadows of trees of the same species often overlapped each other, and estimating the fecundity of trees is difficult. On the basis of the long-term observation of fruiting trees and seed-trap data in Ogawa Forest Reserve, mean per capita seed production of each species was estimated also to vary in a wide range from 7.0×10^2 to 3.4×10^5 (Shibata et al. unpublished; Nakashizuka et al. unpublished).

10.3 Seed Shadows of Different Dispersal Modes

10.3.1 Wind Dispersal

Seed shadows (postmodal) of wind-dispersed tree species are generally expected to follow negative exponential curves (Okubo and Levin 1989; Willson 1993). Five of six representative wind-dispersed species (*Ostrya japonica*, *Carpinus laxiflora*, *C. tschonoskii*, *Acer mono*, *A. amoenum*) in this forest basically followed this form, but one dispersal curve with a long-tail (*Betula grossa*) could not be described by a negative exponential function (Tanaka et al. 1998, Fig. 10.1). To understand dispersal mechanisms, the application of an interpretative model would be more effective than fitting descriptive curves. Actually, seed shadows of all wind-dispersed species fitted better to the dispersal model proposed by Greene and Johnson (1989), which incorporates biological factors such as tree height (H), total seed number dispersed (Q), descent velocity of seeds (F), and the parameters of the wind regime [mean wind speed (u) and variation in wind speed (σ)] (Greene and Johnson 1989; Tanaka et al. 1998; Figs. 10.1 and 10.2).

Among the parameters incorporated into the model, both diaspore type (samara, involucre, scale) and weight (0.6–46.3 mg) and the average number of dispersed seeds per individual in a mast year ($2 \times 10^4 - 1.3 \times 10^6$ tree^{-1}) varied greatly among species (Tanaka et al. 1998). However, variation in descent velocity of seeds (F, 0.83–1.12 m sec^{-1}) and average tree height (H, 15.6–23.7 m) were relatively small. Either increasing the total number of dispersed seeds or increasing the ratio H/F contributed to an increase in the dispersal distance, but there was a tendency for species producing large number of seeds to also have larger H/F ratios, i.e., more efficient architectural and morphological traits for seed dispersal. The variation in the total number of dispersed seeds was larger than that in the H/F ratio, and thus contributed more to the difference in dispersal distance.

In the intensively studied 1-ha area, the area with dispersed seed density > 10 m^{-2} and > 2 m^{-2} in a mast year were 51% to 97% and 87% to 98% of the total area, respectively, for the six wind-dispersed species (Tanaka et al. 1998, Fig. 10.2). The proportion with a given dispersed seed density is affected by both the seed shadow of an individual tree and the density of fruiting trees in the forest community. *Betula*

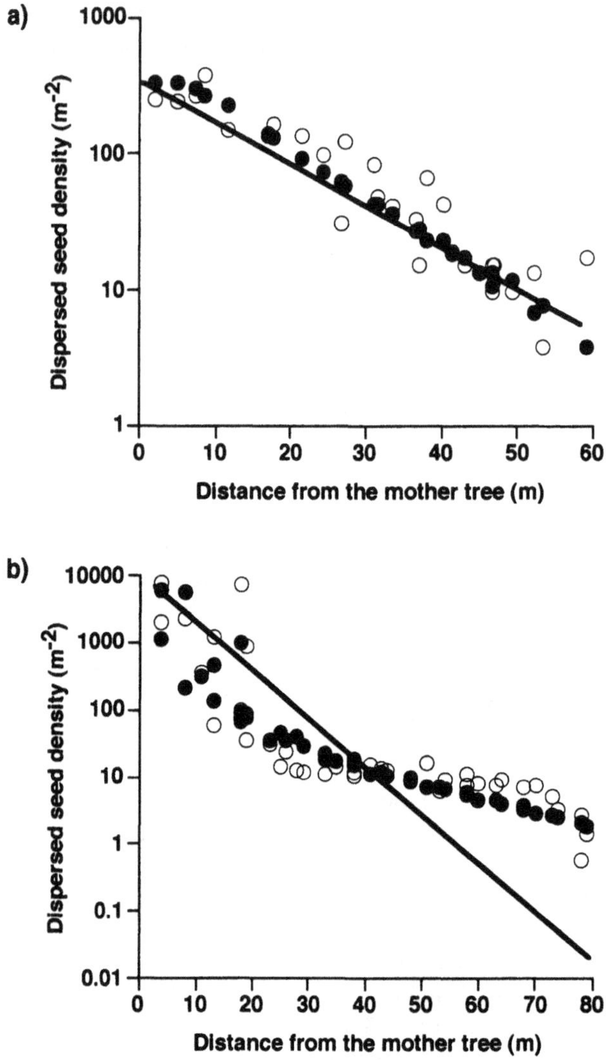

Fig. 10.1. Nonlinear regression of dispersal with distance for a single seed source according to the Greene-Johnson (GJ) model. *Open* and *solid circles* show the actual density and that estimated by the GJ model, respectively. The GJ model takes into account the influence of tree height and topography, and density was estimated at each deposition site. The line shows the results of regression according to a negative exponential model in which only the horizontal distance is considered. **a** *Acer mono*, **b** *Betula grossa* (after Tanaka et al. 1998)

grossa could achieve a seed density > 2 m^{-2} for 97% of the total area, as indicated by the long-tailed dispersal curve, although there was only one fruiting stem in the 1-ha plot and one other near the plot. For these six species, local seed dispersal seemed to be effective to colonize in any part of the forest floor in the community; that is, dispersal was not necessarily a limiting factor for the establishment of the present

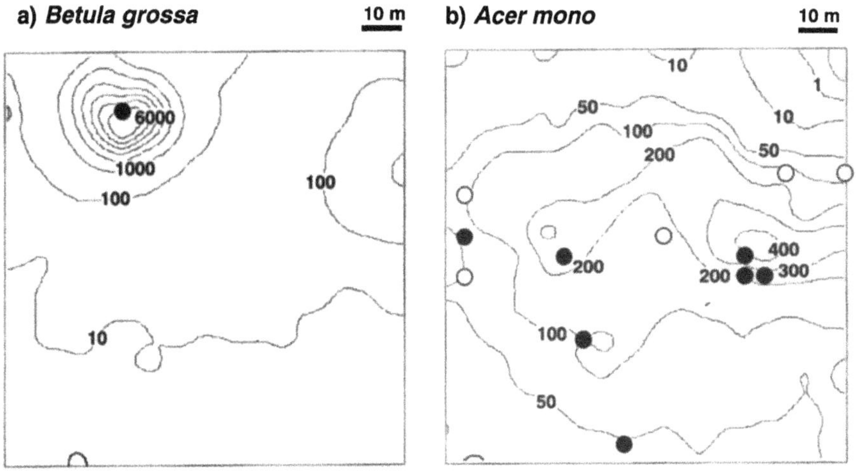

Fig. 10.2. Seed dispersal patterns of *Betula grossa* (**a**) and *Acer mono* (**b**) in a mast seeding year (partly after Tanaka et al. 1998). Position of reproductive trees are shown as *solid* (DBH > 30 cm) and *open* (DBH < 30 cm) *circles*

adult tree density (cf. Clark et al. 1998). For rarer species or species with extremely localized distribution, such as *Acer nikoense* and *A. cissifolium* in this forest, seed dispersal could be critical for local colonization and might be a limiting factor.

10.3.2 Bird Dispersal

Seed shadows of bird-dispersed tree species are actually a mixture of directly fallen (gravity-dispersed) seeds and bird-dispersed seeds. The seeds digested and dispersed by birds are without fleshy pericarp, and the ones that fall directly have the fleshy pericarp intact. In Ogawa Forest, the seed shadows of two bird-dispersed species were obtained by analyzing these dimorphic seed conditions (*Swida controversa*, Masaki et al. 1994; *Kalopanax pictus*, Iida and Nakashizuka 1998). In the case of *K. pictus*, information on the spatial pattern of seedling emergence compensated for the lower detectability (probably caused by their small size) of the bird-dispersed seeds (see Chapter 11).

The dispersal curve of directly fallen seeds declined sharply, almost fitting a log-linear (negative exponential) line for both species. A large number of directly fallen seeds fell under and near the canopy, but no seeds were dispersed more than 35 (*Swida*) and 37 m (*Kalopanax*) from the adult trees (Fig. 10.3). In contrast, the density of bird-dispersed seeds decreased slowly with the distance from an adult tree, and the magnitude of the decrease was not as large as that of the directly fallen seeds. Moreover, the dispersal curve of bird-dispersed seeds even increased at > 40 m from the nearest adult trees. Bird-dispersed seeds could potentially be dispersed to a distance of more than 90 m from the conspecific adult tree in the case of *K. pictus* (Iida and Nakashizuka 1998). The proportion of bird-dispersed seeds to total

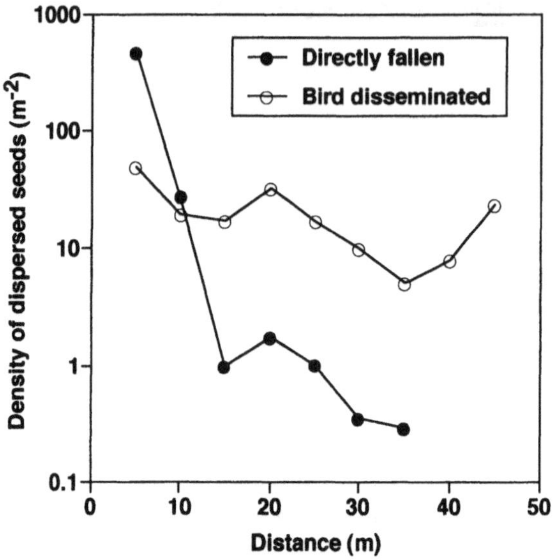

Fig. 10.3. Seed density at 5-m intervals measured from the edge of the nearest mature *Swida controversa* tree crown. Distance class "–5 m" includes a site just under the crown of a conspecific (after the data in Masaki et al. 1994)

seed dispersal was estimated to be only 17% for *Swida controversa* (because of its earlier fruiting phenology than the migration of the main bird-dispersal agent), but birds contributed to the removal of fruits from the vicinity (< 5 m) of the parent trees. More than half (56%) of the bird-dispersed fruits, but only 4% of directly fallen fruits were > 5 m from the parent trees. These patterns of spatial distribution of heteromorphic seeds with different seed dormancy conditions may contribute to the efficient spatiotemporal dispersal of seeds (see Chapter 11).

Swida controversa in OFR showed a heterogeneous spatial pattern of bird-dispersed seed shadows influenced by coexisting plants fruiting in the same season (Fig. 10.4). Since frugivorous birds are usually generalists in temperate forests, the spatial configuration of coexisting conspecific and heterospecific fruiting plants, or the "fruiting environment" (Herrera 1986), is an important factor affecting bird-dispersed seed shadows. For *S. controversa*, the spatial distribution of conspecific trees, heterospecific fruiting trees (two canopy trees and three woody vines), and canopy gaps accounted for 42.9% of the variance of the bird-dissemination pattern. Common frugivorous birds associated *S. controversa* with heterospecific fruiting plants in the seed rain. The heterogeneous spatial pattern of the fruiting environment observed in the bird-dispersed seed shadows of *S. controversa* demonstrated a plant–plant interaction caused by the fruit–frugivore interaction, which is an obvious peculiarity of animal-dispersed seed shadows.

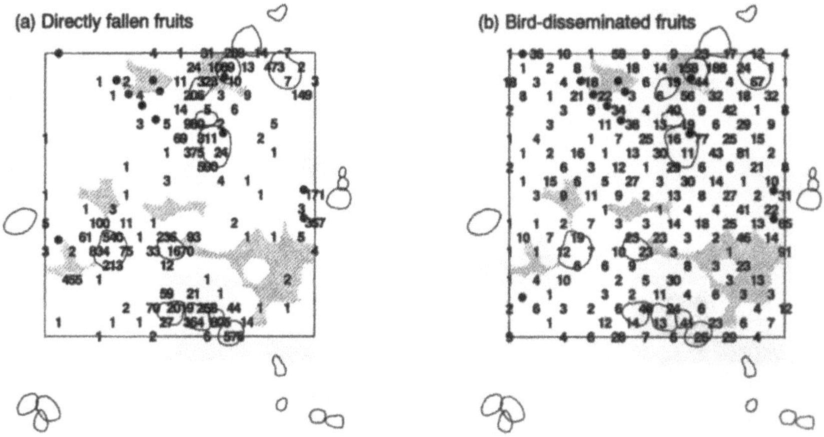

Fig. 10.4. Total numbers and locations of (**a**) directly fallen fruits and (**b**) bird-disseminated fruits of *Swida controversa* collected in traps in the 1-ha plot during 1988. Each *number* shows the number of fruits per trap (0.5 m^2). The *open irregular circles* represent the crowns of mature trees of *Swida controversa* and *shaded areas* represent canopy gaps. *Solid circles* indicate inferred locations of heterospecific fruiting tree crowns (after Masaki et al. 1994)

10.3.3 Dispersal by Seed-Hoarding Animals

Seed shadows of the directly fallen seeds of certain species (*Fagus crenata*, *F. japonica*, *Quercus serrata*, *Q. crispula*, *Castanea japonica*, *Stylax obassia*, and *S. japonica*) were mostly restricted to the space under and near the adult tree crown (Nakashizuka et al. 1995; Iida et al. unpublished data), but the seedling emergence data suggest that transportation of seeds (acorns, nuts) by scatter-hoarding animals actually contributed to enlarge the dispersal distance. Since these acorns and nuts have no period of dormancy, i.e., no temporal dispersal potential (see Chapter 11), dispersal by seed-hoarding animals is important for the successful establishment of these species. From the seed-trap study, it is hard to detect the seed shadow of the species dispersed by seed-hoarding animals. An experimental study of *Quercus serrata* that used magnetic locators (Iida 1996) showed that the average transported distance of acorns by rodents was 21.2 m (maximum = 38.5 m) and the only one acorn survived until the next spring was transported for > 26 m (Fig. 10.5). Although the dispersal probability by vector was low because of the high mortality by predation (see Chapter 11) and the dispersed range was narrow compared with bird-dispersed fleshy fruits, acorn-bearing species had certain dispersal success by food-hoarding small mammals.

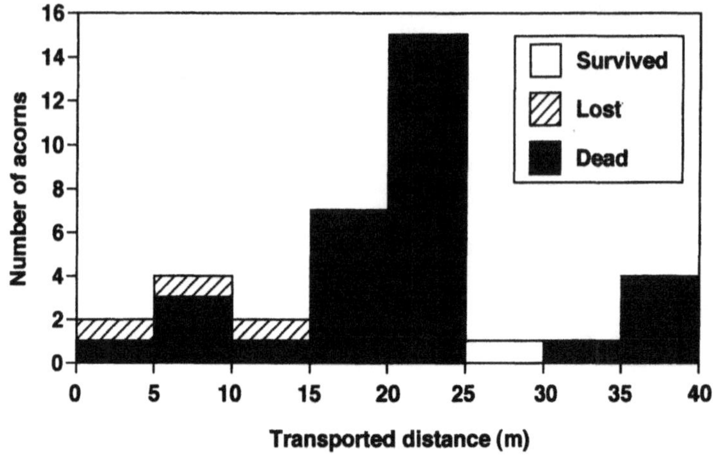

Fig. 10.5. Dispersal of magnet-tagged acorns ($n = 36$: 27 *Quercus serrata* and 9 *Q. acutissima*) by scatter-hoarding animals. The transported distances from the seed source to the places where the acorns were either consumed or found intact in April 1993 are shown. Lost acorns are those which were once detected intact at the place but which could not be found later during the study (after Iida 1996)

10.4 Testing Hypotheses on the Adaptive Significance of Seed Dispersal

10.4.1 Avoiding High Mortality Around Parents

When natural enemies (pathogens, predators, herbivores, parasites, etc.) of the offspring (especially, seeds and seedlings) are concentrated around the parent trees, offspring growing in the vicinity of the parent tree may experience higher mortality than those growing in more distant places (distance-dependent mortality; Janzen 1971). When attack by natural enemies occurs in proportion to the seed or seedling density (density-dependent mortality), mortality patterns will be similar to those for distance-dependent mortality because offspring density is higher around the parent (Connell 1979). Both effects have been recognized as selective pressures favoring the evolution of dispersal (Howe and Smallwood 1982; Willson 1992; Dirzo and Dominguez 1986), and they are known as the Janzen-Connell hypothesis (Clark and Clark 1984).

In the OFR, spatial patterns of seed and seedling mortality suggested that mechanisms such as those of the Janzen-Connell hypothesis operated for some species, but not for others. As is discussed in detail in Chapter 11, high seed mortality was detected for many species in this forest. Three of the four *Carpinus* species in this forest (*C. laxiflora*, *C. tschonoskii*, and *C. cordata*), which have wind-dispersed seeds, showed significant positive correlation between postdispersal seed mortality (estimated from the ratio of seedlings emerged in a quadrat to the number of seeds fallen

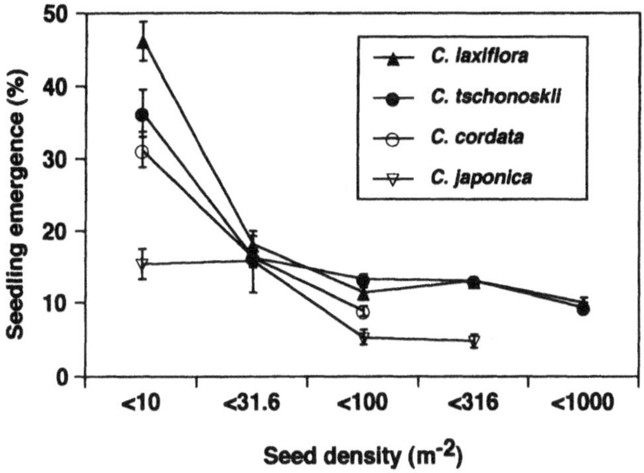

Fig. 10.6. Density-dependent seedling emergence ratios of *Carpinus* species. Three species (except *C. japonica*) had significant partial correlations with seed density ($P < 0.001$) in multiple regression analyses taking into account other factors relating to environmental conditions (after Nakashizuka 1996)

into the trap adjacent to that quadrat) and the density of dispersed seeds (Shibata and Nakashizuka 1995; Nakashizuka et al. 1995; Fig. 10.6). It is likely that predation by wood mice and ground-foraging birds is the major cause of density-dependent mortality of *Carpinus* seeds (see Chapter 11; M. Shibata, personal communication). Density-dependent postdispersal seed mortality was also observed for *Swida controversa* (Masaki et al. 1998) and *Acer mono* (Tanaka 1995), and the highly probable cause of mortality was predation by wood mice (*Apodemus speciosus* and *A. argenteus*). Since the fleshy mesocarp of *Swida* seeds attracts predators, the high density of nondispersed seeds fallen near the parent trees also led to increased mortality of seeds in a distance-dependent manner (see Chapter 11; Masaki et al. 1998). Two *Carpinus* species (*C. japonica, C. cordata*) and *Kalopanax pictus* showed significant positive correlation between the survival rate of first-year seedlings and the distance from the nearest conspecific adults, but the reasons for the high mortality near conspecific adult trees are still unknown (Shibata and Nakashizuka 1995; Nakashizuka et al. 1995; Iida and Nakashizuka 1998). In these cases, postdispersal seed mortality was density-dependent, while seedling mortality was distance-dependent, suggesting different mechanisms operated between the seed and seedling stages. However, density- or distance-dependent causes of mortality (e.g., predation, herbivory, fungal infection) could not, with a few exceptions, be detected clearly. Our intensive observation of seedling survivorship (every two weeks) was not frequent enough to identify the specific mortality factors (Shibata and Nakashizuka 1995). *Quercus serrata*, a relatively shade-intolerant large-seeded species, showed no such spatial mortality patterns of seeds and seedlings (Iida, unpublished data).

Considering the relative contribution to the increase in plant fitness, escaping from the area of high mortality around the parent tree will be more important for species that have relatively higher survival rates under closed canopies, such as *Acer mono* in this forest. The population of offspring of species that require special physical environments to become established (e.g., highly gap-dependent or pioneer species) will decrease very quickly regardless of the distance from the parent or seed/seedling density (see Chapter 12). Thus, for such species, the main selective pressure favoring dispersal seems to be increasing the probability of finding suitable sites for establishment (see Chapter 12; Martinez-Ramos and Alvarez-Buylla 1986; Murray 1988; Shibata and Nakashizuka 1995).

10.4.2 Colonization or Finding Safe Sites

Seed dispersal can increase the chance of reaching a "safe site," or a favorable site for the establishment and later growth of the seedlings, if environmental heterogeneity (spatial and temporal variation of environment) is present (Venable and Brown 1988, 1993). Although this has been accepted as obvious, it is true only in the sense that dispersal contributes to the averaging of the probability of demographic success: escaping the risk of low success (or failure) at one site (near the parent) while sacrificing the chance of having all seeds at one rare high-yield site (near the parent) (Venable and Brown 1988). Safe site variation is produced both by relatively stable physical conditions (such as topography) and by unpredictable disturbances. In a forest community, open sites created by canopy disturbances are especially important safe sites for many species (see Chapter 12; Brokaw 1986; Martinez-Ramos and Alvarez-Buylla 1986). The size and frequency of safe-site patches depend greatly on the disturbance regime of the forest.

The quality of safe sites exists as a continuum, not as a dichotomy (Schupp 1993). The parameters related to establishment (seed germination and seedling survival rate) can be used as indices of safe site quality for a species (Andersen 1989); sites with a high establishment rate are safe sites of high quality. Each species has its own frequency distribution of safe sites in a community (safe site abundance, Fig. 10.7). The increased fitness through dispersal should be evaluated by the establishment success in this heterogeneous environment: the result of the interaction between the size of the seed shadow and safe site distribution in the community.

In the OFR, a major disturbance is a single or multiple tree fall that creates gaps with a mean size of 80 m^2 at a formation rate of 0.42% yr^{-1} (see Chapters 7 and 12; Nakashizuka et al. 1992). These values are comparable to those of other old-growth temperate forests (Runkle 1985; Brokaw 1985; Denslow 1987; Yamamoto 1992). Based on mechanistic analyses of the seed dispersal of six wind-dispersed species (Tanaka et al. 1998), seed shadows of the species in this forest (e.g., the area with seed fall > 1 m^{-2}) covers 10 to 10^2 times the mean gap size. Bird-dispersed species (*Swida* and *Kalopanax*) also had seed shadows of the same size. This may explain why many species seem to have relatively short-range dispersal (Willson 1993). If the forest is under a regime with larger but more infrequent disturbances (e.g., fire or large-scale blowdown), the patch size of the safe site would exceed the scale of these seed shadows, and safe sites would be created rather episodically. If the size of the disturbed patches is much larger than that of the seed shadow, models to evaluate

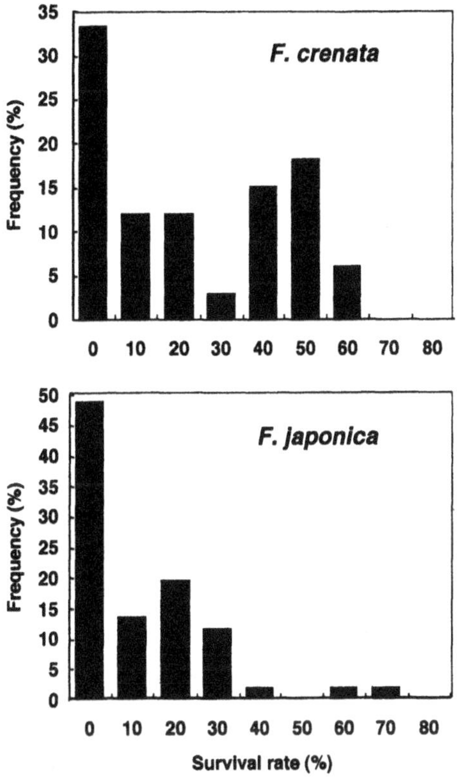

Fig. 10.7. Distribution of safe sites. "Safe sites" (sites with favorable conditions for seedling survival) are evaluated by the seedling survival rate (the ratio of seedlings surviving through the first growing season to the total number of seedlings emerged). Data for *Fagus crenata* (**a**) and *F. japonica* (**b**) from 221 seedling quadrats are presented. On the basis of data from quadrats with > 10 emerged seedlings m^{-2}, the frequency (%) of the quadrats in each survival rate class was calculated

dispersal success should be different from those built for frequent small-scale disturbances; the importance of the tail of the dispersal curve or seed bank should be stressed (cf. Geritz et al. 1984).

Dispersal efficacy seemed to correlate with the abundance of safe sites for the wind-dispersed species in the Ogawa Forest (cf. Green 1983; Willson 1993). A negative correlation between first-year seedling survival and seed-shadow size in the six wind-dispersed species was observed (Fig. 10.8), as was also reported by Augspurger (1986). Among these species, seedling survival had a significantly positive correlation with seed size, and seed size itself had a significantly negative correlation with the number of seeds dispersed per fruiting individual, which is an important parameter for enlarging the seed shadow (Tanaka et al. 1998; Fig. 10.9). As a result, the species with many small seeds tend to have higher dispersal efficacy, but the safe sites for the species are more restricted. Even when the species with other dispersal modes are included in the analysis, safe site abundance for a species might also have a negative correlation with seed shadow size, but a positive one with seed size. The trade-off between dispersal efficacy and safe site abundance supports the colonization hypothesis, but the existence or nonexistence of such a trade-off in this community should be determined after detailed analyses of more species of different dispersal types.

As a temporal aspect of dispersal, seed dormancy has the effect of buffering

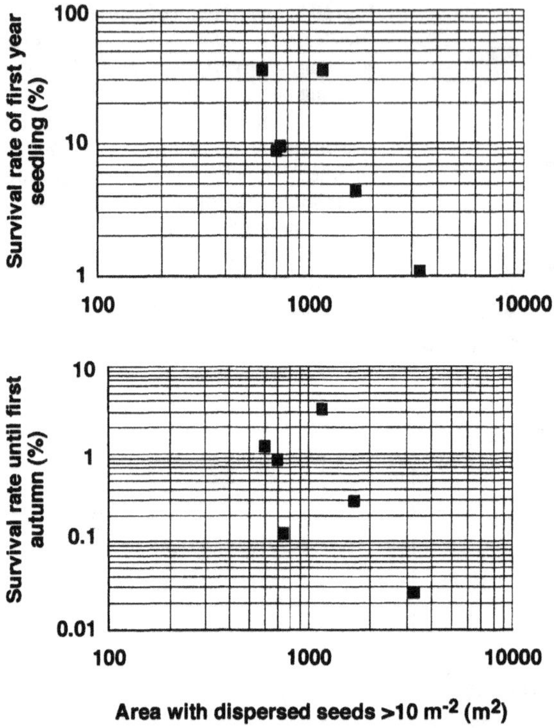

Fig. 10.8. Safe site abundance and dispersal efficacy of six wind-dispersed species (*Acer mono*, *A. amoenum*, *Betula grossa*, *Ostrya japonica*, *Carpinus laxiflora*, and *C. tschonoskii*) in Ogawa Forest Reserve. Safe site abundance is defined here as the mean survival rate of first-year seedlings after emergence (*top panel*) and the survival rate until the first autumn after dispersal (*bottom panel*). The survival rates were averaged among the 221 sets of seed traps and quadrats established regularly in a 1-ha area. The area of seed shadow (more than 10 seeds m^{-2}) per parent tree is considered to be an index of dispersal efficacy. Correlation coefficients were –0.79 (top, $P < 0.05$) and –0.63 (bottom, n.s.) (after Nakashizuka et al. 1995)

temporal variations in viable seed density, and the capacity for extended dormancy might increase the probability of finding safe sites, which are temporally unpredictable (see also Chapter 11). The fluctuation of the density of annual seedling emergence is considerably lower than that of annual seed production for species that have dormant seeds (Shibata and Nakashizuka 1995; Tanaka 1995; Iida and Nakashizuka 1998; Masaki et al. 1998; Abe et al. 1998). Seed dormancy may enhance the temporal probability of finding a safe site, or compensate for the insufficiency of spatial dispersal efficacy. In the OFR, a comparative study of seed and seedling stages of *Carpinus* and *Acer* showed that species with relatively low dispersal (*Carpinus cordata*, *Acer amoenum*) had extended dormancy (one year delayed germination) among congeneric species (see Chapter 11; Shibata and Nakashizuka 1995; Tanaka 1995). In the case of *Kalopanax pictus*, a bird-dispersed species, the nondispersed seeds with fleshy fruits showed extended seed dormancy and the species exhibited a

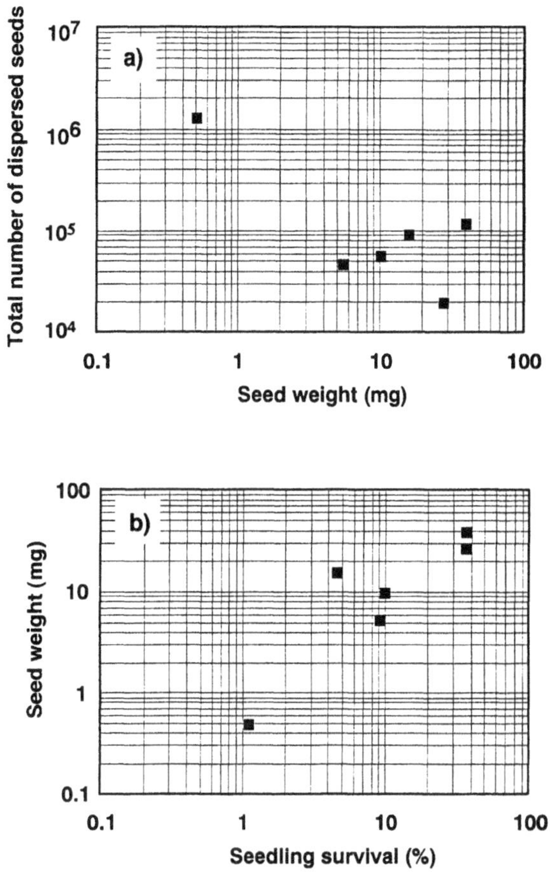

Fig. 10.9. Trade-off relationships between the total number of dispersed seeds and seed size. **a** Negative relationship between the total number of dispersed seed and seed weight ($P < 0.01$), and **b** positive relationship between seed weight and seedling survival ($P < 0.05$) (after Tanaka et al. 1998)

temporally different seedling emergence pattern near and far from the parent trees (Iida and Nakashizuka 1998). In this case, bird dispersal contributed not only to spatial dispersal, but also to the enhancement of temporal variation in germination (see Chapter 11). Seed bank dynamics are regulated not only by seed dormancy or germination habit, but also by many other factors such as predation by vertebrates or insects and fungal attack. The possible formation of a persistent long-term seed bank in the OFR was suggested only for a few species such as *Swida controversa*.

10.4.3 Directional Dispersal

Seed dispersal increases fitness if the agents of the dispersal carry the seeds "directionally" to nonrandom sites favorable for seedling establishment (Venable and Brown 1993). Well-known examples are the ant dispersal of several herbaceous species (e.g., *Carex*, *Viola*) and bird dispersal of mistletoes (see Howe and Smallwood 1982). Since canopy gaps have been considered as safe sites for many tree species in forest communities (Brokaw 1986; Martinez-Ramos and Alvarez-Buylla 1986; Shibata and Nakashizuka 1995), investigations have concentrated mainly on the question of whether the dispersal agents carry the seeds disproportionately to gaps. In some cases, seeds were dispersed disproportionately to gaps (Thompson and Willson 1978; Herrera and Jordano 1981; Hoppes 1988), but not in others (Brokaw 1986; Murray 1988). Burial of scatter-hoarded seeds by birds or rodents has also been suggested to be a kind of directed dispersal providing favorable conditions for germination (Miguchi 1994).

By tracking the fate of individual acorns over a winter with a magnetic locator, Iida (1996) showed that the dispersal pattern of scatter-hoarded *Quercus serrata* acorns partly supports the hypothesis that this mode of dispersal is in effect directed dispersal for this species in the OFR. Many of the dispersed acorns were found in and around the burrows of wood mice (*Apodemus speciosus*), which were concentrated around fallen branches or logs (Iida 1996). Fallen branches or logs covers the forest floor and may provide refuges for rodents from their predators (Rood and Test 1968). Since fallen branches and logs are mainly located in canopy gaps, acorns cached in such a place (if they are not eaten) will be more likely to become established. However, gaps with fallen logs are not always newly created gaps, and old gaps are not safe sites for seeds or seedlings (see Chapter 12). The hoarding depth of surviving acorns should also be estimated to test the effectiveness of scatter-hoarding in dispersing the seeds to favorable germination sites, because some of the acorns were found in a burrow too deep for seedling emergence (Iida 1996).

Seed dispersal of *Swida controversa* by birds was also found to be nonrandom, but bird-dispersed seeds were not necessarily delivered closer to canopy gaps in the OFR (Masaki et al. 1994, Fig. 10.4). Fruiting species were not disproportionately abundant or attractive to birds in gaps compared with the canopy fruiting trees. Gaps with fruiting plants attractive to birds are, if they exist, old gaps and not safe sites. Although heterospecific fruiting plants had a positive effect on the bird-dispersed seed shadows of *S. controversa*, it is also difficult to regard the forest floor under heterospecific fruiting trees as a safe site. An abundance of heterospecific fruiting trees may promote widely scattered dispersal of *Swida* seeds and, thus, simply enlarge the possibility of finding safe sites.

10.5 Conclusion

The role of seed dispersal in tree population dynamics has been emphasized as an important, but often neglected aspect of forest community dynamics. To estimate the increased fitness of tree populations through seed dispersal and to evaluate its contribution to community organization, we need to obtain demographic information on each species throughout its life cycle, since the relative importance of the early stages of a tree life cycle (seed dispersal and seedling establishment) varies among species in a community (Harcombe 1987). Additionally, the spatial scale of disturbance relative to the size of seed shadows and the temporal variation of disturbances relative to the interval between mast years and to the length of seed dormancy are all critical variables for estimating the reproductive success of trees. A long-term study in a large permanent plot is important, because it enables us to obtain enough data not only for demographic analyses but also for studies of spatiotemporal variations in demographic parameters. We have quantitatively analyzed several important aspects of seed dispersal and were able to partly estimate its contribution to the population growth and community organization in the OFR; however, we still need more information about the long-term consequences of seed dispersal. Observing the fate of seedlings germinated from seeds dispersed in different manners (including those not dispersed), at varied distances from the parent trees for different seeding years would be a straightforward approach to the estimation of the contribution of seed dispersal to population recruitment.

References

Abe S, Nakashizuka T, Tanaka H (1998) Effects of canopy gaps on the demography of the sub-canopy tree species *Styrax obassia*. J Veg Sci 9:787–796

Alvarez-Buylla ER, Martinez-Ramos M (1990) Seed bank versus seed rain in the regeneration of a tropical tree. Oecologia 84:314–325

Andersen AN (1989) Effects of seed predation by ants on seedling densities at a woodland site in SE Australia. Oikos 48:171–174

Augspurger CK (1984) Seedling survival among tropical tree species: interactions of dispersal distance, light-gaps, and pathogens. Ecology 65:1705–1712

Augspurger CK (1986) Morphology and dispersal potential of wind-dispersed diaspores of neotropical trees. Am J Bot 73:353–363

Baker HG (1972) Seed mass in relation to environmental conditions in California. Ecology 53:997–1010

Brokaw NVL (1985) Treefalls, regrowth, and community structure in tropical forests. In: Pickett STA, White PS (eds) The ecology of natural disturbance and patch dynamics. Academic, New York, pp 53–69

Brokaw NVL (1986) Seed dispersal, gap colonization, and the case of *Cecropia insignis*. In: Estrada A, Fleming TH (eds) Frugivores and seed dispersal. Dr W Junk, Dordrecht, The Netherlands, pp 323–331

Clark DA, Clark DB (1984) Spacing dynamics of a tropical rain forest tree: evaluation of the Janzen–Connell model. Am Nat 124:769–788

Clark JS, Macklin E, Wood L (1998) Stages and spatial scales of recruitment limitation in southern Appalachian forests. Ecol Monogr 68:213–235

Connell JH (1979) Tropical rain forests and coral reef as open non-equilibrium systems. In: Anderson RM, Turner BD, Taylor LR (eds) Population dynamics. Blackwell, Oxford, pp 141–163

Denslow JS (1987) Tropical rainforest gaps and tree species diversity. Annu Rev Ecol Syst 18:431–451

Dirzo R Dominguez CA (1986) Seed shadows, seed predation and the advantages of dispersal. In: Estrada A, Fleming TH (eds) Frugivores and seed dispersal. Dr W Junk, Dordrecht, The Netherlands, pp 237–249

Foster SA, Jahnson CH (1985) The relationship between seed size and establishment conditions in tropical woody plants. Ecology 66:773–780

Geritz SAH, de Jong TJ, Klinkhamer PGL (1984) The efficacy of dispersal in relation to 'safe site' and seed production. Oecologia 62:219–221

Green DS (1983) The efficacy of dispersal in relation to 'safe site' density. Oecologia 56:356–358

Greene DF, Johnson EA (1989) A model of wind dispersal of winged or plumed seeds. Ecology 70:339–347

Harcombe PA (1987) Tree life table: Simple birth, growth, and death data encapsulate life histories and ecological roles. Bioscience 37:557–568

Herrera CM (1986) Vertebrate-dispersed plants: why they don't behave the way they should. In: Estrada A, Fleming TH (eds) Frugivores and seed dispersal. Dr W Junk, Dordrecht, The Netherlands, pp 5–18

Herrera CM, Jordano P (1981) *Prunus mahaleb* and birds: the high-efficiency seed dispersal system of a temperate fruiting tree. Ecol Monogr 51:203–218

Hoppes WG (1988) Seed fall pattern of several species of bird-dispersed plants in an Illinois woodland. Ecology 69:1092–1095

Houle G (1994) Spatiotemporal patterns in the components of regeneration of four sympatric tree species—*Acer rubrum*, *A. saccharum*, *Betula alleghaniensis* and *Fagus grandifolia*. J Ecol 82:39–53

Howe HF, Smallwood J (1982) Ecology of seed dispersal. Annu Rev Ecol Syst 13:201–228

Howe HF, Schupp EW, Westley LC (1985) Early consequences of seed dispersal for a neotropical tree (*Virola surinamensis*). Ecology 66:781–791

Iida S (1996) Quantitative analysis of acorn transportation by rodents using magnetic locator. Vegetatio 124:39–43

Iida S, Nakashizuka T (1998) Spatial and temporal dispersal of *Kalopanax pictus* seeds in a temperate deciduous forest, central Japan. Plant Ecol 135:243–248

Janzen DH (1971) Seed predation by animals. Annu Rev Ecol Syst 2:465–492

Jordano P (1992) Fruit and frugivory. In: Fenner M (ed) Seeds. The ecology of regeneration in plant communities. CAB International, Wallingford, UK, pp 105–156

Martinez-Ramos M, Alvarez-Buylla ER (1986) Seed dispersal, gap dynamics and tree recruitment: the case of *Cecropia obtusifolia* at Los Tuxtlas. In: Estrada A, Fleming TH (eds) Frugivores and seed dispersal. Dr W Junk, Dordrecht, The Netherlands, pp 333–346

Masaki T, Suzuki W, Niiyama K, Iida S, Tanaka H, Nakashizuka T (1992) Community structure of a species-rich temperate forest, Ogawa Forest Reserve, central Japan. Vegetatio 98:97–111

Masaki T, Kominami Y, Nakashizuka T (1994) Spatial and seasonal patterns of seed dissemination of *Swida controversa* in a temperate forest. Ecology 75:1903–1910

Masaki T, Tanaka H, Shibata M, Nakashizuka T (1998) The seed bank dynamics of *Swida controversa* and their role in regeneration. Seed Sci Res 8:53–63

McPeek MA, Holt RD (1992) The evolution of dispersal in spatially and temporally varying environments. Am Nat 140:1010–1027

Miguchi H (1994) Role of wood mice on the regeneration of cool temperate forest. In: Kobayashi S, Nisikawa K, Danilin IM, Matsuzaki T, Abe N, Kamitani T, Nakashizuka T (eds) Proceedings of NAFRO seminar on sustainable forestry and its biological environment, Japan Society of Forest Planning Press, Tokyo, pp 115–121

Murray KG (1988) Avian seed dispersal of three neotropical gap-dependent plants. Ecol Monogr 58:271–298

Nakashizuka T (1996) Factors maintaining forest tree diversity. In: Turner IM, Diong CH, Lim SSL, Ng PKL (eds) Biodiversity and the dynamics of ecosystems. The international network for DIVERSITAS in Western Pacific and Asia (DIWPA), pp 51–62

Nakashizuka T, Iida S, Tanaka H, Shibata M, Abe S, Masaki T, Niiyama K (1992) Community dynamics of Ogawa Forest Reserve, a species rich deciduous forest, central Japan. Vegetatio 103:105–112

Nakashizuka T, Iida S, Masaki T, Shibata M, Tanaka H (1995) Evaluating increased fitness through dispersal: A comparative study on tree population in a temperate forest, Japan. Ecoscience. 2:245–251

Okubo A, Levin SA (1989) A theoretical framework for data analysis of wind dispersal of seeds and pollen. Ecology 70:329–338

Rood JP, Test FH (1968) Ecology of the spiny rat, *Heteromys anomalus*, at Rancho Grande, Venezuela. Am Midl Nat 79:89–102

Runkle JR (1985) Disturbance regimes in temperate forests. In: Pickett STA, White PS (eds) The ecology of natural disturbance and patch dynamics. Academic. New York, pp 17–33

Salisbury EJ (1942) The reproductive capacity of plants. Bell, London

Schupp EW (1993) Quantity, quality and the effectiveness of seed dispersal by animals. Vegetatio 107/108:15–29

Shibata M, Nakashizuka T (1995) Seed and seedling demography of co-occurring *Carpinus* species in a temperate deciduous forest. Ecology 76:1099–1108

Tanaka H (1995) Seed demography of tree co-occurring *Acer* species in a Japanese temperate deciduous forest. J Veg Sci 6:887–896

Tanaka H, Shibata M, Nakashizuka T (1998) A mechanistic approach for evaluating the role of wind dispersal in tree population dynamics. J Sust For 6:155–174

Thompson JN, Willson MF (1978) Disturbance and the dispersal of fleshy fruits. Science 20:1161–1163

Venable DL, Brown JS (1988) The selective interactions of dispersal, dormancy, and seed size as adaptations for reducing risk in variable environments. Am Nat 131:360–384

Venable DL, Brown JS (1993) The population-dynamic functions of seed dispersal. Vegetatio 107/108:31–55

Willson MF (1992) The ecology of seed dispersal. In: Fenner M (ed) Seeds. The ecology of regeneration in plant communities. CAB International, Wallingford, UK, pp 61–85

Willson MF (1993) Dispersal mode, seed shadows, and colonization patterns. Vegetatio 107/108:261–280

Yamamoto S (1992) Gap characteristics and gap regeneration in primary evergreen broad-leaved forests of western Japan. Bot Mag Tokyo 105:29–45

11 Seed Dynamics

SHIGEO IIDA and TAKASHI MASAKI

11.1 Introduction

Seeds of various plant species exist in forest soil (e.g., Nakagoshi 1985; Garwood 1989; Pickett and McDonnell 1989). Viable buried seeds are regarded as potential or initial plant communities, in contrast to plants growing on the ground (Nakagoshi 1985). Quick plant recruitment from a seed bank takes place even if the vegetation disappears as the result of disturbance.

A seed-bank strategy is considered to be important, especially for species that depend on infrequent disturbances to regenerate (Marks 1974; Howe and Smallwood 1982; Young et al. 1987; Murray 1988; Alvarez-Buylla and Garcia-Barrios 1991; Thompson 1992). In general, pioneer tree species tend to make seed banks (Nakagoshi 1985; Thompson 1992), while shade-tolerant tree species tend to make seedling banks (Silvertown and Lovett Doust 1993). However, these tendencies are not necessarily fixed, and some exceptions exist (Alvarez-Buylla and Martinez-Ramos 1990; Abe 1996). The relative significance of a seed-bank strategy will differ among woody species because other alternatives exist, such as seed rain or a seedling bank.

A negative aspect of a persistent seed bank is that seeds which remain in the soil for a long time may have a high mortality rate due to predation or senescence (Cavers 1983; Howe and Smallwood 1982). Therefore, there is a trade-off between the seed-bank strategy of timely germination but with fewer offspring, and other alternatives which mean less timely germination but result in more offspring.

To evaluate the importance of a seed-bank strategy in the regeneration process of woody species, it is necessary to undertake a quantitative investigation of seed-bank dynamics together with the other regeneration strategies (e.g., seed rain, seedling bank) of each tree species. In this chapter, we document the features of seed-bank dynamics for the main woody species in Ogawa Forest Reserve (OFR), and discuss their ecological significance based on the results obtained.

Ecological Studies, Vol. 158
Nakashizuka, Matsumoto (eds.) Diversity and Interaction in a Temperate Forest Community: Ogawa Forest Reserve of Japan

11.2 Seed Fall and Seedling Emergence

Germination patterns and the duration of seed banks of 17 important woody species in OFR are discussed in relation to the results obtained during 9 years of monitoring seed fall (1987–1995) and seedling emergence (1988–1996). The density of seed fall and seedling emergence was monitored using 263 pairs of seed trap (0.5 m^2) and adjacent quadrat (1 m^2) installed in a 1.2-ha subplot set in the 6-ha plot (see Chapters 5 and 9). Based on these data, the densities of total seed rain and total seedling emergence within 1.2 ha were calculated. These 17 tree species were classified into types A, B, C, and D depending on their differences in germination pattern and the estimated duration of their seed bank (Table 11.1).

The ten tree species of types A and B are significantly correlated in relation to the mean densities of seedling emergence and the sound seed fall in the previous year (Table 11.1). The five type A Fagaceae species (*Castanea crenata, Fagus crenata, F. japonica, Quercus crispula,* and *Q. serrata*) do not generally produce a seed bank which lasts more than 1 year. The seeds of the five type B species (*Acer mono* var. *marmoratum* f. *dissectum, Betula grossa, Carpinus japonica, C. laxiflora,* and *C. tschonoskii*) usually germinate in the spring following the seed fall, although a few new seedlings are found even if there was no seed fall in the previous year. The mean density of seedling emergence sometimes exceeds the mean density of seed fall in the previous year (Table 11.2). Kiyono and Kawahara (1989) reported that 91% of *Betula grossa* seeds artificially buried in soil were viable 1 year after burial. Viable buried seeds of *C. japonica* and *C. tschonoskii* were found in late May in a temperate forest, whereas no viable buried seeds of *A. mono* var. *marmoratum* f. *dissectum*, *C. crenata*, *F. crenata,* or *Q. crispula* were found (Nakagoshi 1985). Small numbers of viable buried seeds of *A. mono* var. *marmoratum* f. *dissectum* were found in soil samples taken in September (Fig. 11.1), and most of a sample of *A. mono* var. *marmoratum* f. *dissectum* seeds packed in stainless steel wire bags set on the forest floor germinated in the following spring (Tanaka 1995). Therefore, although most seeds of type B species germinate in the first germination season, some remain for more than a year to form a seed bank. However, they probably do not remain viable for very long.

The three type C species (*Carpinus cordata*, *Acer rufinerve*, and *A. amoenum*) showed a significant correlation between their mean densities of seedling emergence and their autumn seed fall 2 years earlier (Table 11.1). Most seeds of these species germinate in the second germination season, suggesting that they make a seed bank which remains viable for more than 1 year. The time required for seed germination in *Acer* species was categorized into two types: one which germinates quickly (< 3 months), and one which takes longer to germinate (> 12 months) (Katsuta et al. 1998). From the results of an examination of seed mortality that excluded predation on the forest floor, it was found that *Acer amoenum* germinated in the second spring for the first time, while *Acer rufinerve* germinated in both the first and second springs (Tanaka 1995).

The four type D species (*Swida controversa, Kalopanax pictus, Ostrya*

Table 11.1. Relationship between mean densities of seedling emergence and mean densities of sound seed fall of previous years in a 1.2-ha subplot for each of the 17main tree species, and the ecological characteristics of their seeds

Species	Germination type	Correlation coefficient between mean seedling emergence densities and mean sound seed fall densities N years before			Estimated germination ratio (%)	Seed weight (g)	Seed type	Dispersal type
		$N = 1$ ($n = 9$)	$N = 2$ ($n = 8$)	$N = 3$ ($n = 7$)				
Castanea crenata	A	0.87**	–0.35	–0.27	8.9	2.98	Acorn	Dyszoochory
Fagus crenata	A	1.00**	–0.16	–0.17	4.5	0.166	Acorn	Dyszoochory
Fagus japonica	A	0.87**	–0.29	–0.12	2.4	0.184	Acorn	Dyszoochory
Quercus crispula	A	0.79**	–0.61	0.14	10.4	3.05	Acorn	Dyszoochory
Quercus serrata	A	0.72*	–0.21	–0.57	7.6	1.21	Acorn	Dyszoochory
Acer mono var. *marmoratum* f. *dissectum*	B	0.75*	–0.06	–0.42	11.4	0.041	Samara	Anemochory
Betula grossa	B	0.78**	0.13	–0.44	1.5	0.003	Nutlet	Anemochory
Carpinus japonica	B	0.97***	–0.27	0.06	5.4	0.004	Nutlet	Anemochory
Carpinus laxiflora	B	0.90***	–0.44	0.58	5.9	0.004	Nutlet	Anemochory
Carpinus tschonoskii	B	0.86**	–0.47	0.57	7.8	0.013	Nutlet	Anemochory
Acer amoenum	C	–0.36	0.87**	–0.37	10.4	0.016	Samara	Anemochory
Acer rufinerve	C	–0.04	0.95***	–0.29	31.6	0.019	Samara	Anemochory
Carpinus cordata	C	–0.26	0.96***	–0.45	1.8	0.011	Nutlet	Anemochory
Swida controversa	D	0.12	0.26	0.37	2.3	0.043	Drupe	Endozoochory
Kalopanax pictus	D	0.61	–0.02	–0.37	0.4	0.005	Drupe	Endozoochory
Ostrya japonica	D	–0.10	0.66	0.09	1.0	0.014	Nutlet	Anemochory
Styrax obassia	D	–0.46	0.55	0.27	14.4	0.402	Capsule	Dyszoochory

A, all seeds germinate within 1 year after seed fall; B, most seeds germinate within 1 year after seed fall, but a very small portion of the seeds form seed banks which last more than 1 year; C, form seed banks which last more than 1 year and mainly germinate in the second germination season; D, form seed banks which last more than 1 year without any particular germination pattern.

***$P < 0.001$; **$P < 0.01$; *$P < 0.05$.

Estimated germination rate, proportion of total seedling emergence to total seed fall during a period of 9 years.

Table 11.2. Mean densities of seed fall and seedling emergence in a 1.2-ha subplot for nine tree species with B- or D-type germination

Fruiting year	B-type germination										D-type germination							
	Acer mono var. *marmoratum* f. *dissectum*		*Betula grossa*		*Carpinus japonica*		*Carpinus laxiflora*		*Carpinus tschonoskii*		*Swida controversa*		*Kalopanax pictus*		*Ostrya japonica*		*Styrax obassia*	
	SFD	SED	SFD	SED	SFD	SED	SFD	SED	SFD	SED	SFD	SED	SFD	SED	SFD	SED	SFD	SED
1987	14.08	1.14	241.12	2.12	0.00	0.00	0.00	0.24	0.00	0.24	0.38	0.45	0.00	0.17	0.12	0.14	0.03	0.02
1988	34.35	4.05	10.91	1.27	8.74	0.64	47.36	6.04	39.65	5.52	98.23	0.98	0.09	0.14	38.48	0.20	4.43	0.02
1989	0.26	0.51	0.00	0.03	0.00	0.03	0.00	0.12	0.00	0.09	0.02	1.07	0.00	0.07	0.00	0.20	0.00	0.22
1990	91.16	9.34	102.08	2.35	0.10	0.07	7.16	1.36	11.64	2.01	48.41	1.91	131.81	0.27	15.32	0.51	0.31	0.20
1991	0.73	0.74	8.88	0.26	0.00	0.00	0.00	0.07	0.08	0.07	0.00	0.38	0.02	0.10	0.64	0.08	0.13	0.08
1992	0.64	0.06	1.21	0.01	0.02	0.00	0.02	0.01	0.00	0.02	0.13	0.31	0.00	0.03	0.64	0.01	0.00	0.11
1993	96.63	5.88	37.21	0.46	1.98	0.17	42.55	3.31	67.65	6.03	95.82	0.47	6.76	0.01	60.33	0.03	1.07	0.01
1994	0.18	6.30	33.57	0.57	0.00	0.12	0.00	0.85	0.02	1.08	0.00	1.73	0.02	0.09	0.15	0.92	0.13	0.24
1995	12.07	0.50	139.47	1.32	30.69	1.21	266.12	9.43	162.60	6.79	127.72	1.11	114.13	0.12	131.31	0.29	1.14	0.15

SFD, mean seed fall density (m^{-2}) in autumn; SED, mean seedling emergence density (m^{-2}) in spring the following year.

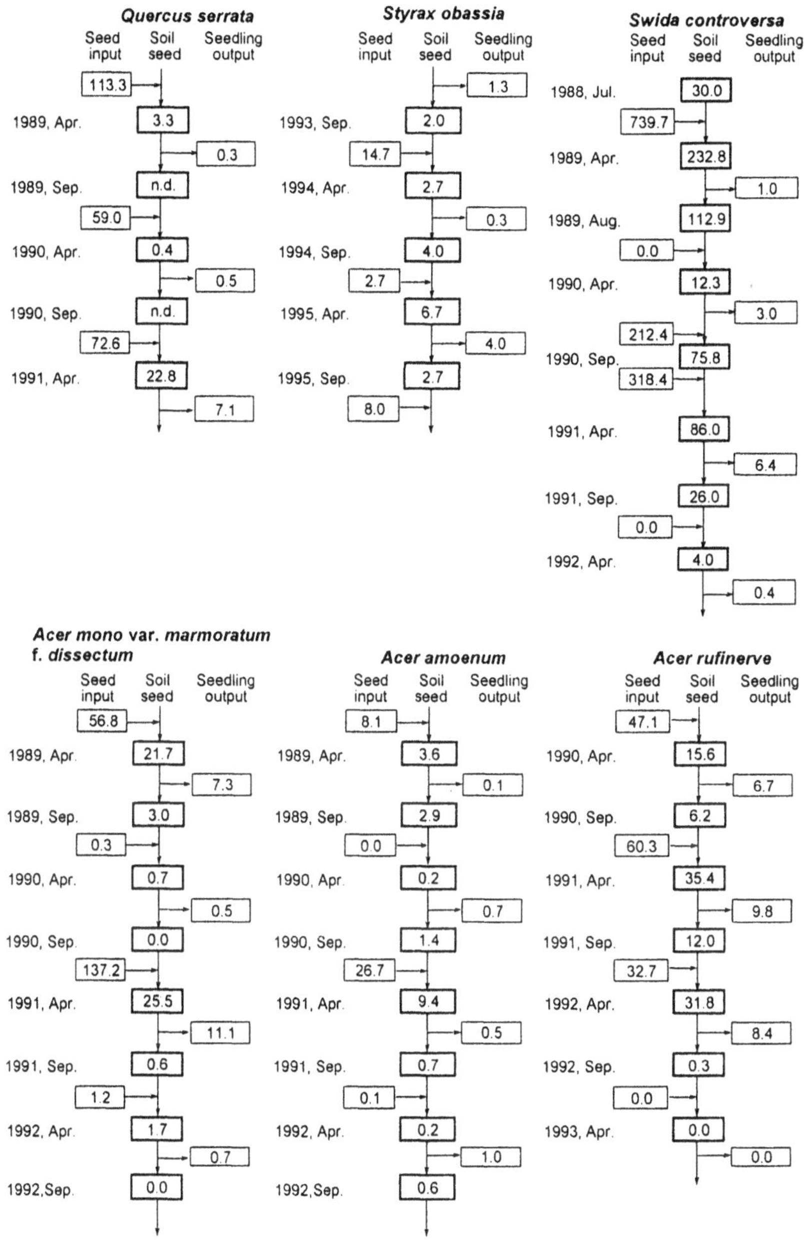

Fig. 11.1. Seed bank dynamics of six tree species under conspecific adult trees. Values are sound seed and seedling densities (m^{-2}). Data are from Iida (unpublished results 1991) for *Quercus serrata*, Abe (1996) for *Styrax obassia*, Masaki et al. (1998) for *Swida controversa*, and Tanaka (1995) for *Acer mono* var. *marmoratum* f. *dissectum*, *A. amoenum* and *A. rufinerve*. *n.d.*, no data

japonica, and *Styrax obassia*) showed no clear relationship between their mean densities of seedling emergence and their seed fall in the previous year (Table 11.1). However, the mean density of seedling emergence of *K. pictus* showed a positive correlation ($P < 0.1$) with the mean density of seed fall 1.5 years earlier. According to the results of a germination test which lasted for 3 years, some *K. pictus* seeds were still dormant after the third germination season (Iida and Nakashizuka 1998). The mean density of seedling emergence of *O. japonica* showed a positive correlation ($P < 0.1$) with the mean density of seed fall 2 years earlier, indicating that *O. japonica* seeds have a physiologically deep endogenous dormancy. *O. japonica* seeds which fall naturally from mother trees can germinate in forests in the first spring, while they germinate in the second spring if they are sown after air-drying (Katsuta et al. 1998). Buried *S. controversa* seeds may be able to maintain their germinative capacity for more than 10 years (Ozawa 1950). *Styrax obassia* seeds maintain their germinative capacity for more than 40 years (Ozawa 1950). Seedling emergence of D-type species was observed even when there was no seed fall in the previous year.

The estimated germination rates from the total seed fall and seedling emergence in forest floor conditions for 9 years showed a large variation (0.4%–31.6%) among the 17 tree species (Table 11.1). However, most fallen seeds (68.4%–99.6%) did not germinate, and these were assumed to have been killed by predation or fungous disease, and/or by lying dormant in the soil.

11.3 Temporal Changes in the Seed Bank

The seed bank dynamics were estimated from the densities of the seed bank in the spring and autumn, together with the seed fall and seedling emergence for six tree species: *Q. serrata* (S. Iida, unpublished results 1991), *S. obassia* (Abe 1996), *S. controversa* (Masaki et al. 1998), and three *Acer* species (Tanaka 1995) (Fig. 11.1). All of the studies were done under conspecific mature tree crowns. In April, before the germination season, the seed bank densities of all species are generally considerably lower than they are in the autumn of the previous year (i.e., the sum of the seed fall and the preexisting seed bank) (Fig. 11.1). The mean disappearance ratio of fallen *Q. serrata* acorns from autumn to the next spring was 88.3%, and their mean germination ratio was only 3.6% (Fig. 11.1). No seed remained after the first germination season.

Styrax obassia had a relatively stable seed-bank population throughout the year under conspecific adults (Fig. 11.1). Seed fall of *Swida controversa* was observed only in 1988 and 1990, but a viable seed bank was found throughout the study period (Fig. 11.1). Masaki et al. (1998) reported that *S. controversa* seeds suffer high mortality immediately after falling. The coefficient of variation of viable seed-bank densities of *S. obassia* (52%) was smaller than that of *S. controversa* (104%). The average mortality of the seed bank from autumn (i.e., the sum of the seed fall and the preexisting seed bank) until the following spring, before the germination season, was 41.9% for *S. obassia* and 80.4% for *S. controversa*. It is thought that the seed bank of *S. obassia* has a lower seed mortality in the soil than that of *S. controversa,* and is therefore more stable. However, at sites far from the crown edge of adult *S. controversa* trees, seed-bank densities were lower than they were under conspecific crowns, although little seed-bank mortality was found there (Masaki et al. 1998).

Germination patterns among the three *Acer* species were different from each other, suggesting that their seed-bank types are also different. The seed density of *A. mono* var. *marmoratum* f. *dissectum* on the forest floor before seed fall was zero or very low (Fig. 11.1). Therefore, this species is considered to have an extremely transient seed-bank, and rarely to make seeds which are viable for more than 1 year. The germination pattern of *A. amoenum* showed a 1-year delayed germination (Fig. 11.1). *A. rufinerve* seeds showed a germination pattern which was intermediate between those of the two previous species (Fig. 11.1). Both *A. amoenum* and *A. rufinerve* form short-term persistent seed banks (sensu Thompson and Grime 1979), but without a supply of a new seeds these seed banks will vanish within a few years. *A. mono* var. *marmoratum* f. *dissectum* and *A. amoenum* are shade-tolerant canopy species, and *A. rufinerve* is a shade-intolerant understory species (Masaki et al. 1992). Saplings of *A. rufinerve* grew more prolifically in canopy gaps, while those of *A. mono* var. *marmoratum* f. *dissectum* were more frequent under a closed canopy, and those of *A. amoenum* had no apparent tendency (see Chapter 12). Therefore, a canopy gap is more important for *A. rufinerve* than for *A. mono* var. *marmoratum* f. *dissectum* and *Acer amoenum*. Even a short-term seed-bank strategy will be advantageous for *A. rufinerve,* since it would level off the temporal variation of seed fall and increase the chances of the seed reaching a suitable site (Tanaka 1995). However, this strategy may relate to the spatial dispersal of seeds, as discussed in Chapter 10.

11.4 Mortality by Predation

In temperate forests, dispersed Fagaceae seeds, which are large, are mostly removed or eaten by small mammals before the germination season begins (Shaw 1968; Kanazawa and Nishikata 1976; Miguchi and Maruyama 1984; Kikuzawa 1988; Iida 1996).

Some *Q. serrata* seeds were experimentally placed on the forest floor under conspecific adults in this stand. It was found that over a period of 4 years, an average of 79.8% were removed or eaten by small mammals, 7.1% died from other causes (disease, desiccation, or insect damage), and 13.1% survived until the following spring. These rates were not significantly different between forest floors under a closed canopy or in canopy gaps (S. Iida, unpublished results 1995). (A similar interaction between acorns and insects is discussed in detail in Chapter 22.)

In this stand, predation pressure on dispersed seeds by small mammals was high not only for Fagaceae, but also for *S. obassia* seeds of relatively large dyszoochory, *S. controversa* seeds of medium-sized endozoochory, and *Acer* species seeds of medium-sized anemochory. Of 3530 *S. obassia* seeds experimentally set under the litter layer in November, 3511 (99.6%) had disappeared by the following May (Abe 1996). Masaki et al. (1998) reported that *S. controversa* seeds that were experimentally set on a forest floor had a high mortality (50%–100%) within 1 year, probably due to rodent predation. The mortality of *S. controversa* seeds tended to be highest during the growing season (April–August). Rodents are assumed to eat the less attractive *S. controversa* seeds because more attractive seeds, such as those of Fagaceae, are not available during this period. The disappearance rates of *S. controversa* seeds showed no specific tendencies in the three sites under or near a conspecific adult patch, in those far away from conspecific crowns, or in canopy gaps away from conspecific adults (see also Chapter 10). This fact indicates that the distance from conspecifics does not affect the mortality of dispersed *S. controversa* seeds. However, the seed disappearance rate does depend on the type of dispersal, i.e., whether the seed falls naturally with the pulp intact (gravity-dispersed) or is dispersed by birds with the pulp removed (bird-dispersed); the disappearance rate of pulpless seeds was significantly lower than that of seeds with pulp. Therefore, the removal of pulp of *S. controversa* seeds during bird dispersal is an effective protection against rodent predation.

Of the seeds of three *Acer* species that were experimentally set on the forest floor in autumn, more than 80% of *A. mono* var. *marmoratum* f. *dissectum* seeds and about half of both *A. amoenum* and *A. rufinerve* seeds had been eaten by the following May (Tanaka 1995). However, the survival rates of seeds of *A. mono* var. *marmoratum* f. *dissectum* and *A. amoenum* which were covered with litter were significantly higher than those of seeds which had been artificially uncovered from litter. This finding suggests that litter is important for seed survival on the forest floor (Jarvis 1964; Thompson 1987), and that the synchronization of seed and litter falls seems to have an adaptive importance for the survival of post-dispersal seeds. Furthermore, a large percentage of small anemochory seeds such as *O. japonica* (T. Nakashizuka, personal communication 1999) and *Carpinus* species (M. Shibata, personal communication 1999) were also eaten by rodents in OFR.

Therefore it is clear that predation by rodents has a large influence on the post-dispersal mortality of seeds almost irrespective of species and size. However, although rodents have a negative effect on dispersed seeds due to predation, they also have a positive effect due to their caching behavior. Therefore, in order to evaluate all the influences of rodents on post-dispersal seed dynamics for various tree species, the effects of both seed predation and seed dispersal should be studied.

11.5 Mortality Other than Predation

If it were not for predation by rodents, how much dispersed seed would survive until the next germination season in this stand? *Q. serrata* acorns packed into wire netting bags to avoid predation by rodents on the forest floor continued to be viable for an average of 5 years; 72.6% survived until the following spring, 21.8% died from drying or disease, and 5.6% died from insect damage (S. Iida, unpublished results 1995).

Abe (1996) reported that *S. obassia* seeds packed in wire netting bags and placed on the forest floor in autumn did not germinate at all in the first germination season 6 months later. In the second autumn, after 1.5 years, 52.0% had germinated, 33.0% were still dormant, 14.1% had decayed, and 0.8% had died from insect damage.

Masaki et al. (1998) buried pulpless *S. controversa* seeds packed in nylon mesh bags 5 cm deep in the soil, and investigated their mortality for 3 years. Most of the seeds were sound 1 year after burial. Subsequently, more than half of the seeds were still sound, but the germination rate was < 5% 3 years after burial. Thus the mortality rate was equivalent to 0.2 $year^{-1}$. Most dead seeds showed symptoms of fungous disease, while the insect damage rate was low (< 2%).

Tanaka (1995) investigated the survival of three varieties of *Acer* seeds packed in wire netting bags and left on the forest floor for 1.5 years. *A. mono* var. *marmoratum* f. *dissectum* seeds had a survival rate of 91% the following April before the next germination season. All the surviving seeds then germinated, and no seeds remained dormant until the next dispersal time. In the case of *A. amoenum* seeds, 97.5% were alive the following April before the next germination season. However, most of these decayed during the following summer without germinating, and only 27% were dormant in the autumn. Their final germination and dormancy rates in the second spring (1.5 years after seed fall) were 7.3% and 0.9%, respectively. Without predation by rodents, most dead *A. amoenum* seeds showed symptoms of decay at the end of the first summer, but there was little further decay during the winter. Since the marked fluctuations in soil moisture and temperature in the growing season make fungi active, seed mortality from fungous disease may increase in the summer, although some seeds decay after death from other causes such as desiccation. In the case of *A. rufinerve* seeds, 97% survived until the following April before the next germination season. Later, 40.4% germinated in the spring, with a large variance among the samples, and most of the remainder survived the growing season without decay. In the second spring, 38.3% germinated and 1.8% were still dormant. Therefore, if the seeds were free from predation by rodents, the estimated mortality rate of *A. rufinerve* seeds after 1.5 years would be 19.5%. It was also found that neither the topographic conditions nor crown closure (gap or closed) significantly affected the survival rates of the seeds of any of these three *Acer* species. Therefore, fungous disease has less effect on the post-dispersal mortality of *A. mono* var. *marmoratum* f. *dissectum* and *A. rufinerve* than predation by rodents.

It is clear that fungous disease of the seeds of *Q. serrata*, *S. obassia*, *S. controversa*, *A. mono* var. *marmoratum* f. *dissectum,* and *A. rufinerve* causes less post-dispersal mortality than rodent predation in OFR, but this is not true for the seeds of *A. amoenum*.

11.6 Response to Disturbance

The most suitable period for seedling establishment after disturbance is 1–5 years, and therefore their arrival time in a disturbed area is critical for shade-intolerant tree species (Canham and Marks 1985). Therefore, in order to succeed in establishing seedlings after a disturbance by quick germination from a seed bank, the seeds need to have a mechanism to detect environmental changes after a disturbance. Mechanisms used by seeds to detect disturbance include a response to changes in heat (Washitani and Takenaka 1987; Vázquez-Yanes and Orozco-Segovia 1993) or light (Vázquez-Yanes and Orozco-Segovia 1990, 1993).

Masaki et al. (1998) reported that while the seed-fall density of *S. controversa* in canopy gaps was low, the density of seedling emergence in the gaps was higher than that in undisturbed forest. Therefore, *S. controversa* seeds seem to have a gap-detection mechanism that responds to conditions such as temperature or light. Since *S. controversa* seedlings in forests require canopy gaps to become established and grow (Cornelissen 1993; Abe et al. 1995), a combination of seed dormancy and a gap-detection mechanism is advantageous to their regeneration.

Abe (1996) packed *S. obassia* seeds in wire netting bags, placed the bags on forest floors in both gaps and closed forest, and investigated their germination rates for 2 years. The germination rate on the floor of the gaps (78.4%) by autumn 2 years later was significantly higher than that on closed forest floors (52.1%). At the same time, the dormancy rate on the floor of the gaps (3.5%) was significantly lower than that on the closed forest floors (33.0%). The peak season of seedling emergence in the closed forest was early June, while seedling emergence in gaps continued until late September. The germination of *S. obassia* seeds may be enhanced by a higher cumulative temperature, which is one mechanism for detecting disturbed sites in deciduous forest. The soil temperature is not different in the gaps and in the closed stands in early spring, but it becomes markedly different when the foliage density in the canopy increases during the growing season. Although *S. obassia* is categorized as a shade-intolerant understory species (Masaki et al. 1992), it can maintain its population in stable stands (Abe 1996). However, both the survival and growth rates of seedlings and saplings in canopy gaps are higher than those in closed forests (Abe 1996). Therefore, the seed dormancy and the gap-detecting mechanism of *S. obassia* may contribute to the maintenance of the population even though it is able to regenerate without gaps. Whether the other tree species (*A. amoenum*, *A. rufinerve*, *O. japonica*, *C. cordata*, and *K. pictus*) that clearly produce a persistent seed bank also have a gap-detecting mechanism is unknown. For the seven species mentioned above, seedling densities in canopy gaps are significantly higher than those on a closed forest floor except for *A. amoenum* and *K. pictus* (see Chapter 12). This does not necessarily indicate that the seeds of the other five species have a gap-detecting mechanism, but simply that canopy gaps are a suitable site for their growth.

11.7 Dispersal and Dormancy

A seed bank with a mechanism for detecting canopy gaps enables a species to colonize safe sites which are temporally unpredictable. This is an equivalent strategy to seed dispersal, which enables a seed to reach a safe site spatially (see Chapter 10).

Kalopanax pictus produces fleshy fruits. In OFR, some seeds of this species are dispersed by frugivorous birds, but other seeds fall naturally with their pulp intact (gravity dispersal). Iida and Nakashizuka (1998) reported that *K. pictus* seeds without pulp (i.e., bird-dispersed seeds) differ from those with a pulp (i.e., gravity-dispersed seeds) in their germination behavior. From the results of germination tests lasting 3 years, the germination of pulpless seeds occurred over a period of 2 years: 30%–35.7% for the first germination season and 11.0%–16.8% for the second germination season. The germination of intact seeds occurred over a period of 3 years: 1.5% for the first, 9.7% for the second, and 0.5% for the third germination season. Furthermore, after the third germination season, very few pulpless seeds were dormant (< 1%), while 6.5% of intact seeds were still dormant. Therefore, seeds without pulp germinate quickly at high germination rates, while intact seeds germinate slowly at low germination rates.

For the same seed source, the dispersal distances of bird-dispersed seeds may be greater than those of gravity-dispersed seeds (see Chapter 10). Seedling emergence patterns of *K. pictus* at sites more than 20 m from conspecific adults showed the same germination patterns as pulpless seeds, while seedling emergence patterns within 20 m from conspecific adults showed the same germination patterns as intact seeds. Therefore, *K. pictus* seeds scattered by birds appear to have an increased possibility of reaching existing safe sites in distant areas where they will germinate quickly (spatial dispersal), while fallen seeds with their pulp appear to form a seed bank near their source (temporal dispersal).

Seed heteromorphism means that individual plants form morphologically and/or characteristically different seeds (Venable 1985). This is common in many plant species (Harper 1977), and generally shows different types of dispersibility and dormancy. With annual desert plants that form heteromorphic seeds, the optimal germination schedule is (1) quick germination for seeds with high dispersibility, and (2) delayed germination for seeds with low dispersibility (Venable and Lawlor 1980). This bet-hedging strategy is advantageous at sites with unpredictable environmental changes. Consequently, the seed behavior of *K. pictus* seems to be analogous to that of annual desert plants (Venable and Lawlor 1980).

The influences of seed pulp on germination are neither uniform nor consistent among plant species (Barnea et al. 1990, 1991). However, removal of the pulp from the seeds of some tree species by bird ingestion interrupts the dormancy of those seeds and enhances their germination (Ridley 1930; Murray 1988; Barnea et al. 1991; Yagihashi et al. 1998). The spatial and temporal performances of their seeds are considered to be similar to those of *K. pictus* seeds.

The colonization of preexisting gaps through a wide-ranging dispersal of seeds may be more advantageous than colonization in temporally unpredictable gaps by buried seeds (Chapter 10). This is because the mortality before germination of the seeds which are dispersed promptly may be lower than that of seeds lying dormant in the soil for a long time. However, wide-ranging seed dispersal is temporally limited, and requires sufficient numbers of seeds to be able to invade disturbed sites successfully (Parker et al. 1989).

11.8 Variations in Seed Fall and Seedling Emergence

Seed banks function as a buffer against the annual fluctuations in seedling emergence and seed fall (Granström 1987). Most of the major tree species in this forest had a large annual fluctuation in seed production (see Chapter 9). Yearly seedling emergence from seed banks is advantageous to tree species that cannot form seedling banks or have infrequent seed production.

The coefficients of annual variation (CV) of seed fall and seedling emergence of 17 tree species were studied for 9 years (Fig. 11.2). For convenience, the ratio of the CV of seedling emergence to the CV of seed fall is defined as the "variation ratio" (VR). A VR of less than 1 indicates that the annual variation in seedling emergence is relatively low compared with that of seed fall, while a VR of more than 1 indicates that the annual variation in seedling emergence is relatively high compared with that of seed fall. The 17 tree species were divided into three groups according to their VR value.

The first group consisted of three species with a CV of seedling emergence which was clearly smaller than that of seed fall (VR < 0.5): *Kalopanax. pictus* (VR = 0.37), *Styrax. obassia* (VR = 0.45), and *Swida controversa* (VR = 0.49). These species seem to have a consistent seed bank strategy since seedlings emerge even in a spring after no seed fall. The second group consisted of eight species with a CV of seedling emergence a little smaller than that of seed fall (0.5 < VR <1): *Carpinus cordata* (VR = 0.57), *Carpinus laxiflora* (VR = 0.65), *Carpinus tschonoskii* (VR= 0.67), *Ostrya japonica* (VR = 0.68), *Castanea crenata* (VR = 0.69), *Betula grossa* (VR =

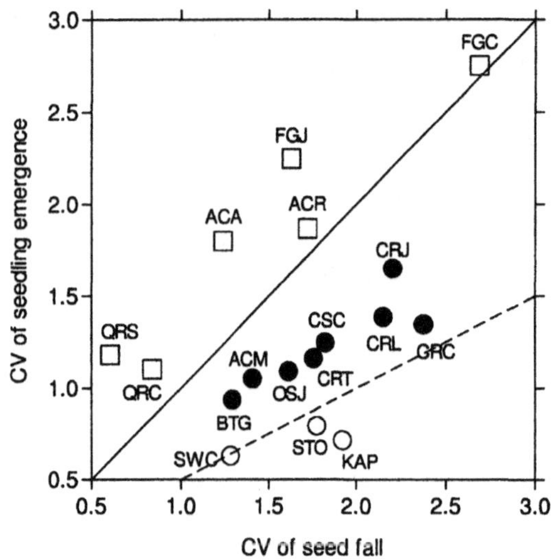

Fig. 11.2. Relationship between the coefficient of variation (CV) of annual seed fall and the CV of annual seedling emergence of the 17 main tree species. *Open squares*, VR (the ratio of the CV of seedling emergence to the CV of seed fall) >1.0; *solid circles*, 0.5 < VR < 1.0; *open circles*, VR < 0.5. *Solid line*, VR = 1; *broken line*, VR = 0.5. Species abbreviations: *ACM*, *Acer mono* var. *marmoratum* f. *dissectum*; *ACA*, *A. amoenum*; *ACR*, *A. rufinerve*; *BTG*, *Betula grossa*; *CRC*, *Carpinus cordata*; *CRJ*, *C. japonica*; *CRL*, *C. laxiflora*; *CRT*, *C. tchonoskii*; *SWC*, *Swida contraversa*; *CSC*, *Castanea crenata*; *FGC*, *Fagus crenata*; *FGJ*, *F. japonica*; *KAP*, *Kalopanax pictus*; *OSJ*, *Ostrya japonica*; *QRC*, *Quercus crispula*; *QRS*, *Q. serrata*; *STO*, *Styrax obassia*

0.73), *Acer mono* var. *marmoratum* f. *dissectum* (VR = 0.75), and *Carpinus japonica* (VR = 0.75). The third group consisted of six species with a CV of seedling emergence which was larger than that of seed fall (VR < 1): *Fagus crenata* (VR = 1.03), *Acer rufinerve* (VR = 1.08), *Quercus crispula* (VR = 1.29), *Fagus japonica* (VR = 1.37), *Acer amoenum* (VR = 1.45), and *Quercus serrata* (VR = 1.94) (Fig. 11.2).

Seed dormancy over 1 year in the species in the second and third groups does not necessarily decrease the annual variation of seedling emergence compared with that of seed production. The short-term seed-bank strategy of *A. amoenum* and *A. rufinerve* does not affect the decrease in their CV of seedling emergence, whereas with *A. mono* var. *marmoratum* f. *dissectum*, which rarely forms a seed bank which lasts longer than 1 year, the CV of seedling emergence is smaller than that of seed fall. *C. cordata* (VR = 0.57), which has a distinct dormancy, has the smallest VR of the *Carpinus* species (*C. laxiflora* 0.65, *C. tschonoskii* 0.67, and *C. japonica* 0.75), and therefore the seed dormancy of *C. cordata* decreases its CV of seedling emergence to some extent. However, the CV of seedling emergence of *C. cordata* is 136%, whereas that of *C. tschonoskii* is 118%, and thus the idea that seed dormancy in *C. cordata* evens out the annual variation in seedling emergence is doubtful. The adaptive significance of this aspect of *C. cordata* seeds is unclear (Shibata and Nakashizuka 1995).

As well as the five Fagaceae trees that form transient seed banks, the VR of *C. crenata* is smaller than 1, while the VRs of the other four trees are greater than 1. This is partly because the seed mortality of these species fluctuates greatly from year to year. These species may represent a complicated web of interaction within the guilds of large seed producers and seed-eating rodents (Hoshizaki et al. 1997).

Consequently, the effect of seed dormancy on the annual variation in seedling emergence is unclear, except for the trees in the first group that can form relatively long-term seed banks.

References

Abe S (1996) Life cycle and estimation of gap dependency for *Styrax obassia*, a sub-canopy tree species (in Japanese). PhD Dissertation, University of Tokyo

Abe S, Masaki T, Nakashizuka T (1995) Factors influencing sapling composition in canopy gaps of a temperate deciduous forest. Vegetatio 120:21–32

Alvarez-Buylla ER, Garcia-Barrios R (1991) Seed and forest dynamics: a theoretical framework and example from the neotropics. Am Nat 137:133–154

Alvarez-Buylla ER, Martinez-Ramos M (1990) Seed bank versus seed rain in the regeneration of a tropical pioneer tree. Oecologia 84:314–325

Barnea A, Yom-Tov Y, Friedman J (1990) Differential germination of two closely related species of *Solanum* in response to bird ingestion. Oikos 57:222–228

Barnea A, Yom-Tov Y, Friedman J (1991) Does ingestion by birds affect seed germination? Funct Ecol 5:394–402

Canham CD, Marks PL (1985) The response of woody plants to disturbance: patterns of establishment and growth. In: Pickett STA, White PS (eds) The ecology of natural disturbance and patch dynamics. Academic, San Diego, pp 197–216

Cavers PB (1983) Seed demography. Can J Bot 61:3578–3590
Cornelissen JHC (1993) Seedling growth and morphology of the deciduous tree *Cornus controversa* in simulated forest gap light environments in subtropical China. Plant Species Biol 8:21–27
Garwood NC (1989) Tropical soil seed banks: a review. In: Leck MA, Parker VT, Simpson RL (eds) Ecology of soil seed banks. Academic, San Diego, pp 149–209
Granström A (1987) Seed viability of fourteen species during five years of storage in a forest soil. J Ecol 75:321–331
Harper JL (1977) Population biology of plants. Academic, London
Hoshizaki K, Suzuki W, Sasaki S (1997) Impacts of secondary seed dispersal and herbivory on seedling survival in *Aesculus turbinata*. J Veg Sci 8:735–742
Howe HF, Smallwood J (1982) Ecology of seed dispersal. Ann Rev Ecol Syst 13:201–218
Iida S (1996) Quantitative analysis of acorn transportation by rodents using magnetic locator. Vegetatio 142:39–43
Iida S, Nakashizuka T (1998) Spatial and temporal dispersal of *Kalopanax pictus* seeds in a temperate deciduous forest, central Japan. Plant Ecol 135:243–248
Jarvis PG (1964) Interference by *Deschampsia flexuosa* (L.) Trin. Oikos 15:56–78
Kanazawa Y, Nishikata S (1976) Disappearance of acorns from the floor in *Quercus crispula* forest. J Jpn For Soc 58:52–56
Katsuta M, Mori T, Yokoyama T (1998) Seeds of woody plants in Japan. Angiospermae (in Japanese). Japan Forest Tree Breeding Association, Tokyo
Kikuzawa K (1988) Dispersal of *Quercus mongolica* acorns in a broadleaved deciduous forest. 1. Disappearance. For Ecol Manage 25:1–8
Kiyono Y, Kawahara T (1989) Distribution of buried seeds of *Betula grossa* around the seed trees (in Japanese). Trans Jpn For Soc 100:327–328
Marks PL (1974) The role of pin cherry (*Prunus pensylvanica* L.) in the maintenance of stability in northern hardwood ecosystems. Ecol Monogr 44:73–88
Masaki T, Suzuki W, Niiyama K, Iida S, Tanaka H, Nakashizuka T (1992) Community structure of a species-rich temperate forest, Ogawa Forest Reserve. Vegetatio 98:97–111
Masaki T, Tanaka H, Shibata M, Nakashizuka T (1998) The seed bank dynamics of *Cornus controversa* and their role in regeneration. Seed Sci Res 8:53–63
Miguchi H, Maruyama K (1984) Ecological studies on natural beech forest. XXXVI. Development and dynamics of beech nuts in a mast year (in Japanese with English summary). Jpn J Ecol 66:320–327
Murray KG (1988) Avian seed dispersal of three neotropical gap-dependent plants. Ecol Monogr 58:271–298
Nakagoshi N (1985) Buried viable seeds in temperate forests. In: White J (ed) Handbook of vegetation science: the population structure of vegetation. Junk, Dordrecht, pp 551–570
Ozawa J (1950) On the durability of germination power of tree seeds buried in soil (in Japanese). Rep Gov For Exp Stn 58:25–43
Parker VT, Simpson RL, Leck MA (1989) Pattern and process in the dynamics of seed banks. In: Leck MA, Parker VT, Simpson RL (eds) Ecology of soil seed banks. Academic, San Diego, pp 367–384
Pickett STA, McDonnell MJ (1989) Seed bank dynamics in temperate deciduous forest. In: Leck MA, Parker VT, Simpson RL (eds) Ecology of soil seed banks. Academic, San Diego, pp 123–147
Ridley HN (1930) The dispersal of plants throughout the world. Reeve, Ashford
Shaw MW (1968) Factors affecting the natural regeneration of sessile oak (*Quercus petraea*) in north Wales. II. Acorn losses and germination under field conditions. J Ecol 56:647–660

Shibata M, Nakashizuka T (1995) Seed and seedling demography of four co-occurring *Carpinus* species in a temperate deciduous forest. Ecology 76:1099–1108

Silvertown JW, Lovett Doust J (1993) Introduction to plant population biology, 3rd edn. Blackwell, Oxford

Tanaka H (1995) Seed demography of three co-occurring *Acer* species in a Japanese temperate deciduous forest. J Veg Sci 6:887–896

Thompson K (1987) Seed and seed bank. New Phytol 106:23–34

Thompson K (1992) The functional ecology of seed banks. In: Fenner M (ed) Seeds. The ecology of regeneration in plant community. CABI, Wallingford, pp 231–258

Thompson K, Grime JP (1979) Seasonal variation in the seed banks of herbaceous species in ten contrasting habitats. J Ecol 67:893–921

Vázquez-Yanes C, Orozco-Segovia A (1990) Seed dormancy in the tropical rain forest. In: Bawa KS, Hadley M (eds) Reproductive ecology of tropical forest plants. Parthenon, New Jersey, pp 247–259

Vázquez-Yanes C, Orozco-Segovia A (1993) Patterns of seed longevity and germination in the tropical rainforest. Annu Rev Ecol Syst 24:69–87

Venable DL (1985) The evolution of ecology of seed heteromorphism. Am Nat 126:577–595

Venable DL, Lawlor L (1980) Delayed germination and dispersal in desert annuals: escape in space and time. Oecologia 46:272–282

Washitani I, Takenaka A (1987) Gap-detecting mechanism in the seed germination of *Mallotus japonicus* (Thunb.) Muell. Arg., a common pioneer tree of secondary succession in temperate Japan. Ecol Res 2:191–201

Yagihashi T, Hayashida M, Miyamoto T (1998) Effects of bird ingestion on seed germination of *Sorbus commixta*. Oecologia 114:209–212

Young KR, Ewel JJ, Brown BJ (1987) Seed dynamics during forest succession in Costa Rica. Vegetatio 71:157–173

12 Seedling/Sapling Banks and Their Responses to Forest Disturbance

Shin Abe

12.1 Introduction

Although all the trees that compose a mature forest originate as seedlings on the forest floor, the composition of a seedling or sapling community often differs from that of the adult trees. While this discrepancy may signal a shift or fluctuation in the forest (see Chapter 6), it can be attributed to demographic behaviors of each species (e.g., Harcombe 1987; Aiba and Kohyama 1997). To understand the maintenance mechanisms of a forest community, or to predict the forest dynamics, we must consider the seedling and sapling stages, since they, as well as the seed stage, are the stages subject to drastic demographic changes (see Chapter 1). Following on the previous chapters of this volume on seed production (Chapter 9), dispersal (Chapter 10), and dormancy and germination (Chapter 11), I describe the young tree community in Ogawa Forest Reserve (OFR) in this chapter, especially as it relates to forest disturbances.

Forests commonly experience disturbances, and researchers have come to notice that disturbances closely relate to the community structure and dynamics (White and Pickett 1985; van der Maarel 1996; also see Chapter 7). Disturbances and the subsequent recovery stages produce various conditions in a forest, including sites that are safe for many species to reproduce or grow (e.g., Harcombe 1987; Spies and Franklin 1989; Grey and Spies 1997). This chapter is concerned with how disturbances affect the juvenile stages in forests.

Since natural disturbances are largely unpredictable and trees cannot move from place to place after their establishment, trees must reach safe sites by means of strategies such as seed rains, seed banks, and seedling/sapling banks. In the seed stage, we can recognize two adaptations in some species: wide spatial dispersal (the production of seeds that are easily spread by wind or animals to safe sites); and temporal dispersal (seed dormancy until circumstances are favorable). Species in the OFR follow both strategies (see Chapters 10 and 11). In the stages after germination, we can recognize another adaptation; under shaded conditions, seedlings of some tree species begin to grow in readiness for a canopy opening (e.g., Uhl et al. 1988; Abe et al. 1995). These strategies are closely related to features that are characteristic of

Ecological Studies, Vol. 158
Nakashizuka, Matsumoto (eds.) Diversity and Interaction in a Temperate Forest Community: Ogawa Forest Reserve of Japan

either pioneer or climax (Swaine and Whitmore 1988; Whitmore 1989) or shade-tolerant or intolerant species (see Chapter 6). Seed rain and seed bank strategies are favored by pioneer or light-demanding species, while seedling/sapling banks are favored by climax or shade-tolerant species.

Canopy gaps are the most important type of disturbance influencing the structure and organization of forest communities in the present OFR, as well as in other mature forests (e.g., Whitmore 1982; Veblen 1989; Nakashizuka et al. 1992; Kupfer and Runkle 1996). However, the responses of trees to gap formation vary among species. Some plants respond by increasing their rate of growth, and others show more recruitments under the new, favorable conditions (Brokaw 1985a; Orwig and Abrams 1995). Gap size is one of the important factors that affects the amount of light in the gap, and thus the seedling/sapling demography. This factor may also act in conjunction with other factors such as topography and nutrient gradients. Other factors possibly important for seedling/sapling demography are gap age and the seed supply from seed-bearing trees around the gap.

I observed sapling demography within gaps during samplings of 651 quadrats (2 m^2) regularly scattered within the 6-ha plot (for the background sampling census, see Chapter 5). In 1990, we set up a total of 115 sampling quadrats (5 m^2) in 55 gaps within the OFR (gap sampling census, Abe et al. 1995). Gaps were randomly sampled by perambulation and were defined as patches ≥ 5 m^2 with no canopy exceeding 10 m in height. Recruitment, growth, and survival of saplings of height ≥ 30 cm in the gap quadrats were surveyed periodically until 1994. The relationship between the regeneration of tree species and the canopy gaps in the OFR is described below.

12.2 Sapling Composition

12.2.1 Sapling Compositions Under the Closed Canopy and in Canopy Gaps

In 1990, of the 68 woody (tree and shrub) species counted by the background sampling census and the gap sampling census, 12 species appeared only in gaps. The canopy gaps may contribute to the maintenance of species diversity in the OFR by enabling at least these 12 species to co-occur.

The overall gap sapling composition correlated highly and significantly with the composition under a closed canopy (Fig. 12.1), suggesting that most saplings growing in gaps are recruited before the gaps appear. The total sapling densities in gaps (8,793 ha^{-1}) and under the closed canopy (8,455 ha^{-1}) did not greatly differ in 1990. That is, advance regeneration by sapling banks is more important than new regeneration in gaps as a strategy to maintain sapling composition in this forest. This seems to be the case in many other temperate and tropical forests as well (e.g., Hara 1987; Uhl et al. 1988).

Nine species were significantly biased to gaps, and 4 other species were biased to the closed canopy (Table 12.1, Fig. 12.1). Dependence on sapling banks varied by species (Tanaka 1995), and shade-intolerant species (those requiring much light)

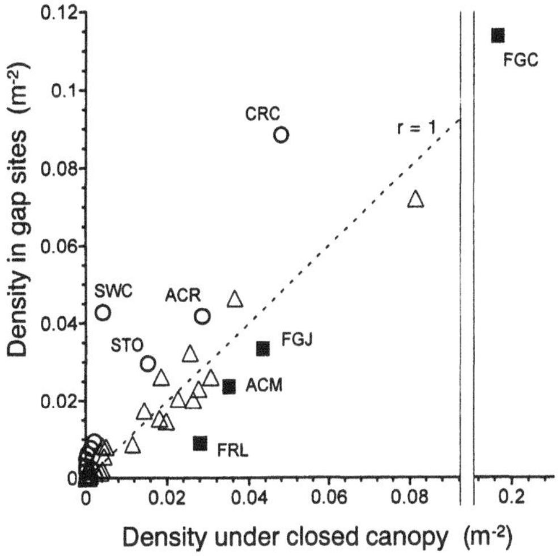

Fig. 12.1. Sapling densities under the closed canopy and in gaps. The most abundant 37 tree species in gaps in 1990 are plotted. A significantly high correlation ($r = 0.88$, $P < 0.001$) was found between densities in gaps and those under the closed canopy. Species with significant ($P < 0.05$) bias are distinguished by *open circles* (bias to gaps) and *solid squares* (bias to closed canopy). *ACM, Acer mono* var. *marmoratum* f. *dissectum*; *ACR, Acer rufinerve*; *SWC, Swida controversa*; *CRC, Carpinus cordata*; *FGC, Fagus crenata*; *FGJ, Fagus japonica*; *FRL, Fraxinus lanuginosa* f. *serrata*; *STO, Styrax obassia*

should tend to appear mostly in gaps in a mature forest. Species biased to gaps were consistent with our evaluation of shade-intolerant species in this region, except for two species: *Carpinus cordata*, which had been considered a shade-tolerant species according to the population structure (Masaki et al. 1992; see Chapter 6); and *Styrax obassia*, a typical shade-tolerant subcanopy species. However, shade-tolerant species do not necessarily avoid abundant light, because abundant light increases the photosynthetic activity of some shade-tolerant species that are physiologically flexible (Lee et al. 1997). *S. obassia* shows high shade tolerance and a high degree of adaptability to growth and regeneration in gaps (Abe 1996; Abe et al. 1998).

12.2.2 The Influence of Topography

Sapling composition differs from gap to gap, because the responses of trees vary according to the characteristics of the gap, such as topographic location, size, age, and morphology of the gap, or the mode of its formation (e.g., by trees being uprooted or snapping) and canopy height (e.g., Brokaw 1985b; Poulson and Platt 1989; Brokaw and Scheiner 1989; Welden et al. 1991; Goldblum 1997). In the OFR, we have noticed some species-specific relationships between gap characteristics and tree responses, suggesting a guild structure of species in response to gap characteristics (Abe et al. 1995).

Because trees in the 6-ha plot varied in distribution according to the topography (Masaki et al. 1992, see Chapter 6), I grouped sampled gaps into upper- and lower-slope gaps by analyzing tree composition surrounding the gaps (Abe et al. 1995). The upper-slope gaps had higher sapling density and greater species richness than the lower-slope gaps (Table 12.2). Several species had a significant distributional bias to either upper- or lower-slope gaps (Table 12.1). This classification of gaps, however, could not remove the biological influences of neighboring trees on sap-

Table 12.1. Sapling distributions for the most abundant 37 tree species in gaps

		Bias to			Topographic locations			Gap size			Adults adjacent	Seed dispersal
Species	Abbr.	Gap		Closed	Lower		Upper	< 70 m^2		≥ 70 m^2	to the gap	agent
Acanthopanax sciadophylloides	APS		ns			ns		+				B
Acer cissifolium	ACC		ns		++					+		W
Acer mono var. *marmoratum* f. *dissectum*	ACA			+		ns				+++		W
Acer amoenum	ACP		ns			ns			ns			W
Acer rufinerve	ACR	+					++			+++		W
Acer sieboldianum	ACS		ns			ns			ns			W
Acer tenuifolium	ACT		ns		+++				ns	++		W
Betula grossa	BTG	+++				ns				+		W
Carpinus cordata	CRC	+++			+++					+++	++	W
Carpinus japonica	CRJ		ns			ns			ns		+	W
Carpinus laxiflora	CRL		ns				+++	+++			++	W
Carpinus tschonoskii	CRT		ns				+		ns		++	W
Castanea crenata	CSC		ns			ns				+	++	O
Clethra barbinervis	CLB		ns				+++	++				O
Swida controversa	SWC	+++					+++			+++		B
Benthamidia japonica	BNJ		ns				++		ns			B
Fagus crenata	FGC			+++			+++	+++				O
Fagus japonica	FGJ			+	+++				ns			O

Fraxinus lanuginosa f. *serrata*	FRL			+++			+		ns		+	W
Hamamelis japonica	HAJ		ns			ns		+++				B
Ilex macropoda	ILM		ns				+	+++				W
Kalopanax pictus	KAP		ns			ns				+		B
Magnolia obovata	MGO	+++				ns			ns			B
Meliosma myriantha	MLM		ns		+++				ns		+	B
Morus australis	MRA		ns			ns			ns			B
Ostrya japonica	OSJ		ns			ns			ns			W
Phellodendron amurense	PDA	++				ns			ns			B
Prunus buergeriana	PRB		ns			ns			ns			B
Prunus grayana	PRG		ns		+++				ns			B
Prunus verecunda	PRV		ns			ns			ns			B
Quercus crispula	QRC		ns				+			+	+	O
Quercus serrata	QRS	+++					+++			++		O
Rhus trichocarpa	RHT	+++				ns			ns			B
Sorbus alnifolia	SRA		ns			ns		+				B
Sorbus japonica	SRJ		ns				+		ns			B
Styrax japonica	STJ		ns			ns			ns			O
Styrax obassia	STO	++				ns				+	+	O

Significance by *t*-test for the presence of adults adjacent to the gap, and by χ squared test for the others.

Seed dispersal agent: B, bird (endozoochore); W, wind (anemochore); O, others. Others include dispersal by caching animals (dyszoochore) Lowercase ns, non significant.

Symbols, +, ++, and +++ indicate both bias and a significant level of $P < 0.05$, 0.01, and 0.001, respectively.

Table 12.2. Density and species diversity of saplings and the attributes of gaps

Attributes of gaps		No. of gaps	Gap size (m^2)	Sapling density (m^{-2})	Spp. Diversity H'
Topographic location	Upper slope	29	55.8 ± 67.3^a	1.04 ± 0.79^a	1.58 ± 0.57^a
	Lower slope	23	87.7 ± 67.4^b	0.54 ± 0.37^b	1.36 ± 0.70^b
Size (m^2)	5 – 25	17	14.1 ± 5.7^a	0.86 ± 0.59^a	1.52 ± 0.50^a
	< 50	11	35.2 ± 5.3^a	0.83 ± 0.56^a	1.55 ± 0.51^a
	< 100	10	78.8 ± 15.1^a	0.76 ± 0.66^a	1.50 ± 0.75^a
	= 100	14	158.6 ± 67.7^a	0.73 ± 0.70^a	1.40 ± 0.69^a
Age (year)	< 5	12	61.2 ± 78.2^{ab}	0.99 ± 0.68^a	1.76 ± 0.41^a
	5–10	13	84.1 ± 60.6^a	0.54 ± 0.55^b	1.21 ± 0.75^c
	10–15	13	106.1 ± 79.2^a	0.93 ± 0.76^a	1.61 ± 0.65^{ab}
	= 15	13	39.2 ± 35.6^b	0.64 ± 0.30^b	1.37 ± 0.49^b
Gap maker	*Quercus* spp.	30	81.5 ± 73.7^a	0.72 ± 0.66^a	1.43 ± 0.68^a
	Fagus crenata	5	169.6 ± 125.4^a	1.28 ± 0.86^b	1.71 ± 0.49^b
	Fagus japonica	5	87.0 ± 43.5^a	0.53 ± 0.33^{ac}	1.23 ± 0.71^a
	Acer spp.	6	34.7 ± 26.5^b	1.09 ± 0.64^{ab}	1.55 ± 0.39^{ab}
	Carpinus spp.	6	88.8 ± 62.8^a	0.66 ± 0.34^{ac}	1.57 ± 0.45^{ac}
	others	46	59.2 ± 56.6^{ab}	0.62 ± 0.61^{abc}	1.08 ± 0.75^{ab}

Data are mean values ± standard deviation.

Symbols with the same letter (a, b, and c) indicate nonsignificant ($P > 0.05$) groups among the categories of each gap attribute

H', Shannon function.

lings in a gap, such as seed source (to be described below) or proximity-related mortality, because conspecific adult trees tended to have the same topographic preference (Clark and Clark 1984; see Chapter 10).

12.2.3 Influence of Seed-Bearing Trees Around the Gaps

For some species in the OFR, the presence of conspecific adults adjacent to the gap was related to sapling density in that gap. Sapling densities of ten species were significantly higher in the gaps where conspecific adults (DBH $\geq$ 10 cm) were within 15 m from the center of the gap compared with other gaps (Table 12.1). We can attribute this relationship to the adults' status as a nearby seed source. Six of these species disperse their seeds by wind (anemochore), one species by birds (endozoochore), and three species by other means including rodents (dyszoochore). Correlations between seed dispersal type and sapling density in gaps were not clear in this analysis.

12.2.4 Gap Size and Sapling Composition in Gaps

Gaps in this forest are relatively small, as is true for many other temperate forests (see Chapter 7; Nakashizuka and Iida 1995), and rapid closing of small gaps by lateral branch extension of neighboring canopy trees suggests that most trees must survive several shady periods between gap events before they reach the canopy (Runkle 1985; Rebertus and Veblen 1993). Therefore, gap dependence and shade tolerance of a species need not always be consistent (Canham 1989). However, single-factor analyses (Table 12.1) showed that the gap-size dependence with respect to the abundance of some species was consistent with their classification with respect to shade tolerance as estimated by population structure (see Chapter 6; Masaki et al. 1992). In large gaps ($\geq$ 70 m^2), sapling densities increased for *Acer rufinerve*, *Swida controversa*, and other species considered shade-intolerant. Saplings of *Fagus crenata*, *Carpinus laxiflora*, and other species that have been characterized as shade-tolerant were abundant in small gaps (< 70 m^2). Sapling density and species diversity (H': Shannon function) were higher in small gaps than in large ones.

Some species that seemed shade-tolerant were sparse in larger gaps, suggesting that the environment of a large gap is not favorable for the survival of these species. In addition to their ecophysiological traits, root and understory competition with forest floor plants seem to be relevant factors (Ehrenfeld 1980; Putz and Canham 1992), and they may be more important factors in large gaps. As mentioned above, some saplings may lose their physiological flexibility when they are exposed to more drastic changes by the formation of large gaps. Physical disturbance by uprooting or from debris on the forest floor associated with the formation of large gaps tends to be severe and may result in damage to the sapling banks of these species.

In a finer categorization of gap size, clear patterns in sapling densities and species diversity were not observed (Table 12.2). In upper-slope gaps, we recognized significant trends in sapling density according to gap size, but not in lower-slope

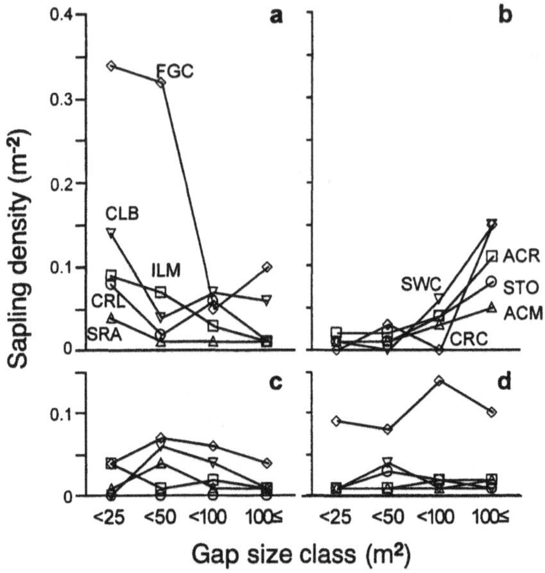

Fig. 12.2. Sapling distribution according to gap size in different topographic locations. Sapling densities of several species are displayed according to four gap size classes. **a**,**b** upper slope; **c**,**d** lower slope. Symbols for species are the same in **c** and **d** as indicated in **a** and **b**, respectively. Abbreviations of species names are as shown in Table 12.1

gaps (Fig. 12.2). That is, the effect of gap size on saplings was not clear in gaps on the lower slopes or in the valley.

12.2.5 The Influence of the Gap Maker and the Mode of Formation

Does the "gap maker," i.e., the species of fallen tree, affect sapling composition in a gap? Principal component analysis showed some variation, suggesting that gap size correlated with the gap-maker species (Abe et al. 1995). In gaps created by tree falls of *F. japonica*, sapling density and species richness were significantly lower than in gaps made by other species (Table 12.2), because *F. japonica* usually sprouts profusely from the base of the gap maker before other species can become established. Thus, the succeeding generation of *F. japonica* tends to fill its own gaps.

I observed no marked differences in sapling composition among gaps created by different causes. Of the 80 gap makers in the 55 sampled gaps, frequently observed gap causes were trunk breaking (38%) and uprooting (34%). Standing death (without toppling) (16%) and falling branches (13 %) were relatively infrequent (see Chapter 7). However, the proportion of gaps created by standing dead trees was most likely underestimated, because most gap makers that were uprooted were small, incompletely uprooted trees, or seemed to have been uprooted after death. The gradual deterioration and death of canopy trees causes the gradual formation of gaps, and different responses by saplings of different species to such gaps may influence sapling composition (Kransy and Whitmore 1991).

Consequently, we identified topography and gap size as two important factors

Table 12.3. Guild structure of saplings according to gap attributes

a Single-factor analyses

	Small gap $< 70m^2$	Large gap $\geq 70m^2$	No relation with gap size
Upper-slope gap	CLB CRL FGC^{-} ILM	ACR^{+} SWC^{+} CRT QRC QRS^{+}	BNJ FRL^{-} SRJ
Lower-slope gap		ACC CRC^{+}	ACT FGJ MLM PRG
No relation with topography	APS HAJ SRA	ACM^{-} BTG^{+} CSC KAP	ACA ACS CRJ MGO^{+} MRA OSJ
		STO^{+}	PDA^{+} PRB PRV RHT^{+} STJ

Species that showed significant relationships ($P < 0.05$) between sapling density and gap size or topographic location are shown.
Abbreviations of species names are as shown in Table 12.1.
$^{+}$, Significant density bias to gaps; $^{-}$, significant density bias to closed canopy.

b MANOVA (multivariate analysis of variance)

	Small gap $< 70m^2$	Large gap $\geq 70m^2$	No relation with gap size
Upper-slope gap	CRL FGC	SWC	ACA ACR
Lower-slope gap			*MLM*
No relation with topography	APS ILM	ACM *CRC* STO	

Species with significant relationships ($P < 0.05$) with the explanatory variables topographic location, gap size, and adjacent conspecific adults are shown.
The italicized species was related to the presence of conspecific adults.

accounting for the present sapling composition in gaps. Single-factor analyses and multivariate analysis of variance (MANOVA) suggested that the guild structure varies among gaps (Table 12.3).

12.3 Sapling Dynamics in Gaps

Over time, disturbed conditions will gradually recover as the saplings in the gaps develop. Because of this process, gap age may be an important factor in sapling dynamics. However, gap age as estimated dendrochronologically by taking samples of tree cores (see Fig. 7.6 in Chapter 7) showed no relationship to the present species composition in the observed gaps. Sapling density and species diversity also did not differ according to gap age (Table 12.2). Topography and other environmental conditions might have concealed the effects of gap age on the present sapling composition.

However, we did find that some tree species in the OFR, particularly shade-intolerant species (those with bell-shaped DBH distributions, see Chapter 6), showed some gap-age dependence in their sapling dynamics. These species had a high recruitment rate and a high mortality in young gaps, although their growth rates were not affected by gap age (Table 12.4). With regard to saplings of shade-tolerant species (those with L-shaped DBH distributions, see Chapter 6), mortality and recruitment did not show any gap-age dependence, but their growth rates did vary according to gap age. Thus, the gap age was important with respect to the recruitment and mortality of shade-intolerant species, and with respect to the growth rate of shade-tolerant species. The shade-intolerant species probably prefer well-lighted sites during the establishing phase when gaps are young, and few become established afterward. In contrast, most saplings of shade-tolerant species seem to have regenerated in advance of gap formation (Fig. 12.1), and these increase their growth rate in young, well-lighted, gaps.

Table 12.4. Demography of saplings in gaps over a two-year period

Spp. group	Gap age young >> old	Gap size small >> large	Topographic location ridge >> valley
Bell-shape[c]			
Mortality	increase	decrease	ns
Recruitment	decrease	ns	decrease
RGR of height	ns	ns	ns
RGR of DBH[a]	ns	ns	ns
L-shape[c]			
Mortality	ns	ns	ns
Recruitment	ns	increase	ns
RGR of height	decrease	increase	ns[b]
RGR of DBH	decrease	ns	ns
All species			
Mortality	ns	ns	ns
Recruitment	ns	increase	ns
RGR of height	decrease	increase	decrease
RGR of DBH	decrease	ns	ns

Significant ($P < 0.05$) fluctuations of species groups according to the structure of adult populations are shown, using MANOVA with the explanatory variables gap age, gap size, and topography.

RGR, relative growth rate; DBH, diameter at breast height; ns, not significant.

[a] DBH of saplings was measured only for individuals with height = 2 m.

[b] Decrease when considered in combination with gap age.

[c] Definitions and species as described in Chapter 6.

The effect of gap size on sapling dynamics was smaller than its effect on species composition (Table 12.4). It had significant effects on the mortality of species with bell-shaped distributions of DBH, and on recruitment and relative growth rates (RGR) of species with L-shaped DBH distributions. Topography had only a small effect, and only on the recruitment of species with bell-shaped DBH distributions.

12.4 Conclusion

Tree regeneration is largely dependent on advance regeneration in this forest. This may be the case under the present disturbance regime, in which fire disturbances no longer occur (see Chapter 6). If the forest were still subject fire disturbance at present, regeneration would more complete after the disturbance. Under the present disturbance regime, the gap characteristics had a strong effect both on species composition and the dynamics of the sapling communities in the gap. The species composition of the sapling community in a gap was largely affected by the gap size and its location along the topographic gradient, but not significantly by gap age. Gap age, however, had greater effects on the dynamics of the sapling community; such as the mortality and recruitment of shade-intolerant species and the growth rate of shade-tolerant species.

References

Abe S (1996) Life cycle and estimation of gap dependency for *Styrax obassia*, a sub-canopy tree species (in Japanese). Ph.D. dissertation, The University of Tokyo, Tokyo

Abe S, Nakashizuka T, Masaki T (1995) Factors influencing sapling composition in canopy gaps of a temperate deciduous forest. Vegetatio 120:21–32

Abe S, Nakashizuka T, Tanaka H (1998) Effects of canopy gaps on the demography of the subcanopy tree *Styrax obassia*. J Veg Sci 9:787–796

Aiba I, Kohyama T (1997) Crown architecture and life-history traits of 14 tree species in a warm-temperate rain forest: significance of spatial heterogeneity. J Ecol 85:611–624

Brokaw NVL (1985a) Treefalls, regrowth, and community structure in tropical forests. In: Pickett STA, White PS (eds) The ecology of natural disturbance and patch dynamics. Academic, Orlando, Florida, pp 53–69

Brokaw NVL (1985b) Gap-phase regeneration in a tropical forest. Ecology 66:682–687

Brokaw NVL, Scheiner SM (1989) Species composition in gaps and structure of a tropical forest. Ecology 70:538–541

Canham CD (1989) Different responses to gaps among shade-tolerant tree species. Ecology 70:548–550

Clark DA, Clark DB (1984) Spacing dynamics of a tropical rain forest tree: evaluation of the Janzen-Connel model. Am Nat 124:769–788

Ehrenfeld JG (1980) Understory response to canopy gaps of varying size in a mature oak forest. Bull Torrey Bot Club 107:29–41

Goldblum D (1997) The effect of treefall gaps on understory vegetation in New York State. J Veg Sci 8:125–132

Grey AN, Spies TA (1997) Microsite controls on tree seedling establishment in conifer forest canopy gaps. Ecology 78:2458–2473

Hara M (1987) Analysis of seedling banks of a climax beech forest: ecological importance of seedling sprouts. Vegetatio 71:67–74

Harcombe PA (1987) Tree life table. Simple birth, growth, and death data encapsulate life histories and ecological roles. Bioscience 37:557–568

Kransy ME, Whitmore MC (1991) Gradual and sudden forest canopy gaps in Allegheny northern hardwood forest. Can J For Res 20:659–667

Kupfer JA, Runkle JR (1996) Early gap successional pathways in a *Fagus-Acer* forest preserve: pattern and determinants. J Veg Sci 7:247–256

Lee DW, Oberbauer SF, Krishnapilay B, Mansor M, Mohamad H, Yap SK (1997) Effects of irradiance and spectral quality on seedling development of two Southeast Asian *Hopea* species. Oecologia 110:1–9

Masaki T, Suzuki W, Niiyama K, Iida S, Tanaka H, Nakashizuka T (1992) Community structure of a species-rich temperate forest, Ogawa Forest Reserve, central Japan. Vegetatio 98:97–111

Nakashizuka T, Iida S (1995) Composition, dynamics and disturbance regime of temperate deciduous forests in Monsoon Asia. Vegetatio 121:23–30

Nakashizuka T, Iida S, Tanaka H, Shibata M, Abe S, Masaki T, Niiyama K (1992) Community dynamics of Ogawa Forest Reserve, a species-rich deciduous forest, central Japan. Vegetatio 103:105–112

Orwig DA, Abrams MD (1995) Dendroecological and ecophysiological analysis of gap environments in mixed-oak understoreys of northern Virginia. Funct Ecol 9:799–806

Poulson TL, Platt WJ (1989) Gap light regimes influence canopy tree diversity. Ecology 70:553–555

Putz FE, Canham CD (1992) Mechanisms of arrested succession in shrublands: root and shoot competition between shrubs and tree seedlings. For Ecol Manag 49:267–275

Rebertus AJ, Veblen TT (1993) Structure and tree-fall gap dynamics of old-growth *Nothofagus* forests in Tierra del Fuego, Argentina. J Veg Sci 4:641–654

Runkle JR (1985) Disturbance regimes in temperate forests. In: Pickett STA, White PS (eds) The ecology of natural disturbance and patch dynamics. Academic, Orlando, Florida, pp 17–33

Spies TA, Franklin JF (1989) Gap characteristics and vegetation response in coniferous forests of the pacific northwest. Ecology 70:543–545

Swaine MD, Whitmore TC (1988) On the definition of ecological species groups in tropical rain forests. Vegetatio 75:81–86

Tanaka H (1995) Seed demography of three co-occurring *Acer* species in a Japanese temperate deciduous forest. J Veg Sci 6:887–896

Uhl C, Clark K, Dezzeo N, Maquino P (1988) Vegetation dynamics in Amazonian treefall gaps. Ecology 69:751–763

van der Maarel E (1996) Pattern and process in the plant community: Fifty years after A.S. Watt. J Veg Sci 7:19–28

Veblen TT (1989) Tree regeneration response to gaps along a transandean gradient. Ecology 70:541–543

Welden CW, Hewett SW, Hubbell SP, Foster RB (1991) Sapling survival, growth, and recruitment: relationship to canopy height in a neotropical forest. Ecology 72:35–50

White PS, Pickett STA (1985) Natural disturbance and patch dynamics: an introduction. In: Pickett STA, White PS (eds) The ecology of natural disturbance and patch dynamics. Academic, Orlando, Florida, pp 3–16

Whitmore TC (1982) On pattern and process in forests. In: Newman EI (ed) The plant community as a working mechanism. Blackwell, Oxford, pp 45–59

Whitmore TC (1989) Canopy gaps and the two major groups of forest trees. Ecology 70:536–538

13 Tree Demography Throughout the Tree Life Cycle

KAORU NIIYAMA and SHIN ABE

13.1 Demographic Studies in Plants

Demographic studies and comparative methods have been the basis of a number of ecological studies of plant populations. The demographic consequences of the life cycle of plants have been analyzed with matrix population models. The matrix model was first developed as an age-structured model to describe the dynamics of human populations (Leslie 1945). Later, a size- or stage-structured matrix model was used to analyze insect populations (Lefkovitch 1965). Werner and Caswell (1977) applied these two types of matrix models to plant populations and reported that the size-structured model had higher predictability in plant population dynamics than the age-structured matrix model. Recently size- or stage-structured matrix population models have become common and effective tools for the analysis of plant populations (Caswell 1989; Silvertown and Lovett Doust 1993).

In studies of herbaceous plant populations, size-classified matrix models based on data from the entire life cycle are common (Silvertown et al. 1993). However, only a few studies have applied this type of model to tree species (Enright and Ogden 1979; Enright, 1982; Harcombe 1987; Alvarez-Buylla 1994) because repeated censuses and a long-term study period are required to obtain accurately the transition probabilities of entire tree populations. Furthermore, most tree species show mast seeding phenomena: an exceptionally large seed production year followed by several years during which few seeds are produced (see also Chapter 9). Evaluation of the fecundity of trees is fairly difficult. Thus, many studies on tree populations lack demographic data for the seed and seedling stages, which suffer the highest mortality and have large spatial and temporal variations in the life cycle of tree species (Platt et al. 1988; Nakashizuka 1991; Abe et al. 1998; Batista et al. 1998). To investigate these reproductive aspects, our study group has been studying the demography of tree species throughout their entire life cycle, incorporating seed and seedling stages, in a temperate deciduous forest, the Ogawa Forest Reserve (OFR), since 1987 (Masaki et al. 1992; Nakashizuka et al. 1992; Shibata and Nakashizuka 1995; Tanaka 1995; also see Chapters 9, 10, 11, and 12).

Ecological Studies, Vol. 158
Nakashizuka, Matsumoto (eds.) Diversity and Interaction in a Temperate Forest Community: Ogawa Forest Reserve of Japan

In this chapter, we first compare the demography of the canopy tree species *Fagus crenata* and *Fagus japonica* and the subcanopy species *Styrax obassia* as an example of effective demographic studies of trees using data on the whole life cycle. Second, the gap dependency of *Styrax obassia* is analyzed by a matrix population model incorporating the effect of canopy gap dynamics (see Chapter 7).

13.2 Matrix Model

Demographic studies of plant populations using matrix models can be categorized into four types as follows:

1. comparative studies of the demography of ecologically or taxonomically related species. The evolution of life-history traits such as the growth rate, vegetative reproduction, and fecundity is analyzed through interspecies demographic comparisons among related species (Sarukhán and Gadgil 1974; Kawano et al. 1987; Silvertown et al. 1993);
2. demographic studies of a single species, analyzing intraspecific differences of spatio-temporal demographic variations (Piñero et al. 1984; van Groenendael and Slim 1988; Abe et al. 1998; Horvitz and Schemske 1995);
3. applied studies to preserve endangered plant species (Menges 1990; Schemske et al. 1994) and to manage plant populations as resources (Nault and Gagnon

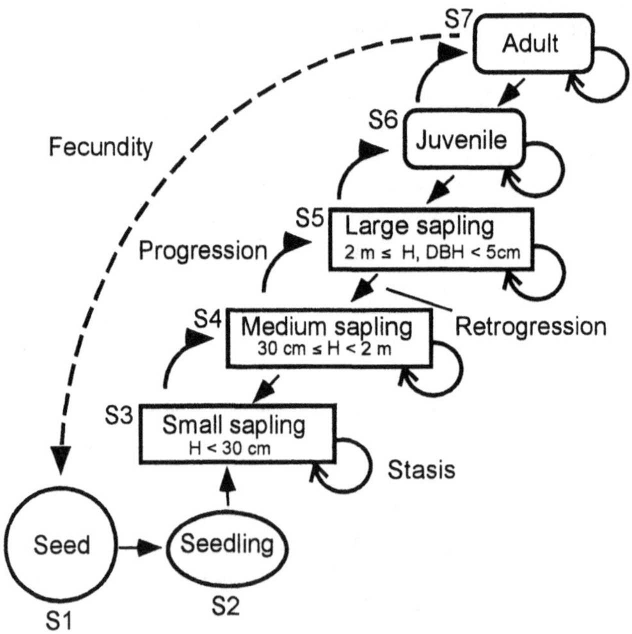

Fig. 13.1. Seven life stages (S1–S7) for the matrix population model. *Frames* show each life stage. *Arrows* show four types of transition: fecundity, progression, retrogression, and stasis

1993; Usher 1966) using matrix models as a simulation tool;
4. theoretical studies using matrix models or mathematical modeling to analyze the evolution of life-history traits (Takada and Nakajima 1992) and the dynamics of forests, using a density-dependent matrix model (Takada and Nakashizuka 1996).

This chapter focuses on types (1) and (2); the comparative demographic study of the two Fagus species and *S. obassia* belongs to type (1), and the gap dependency study of *S. obassia* belongs to type (2). The life cycles of *F. japonica* and *F. crenata* were classified into seven stages (Fig. 13.1): sound seed (S1); current year seedling (S2); small sapling (S3); medium sapling (S3); large sapling (S5); juvenile tree (S6); and adult tree (S7). We constructed a matrix population model using demographic data from 1987 to 1992 for S1 and S2, from 1989 to 1994 for S3, S4, and S5, and from 1987 to 1997 for S6 and S7. These different census periods were chosen on the basis of the expected mortality during each stage. Stages having lower mortality need longer census periods to obtain accurate demographic data. Our model is a genet-based, stage-classified projection matrix model, which gives results different from those obtained with a stem-based (ramet-based) matrix model in studies of multi-stemmed species such as *Fagus japonica* (Niiyama, unpublished data). Thus, the stage of an individual is based on the size of the largest stem among multiple stems of an individual (genet).

The population projection matrix A is composed of elements a_{ij} that describe transition probabilities between the life stages from stage j to i during a time interval from t to $t + 1$. The abundance of each stage is described by a vector $\mathbf{n}_t$. The stage structures of the population after one time interval are given as

$$\mathbf{n}_{t+1} = A\mathbf{n}_t \qquad (1)$$

The column vector will converge after intervals as

$$\mathbf{n}_{t+1} = \lambda\mathbf{n}_t \qquad (2)$$

where λ is a population growth rate in equilibrium and is given by the dominant eigenvalue of the matrix A (Caswell 1989).

The sensitivity of the population growth to a small change in each transition probability (i.e., $\delta\lambda/\delta a_{ij}$) is calculated by the product of the j-th element of the stable size distribution vector ($\mathbf{n}$) and the i-th element of the reproductive value vector. The elasticity (e) that de Kroon et al. (1986) introduced is defined as:

$$e_{ij} = (a_{ij} / \lambda) \cdot (\delta\lambda/\delta a_{ij}) \qquad (3)$$

The elasticity e_{ij} is used to estimate the relative contribution of each stage to the population growth rate of a tree species. Elasticity elements are summarized into four ecologically meaningful components based on Silvertown et al. (1993), and these components can reveal the differences in the life-history traits among the species. Elements on the diagonal of the matrix represent the probability of staying at the same stage, "stasis"; elements below the diagonal represent the probability of growth to larger stages, "progression"; and elements above the diagonal represent the probability of downgrading to smaller stages, "retrogression." This retrogression component is a special characteristic of woody species that can sprout from stem or root collar. Elements of seed production from the adult stage to the beginning stages represent "fecundity." Although clonal growth is an important life-history trait in herbaceous plants, the three species of trees studied here do not exhibit clonal growth.

To analyze the gap dependency of tree species, four types of projection matrix models can be constructed from the demographic data in a plot (Abe et al. 1998) as follows:

1. Whole-population matrix model: The annual transition probabilities are calculated as an average of all individuals, both those growing in gaps and those growing in the shade.
2. Shaded-population matrix model: The transition probabilities are calculated only for the individuals growing under the closed canopy.
3. Gap-site population matrix model: The transition probabilities are calculated only for those individuals growing in gaps.
4. Compound-population matrix model composed of the shaded and gap-site subpopulations: The transition of individuals between shaded sites and gap sites are incorporated into this model.

A whole-population matrix model was constructed to compare the life-history traits among the two *Fagus* species and *S. obassia*. Shaded-, gap-site, and compound-population matrix models were constructed to examine the gap dependency of *S. obassia*.

13.3 Elasticity Analysis in a Whole-Population Matrix Model

First, we constructed a whole-population matrix model in which populations under canopy gaps and under the closed canopy are not separated. Although *F. crenata* and *F. japonica* are shade-tolerant canopy species, they have contrasting growth forms: *F. crenata* has a single stem, whereas *F. japonica* naturally develops a multi-stem stool by vigorous root-collar sprouts (Ohkubo 1992). On the other hand, *Styrax obassia* is a typical shade-tolerant subcanopy species such as dominates in the OFR. Whole-population matrix models are constructed through the seven life-stages for the *Fagus* species (Fig. 13.1). For the *Styrax obassia* population, the medium- and large-sapling stages are combined, the seed stage is omitted, and recruitment of seedlings is estimated as fecundity, because we do not have data on seed dormancy for this species.

13.3.1 Contribution of Life Stages to Population Growth

Transition probabilities are different among the three species (Tables 13.1, 13.2, and 13.3). *F. japonica* produces tenfold more seeds than *F. crenata*, but has much lower transition probabilities from seed (S1) to seedling (S2), and from seedling (S2) to

Table 13.1. Transition probability and elasticity matrices for the *F. crenata* population based on a whole-population matrix model

a) Transition probability λ = 1.0033

	S1	S2	S3	S4	S5	S6	S7
S1							1438
S2	0.0066						
S3		0.3330	0.8992	0.0047			
S4			0.0328	0.9760	0.0019		
S5				0.0060	0.9945		
S6					0.0017	0.9958	
S7						0.0009	0.9977

b) Elasticity matrix

	S1	S2	S3	S4	S5	S6	S7	
S1							0.0021	
S2	0.0021							
S3		0.0021	0.0188	0.0001				
S4			0.0022	0.0818	0.0001			
S5				0.0022	0.2462			
S6					0.0021	0.2717		
S7						0.0021	0.3666	
Total	0.0021	0.0021	0.0210	0.0841	0.2484	0.2738	0.3687	1.0000

Table 13.2. Transition probability and elasticity matrices for the *F. japonica* population based on a whole-population matrix model

a) Transition probability λ = 0.9947

	S1	S2	S3	S4	S5	S6	S7
S1							1438
S2	0.0003						
S3		0.1000	0.8906	0.0174			
S4			0.0139	0.9419	0.0620	0.0020	
S5				0.0174	0.9342	0.0101	0.0007
S6					0.0038	0.9797	0.0049
S7						0.0020	0.9937

b) Elasticity matrix

	S1	S2	S3	S4	S5	S6	S7	
S1							0.0002	
S2	0.0002							
S3		0.0002	0.0023	0.0000				
S4			0.0003	0.0081	0.0002	0.0000		
S5				0.0004	0.0079	0.0001	0.0000	
S6					0.0003	0.0651	0.0007	
S7						0.0009	0.9130	
Total	0.0002	0.0002	0.0026	0.0086	0.0084	0.0661	0.9139	1.0000

Table 13.3. Transition probability and elasticity matrices for the *Styrax obassia* population based on a whole-population matrix model

a) Transition probability $\lambda = 1.0077$

	S2	S3	S4 & S5	S6	S7
S2					37.59
S3	0.2380	0.5960	0.0480		
S4 & S5		0.0380	0.8580	0.0010	
S6			0.0070	0.9780	
S7				0.0110	0.9930

b) Elasticity matrix

	S2	S3	S4 & S5	S6	S7	Total
S2					0.0088	
S3	0.0088	0.0132	0.0003			
S4 & S5		0.0091	0.0523			
S6			0.0088	0.2917		
S7				0.0088	0.5981	
Total	0.0088	0.0223	0.0614	0.3005	0.6069	1.0000

the small-sapling stage (S3). As a result of these low recruitment rates to advanced stages, *F. japonica* has a smaller population growth rate ($\lambda = 0.9947$) than *F. crenata* ($\lambda = 1.0033$). Another type of simulation model (Table 14.3. in Chapter 14) also predicts a decreasing *F. japonica* population and an increasing *F. crenata* population. These results suggest that large seed production (high fecundity) does not necessarily mean a high population growth rate in long-lived tree species. *F. crenata* shows higher probabilities of staying at the same stage (stasis). On the other hand, *F. japonica* could pass into the lower stages from juvenile (S6) and adult (S7) stages. As a result of this retrogression probability from the adult stage to lower stages, adult individuals of *F. japonica* show lower mortality than those of *F. crenata*. The stasis probability of *S .obassia* is similar to that of *F. japonica* at stages 6 and 7, but lower than that of the two *Fagus* species at stages 3, 4, and 5, suggesting higher mortality at these smaller stages.

13.3.2 Elasticity Components

The contribution of each stage to population growth is more clearly revealed in the elasticity matrices (Tables 13.1, 13.2, and 13.3). All of the species have the largest elasticity at stage 7. The elasticity of *F. crenata* and *S. obassia* at stages 5 and 6 is also large, but the elasticity of *F. japonica* strongly concentrates at stage 7. These results suggest that the mortality of advanced stages, in particular adult stages, strongly affects the population growth of these tree species. The disturbance regime (see Chapter 7), which primarily affects the mortality at advanced stages, is an important factor that determines the population persistence of tree species in a community.

Table 13.4. Comparison of the four components of elasticity

	F. crenata	*F. japonica*	*S. obassia*
Fecundity	0.0021	0.0002	0.0088
Stasis	0.9852	0.9964	0.9553
Progression	0.0126	0.0024	0.0356
Retrogression	0.0002	0.0009	0.0003
Total	1.0000	1.0000	1.0000

Disturbances also increase sapling mortality, but improved light conditions in a gap site accelerate sapling growth and increase the transition probability to the next stages. Thus, the effect of disturbance on the population growth rate is complex.

These elasticities can be classified into four components (Table 13.4) after Silvertown et al. (1993). Stasis is the largest component for the three species. This may be a common pattern among most tree species (Silvertown et al. 1993). *Fagus* crenata has larger fecundity and progression elasticity components than *F. japonica*, whereas the sprouting beech, *F. japonica*, has large stasis and retrogression components. Although both *Fagus* species are shade-tolerant canopy tree species in the OFR, their life-history traits as represented by the elasticity components are different. Regeneration of *F. crenata* depends on recruitment by seed released in the canopy gaps, whereas *F. japonica* individuals persist and await rare recruitment by seed through their resprouting ability. *S. obassia*, which dominates in the subcanopy, has large fecundity and progression elasticity components. This result predicts that the population growth of *S. obassia* will accelerate in canopy gaps, where large seed production and rapid growth into the adult stage are expected.

13.4 Gap Dependency of *Styrax obassia*

The responses of trees to canopy gaps vary among species (see Chapter 12), and these variations contribute to the coexistence of many species in the forest (Whitmore 1982; Abe et al. 1995; Goldblum 1997; Midgley et al. 1995). The discussion of gap dependency should be based on quantitative demographic analyses taking into account the differences between gap and nongap environments. The transition probabilities of *S. obassia* under the closed canopy and in canopy gaps were different (Tables 13.5 and 13.6). Seed production, seedling survivorship, and transition probability to the next stage increased in gap-site populations (Table 13.6). In addition, *S. obassia* saplings showed a significant gap-biased distribution (Table 12.1. in Chapter 12). However, the population growth rate was larger than 1.00 both in gap-site (1.0880) and shaded populations (1.0045). These results show that the S. obassia population has the potential to increase even under a closed canopy. Such a positive population growth under shaded conditions is a very important ecological trait for a subcanopy species. A simulation using an individual-based model (see Chapter 14), however, predicted a slightly decreasing population of *S. obassia*. This different result probably can be explained by differences in the mortality estimation and the

Table 13.5. Transition probability and elasticity matrices for the *Styrax obassia* population based on a shaded-population matrix model

a) Transition probability $\lambda = 1.0045$

	S2	S3	S4 & S5	S6	S7
S2					35.15
S3	0.2120	0.6900	0.0300		
S4 & S5		0.0300	0.8660	0.0010	
S6			0.0060	0.9800	
S7				0.0090	0.9930

b) Elasticity matrix

	S2	S3	S4 & S5	S6	S7	Total
S2					0.0071	
S3	0.0071	0.0160	0.0002			
S4 & S5		0.0073	0.0457			
S6			0.0072	0.2861		
S7				0.0071	0.6162	
Total	0.0071	0.0233	0.0530	0.2933	0.6233	1.0000

Table 13.6. Transition probability and elasticity matrices for the *Styrax obassia* population based on a gap-site population matrix model

a) Transition probability $\lambda = 1.0880$

	S2	S3	S4 & S5	S6	S7
S2					53.04
S3	0.5020	0.6820			
S4 & S5		0.0930	0.7890		
S6			0.0270	0.9580	
S7				0.0310	0.9930

b) Elasticity matrix

	S2	S3	S4 & S5	S6	S7	Total
S2					0.0392	
S3	0.0392	0.0643				
S4 & S5		0.0392	0.1001			
S6			0.0392	0.2684		
S7				0.0392	0.3710	
Total	0.0392	0.1035	0.1394	0.3077	0.4103	1.0000

Table 13.7. Transition probability and elasticity matrices for the *Styrax obassia* population based on a compound-population matrix model

a) Transition probability $\lambda = 1.00723$

	Shaded					Gap-site				
	S2	S3	S4 & S5	S6	S7	S2	S3	S4 & S5	S6	S7
Shaded										
S2					35.00					27.37
S3	0.2111	0.6872	0.0299			0.3239	0.4400			
S4 & S5		0.0299	0.8623				0.0600	0.0508		
S6			0.0060	0.9758				0.0017	0.0618	
S7				0.0090	0.9888				0.0020	0.0641
Gap-site										
S2					0.1499					25.6671
S3	0.0009	0.0028	0.0001			0.1781	0.2420			
S4 & S5		0.0001	0.0037				0.0330	0.7381		
S6				0.0042				0.0253	0.8962	
S7				0.0000	0.0042				0.0290	0.9289

b) Elasticity matrix

	Shaded					Gap-site					
	S2	S3	S4 & S5	S6	S7	S2	S3	S4 & S5	S6	S7	
Shaded											
S2					0.0065					0.0005	
S3	0.0069	0.0172	0.0002			0.0007	0.0003				
S4 & S5		0.0077	0.0488				0.0004	0.0001			
S6			0.0072	0.2626				0.0001	0.0012		
S7				0.0066	0.5226				0.0001	0.0031	
Gap-site											
S2					0.0001					0.0017	
S3	0.0001	0.0002				0.0011	0.0005				
S4 & S5		0.0001	0.0007				0.0008	0.0044			
S6			0.0000	0.0018				0.0015	0.0268		
S7				0.0000	0.0031				0.0021	0.0622	
Total	0.0070	0.0253	0.0570	0.2711	0.5323	0.0018	0.0019	0.0060	0.0301	0.0675	1.0000

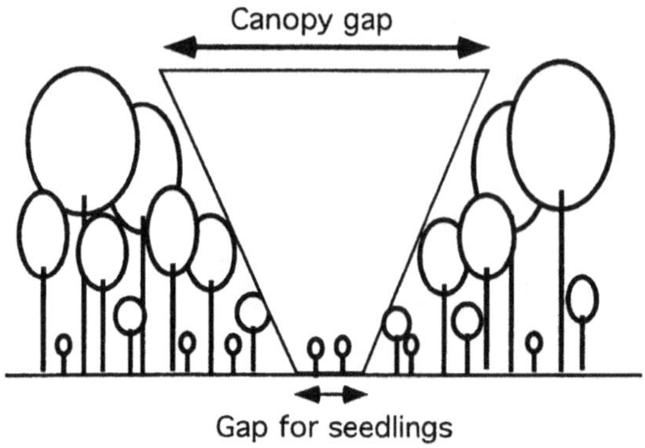

Fig. 13.2. Diagrammatic representation of "conical" gap architecture. Gap area for seedlings is smaller than the total canopy gap area even for the whole forest

size limit (DBH > 5 cm) used between the matrix model and the individual-based model (see Chapter 14).

The compound-population model incorporates mutual transitions between a shaded subpopulation and a gap-site subpopulation into the matrix (Table 13.7). For example, the probability of transition from a shaded to a gap-site at stage 7 is 0.0042 and the reverse probability is 0.0641 (Abe et al. 1995). The value 0.0042 is the gap formation rate, 0.42 %/yr, in the plot (see Table 7.1, Chapter 7). The shape of the gap is assumed to be conical (Fig. 13.2). The proportion of the gap area on the forest floor where it is available for the seedling stages is calculated as follows:

$$p = (\text{gap area for seedlings}) / (\text{canopy gap area}) \qquad (4)$$

where p is the degree to which the gaps in the plot are cone-shaped. A larger p indicates that a larger gap area is available for seedlings, where the probability of growth and survivorship of seedlings would be improved. The stage class distribution in an equilibrium population in which the gaps are assumed to be conical and p is 0.1 ($\lambda = 1.00723$) is close to that found in the OFR at present.

13.4.1 Simulation with a Different Gap Formation Regime

We adopted the compound-population matrix model to simulate the effect of the disturbance regime of the forest on population growth, because the model with conical gaps could simulate a population structure and population growth rate very close to those found at present in the OFR. The size distribution of gaps according to canopy height is shown in Fig. 7.4 in Chapter 7, which shows that the gaps are indeed conical. The value p was arbitrarily chosen as 0.1 for simplicity. Results of the simulation show that both the proportions of canopy gap area and the gap closure rate greatly affect the population growth rate. A larger proportion of gap area or a longer period required for canopy closure (i.e., slower canopy recovery) results in a larger population growth rate (Fig. 13.3). The growth rate of the present population is very low compared to the range of variation along the possible gradient of disturbance regimes. This low population growth rate suggests a relatively low disturbance intensity during recent decades in the OFR (see Chapter 7).

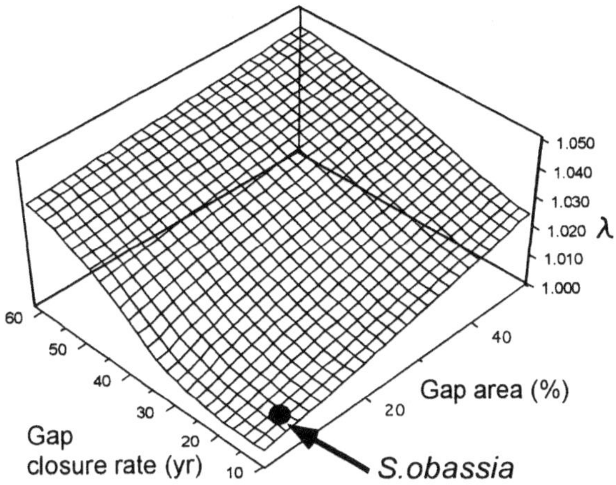

Fig. 13.3. Simulation from a compound-population matrix model assuming conical gaps showing the dependence of the population growth rate (λ) on disturbance intensity and gap closure rate

Life-history traits of tree species can be effectively described and analyzed using a matrix population model. This model can also simulate the response of a population to variations in the disturbance regime. The contribution of each life-history characteristic, such as fecundity, mast seeding, seedling mortality, gap dependency, and sprouting ability, on tree population growth should be evaluated quantitatively using a matrix population model that incorporates canopy gap effects.

References

Abe S, Nakashizuka T, Masaki T (1995) Factors influencing sapling composition in canopy gaps of a temperate deciduous forest. Vegetatio 120:21–32

Abe S, Nakashizuka T, Tanaka H (1998) Effects of gaps on the demography of the subcanopy tree Styrax obassia. J Veg Sci 9:787–796

Alvarez-Buylla ER (1994) Density dependence and patch dynamics in tropical rain forests: matrix models and applications to a tree species. Am Nat 143:155–191

Batista WB, Platt WJ, Macchiavelli RE (1998) Demography of a shade-tolerant tree (*Fagus grandifolia*) in a hurricane-disturbed forest. Ecology 79:38–53

Caswell H (1989) Matrix population model. Sinauer Associates, Sunderland, Massachusetts

de Kroon H, Plaisier A, van Groenendael, Caswell H (1986) Elasticity: the relative contribution of demographic parameters to population growth rate. Ecology 67:1427–1431

Enright NJ (1982) The ecology of Araucaria species in New Guinea. III. Population dynamics of sample stands. Aust J Ecol 7:227–237

Enright NJ, Ogden J (1979) Applications of transition matrix models in forest dynamics: *Araucaria* in Papua New Guinea and *Nothofagus* in New Zealand. Aust J Ecol 4:3–23

Goldblum D (1997) The effect of treefall gaps on understory vegetation in New York State. J Veg Sci 8:125–132

Harcombe PA (1987) Tree life tables. Simple birth, growth, and death data encapsulate life histories and ecological roles. Bioscience 37:557–568

Horvitz CL, Schemske DW (1995) Spatiotemporal variation in demographic transitions of a tropical understory herb: projection matrix analysis. Ecol Monogr 65:155–192

Kawano S, Takada T, Nakayama S, Hiratsuka A (1987) Demographic differentiation and life history evolution in temperate woodland plants. In: Urbanska KM (ed) Differentiation patterns in higher plants. Academic, London, pp 153–181

Lefkovitch LP (1965) The study of population growth in organisms grouped by stages. Biometrics 21:1–18

Leslie PH (1945) On the use of matrices in certain population mathematics. Biometrika 33:183–212

Masaki T, Suzuki W, Niiyama K, Iida S, Tanaka H, Nakashizuka T (1992) Community structure of a species-rich temperate forest, Ogawa Forest Reserve, central Japan. Vegetatio 98:97–111

Menges ES (1990) Population viability analysis for an endangered plant. Conserv Biol 4:52–62

Midgley JJ, Cameron MC, Bond WJ (1995). Gap characteristics and replacement patterns in the Knysna Forest, South Africa. J Veg Sci 6:29–36

Nakashizuka T (1991) Population dynamics of coniferous and broad-leaved trees in a Japanese temperate mixed forest. J Veg Sci 2:413–418

Nakashizuka T, Iida S, Tanaka H, Shibata M, Abe S, Niiyama K (1992) Community dynamics of Ogawa Forest Reserve, a species-rich deciduous forest, central Japan. Vegetatio 103:105–112

Nault A, Gagnon D (1993) Ramet demography of *Allium tricoccum*, a spring ephemeral, perennial forest herb. J Ecol 81:101–119

Ohkubo T (1992) Structure and dynamics of Japanese beech (*Fagus japonica* Maxim.) stools and sprouts in the regeneration of the natural forests. Vegetatio 101:65–80

Piñero D, Martínez Ramos M, Sarukhán J (1984) A population model of *Astrocaryum mexicanum* and a sensitivity analysis of its finite rate of increase. J Ecol 72:977–991

Platt WJ, Evans GW, Rathburn SL (1988) The population dynamics of a long-lived conifer (*Pinus palustris*). Am Nat 131:491–525

Sarukhán J, Gadgil M (1974) Studies on plant demography; *Ranunculus repens* L., *R.. bulbosus* L. and *R.. acris* L. III. A mathematical model incorporating multiple modes of reproduction. J Ecol 62:921–936

Schemske DW, Husband BC, Ruckelshaus MH, Goodwillie C, Parker IM, Bishop G (1994) Evaluating approaches to the conservation of rare and endangered plants. Ecology 75:584–606

Shibata M, Nakashizuka T (1995) Seed and seedling demography of four co-occurring *Carpinus* species in a temperate deciduous forest. Ecology 76:1099–1108

Silvertown J, Lovett Doust J (1993) Introduction to plant population biology. Blackwell, Oxford, pp 96–100

Silvertown J, Franco M, Pisanty I, Mendoza A (1993) Comparative plant demography: relative importance of life-cycle components to the finite rate of increase in woody and herbaceous perennials. J Ecol 81:465–476

Takada T, Nakajima H (1992) An analysis of life history evolution in terms of the density-dependent Lefkovitch matrix model. Math Biosci 112:155–176

Takada T, Nakashizuka T (1996) Density-dependent demography in a Japanese temperate broad-leaved forest. Vegetatio 124:211–221

Tanaka H (1995) Seed demography of three co-occurring *Acer* species in a Japanese temperate deciduous forest. J Veg Sci 6:887–896

Usher MB (1966) A matrix approach to the management of renewable resources, with reference to selection forests. J Appl Ecol 3:355–367

van Groenendael JM, Slim P (1988) The contrasting dynamics of two populations of *Plantago lanceolata* classified by age and size. J Ecol 76:585–599

Werner PA, Caswell H (1977) Population growth rates and age- versus stage-distribution models for teasel (*Dipsacus sylvestris* Huds.). Ecology 58:1103–1111

Whitmore TC (1982) On pattern and process in forests. In: Newman EI (ed) The plant community as a working mechanism. Blackwell, Oxford, pp 45–59

14 Individual-Based Model of Forest Dynamics

TAKUYA KUBO

We constructed and analyzed a computer simulator for forest dynamics based on the newly obtained information of the long-term and large-area census in the Ogawa Forest Reserve (OFR) plot during the period from 1987 to 1993. Our main objective was the development of new methods to estimate values for demographic parameters defined by field measurements.

14.1 Density Model vs. Individual-Based Model

One of the most popular ways of modeling forest dynamics is the "density model," which takes into account time changes of a single statistic, tree density. Such a model has a simple structure and is calculated by well-known methods such as a transition matrix (Takada and Nakashizuka 1996) or partial differential equations. On the other hand, density models are often too simplified to generate many interesting patterns, such as the spatial structures observed in a real forest. Moreover, such density models may be inadequate to treat dynamics that include interactions between individuals (e.g., Kubo and Ida 1998, a justification for adopting an individual-based model).

In order to avoid these disadvantages, several kinds of forest dynamics models based on the behavior of individual trees have been developed (e.g., Botkin et al. 1972; Shugart 1984; Pacala et al. 1993, 1996; Kubo and Ida 1998).

14.2 The OFR Simulation: Grand Design and Species Screening

We decided to develop our forest simulator as an individual-based model because of the benefits mentioned above and to be consistent with the OFR census data. Such consistency is essential not only for comparisons with global patterns, but also for detailed modeling such as parameterization. Accordingly, our forest simulator consists of a grid of 60 × 40 5-m quadrats (6 ha total), with an altitude specified for each

Ecological Studies, Vol. 158
Nakashizuka, Matsumoto (eds.) Diversity and Interaction in a Temperate Forest Community: Ogawa Forest Reserve of Japan

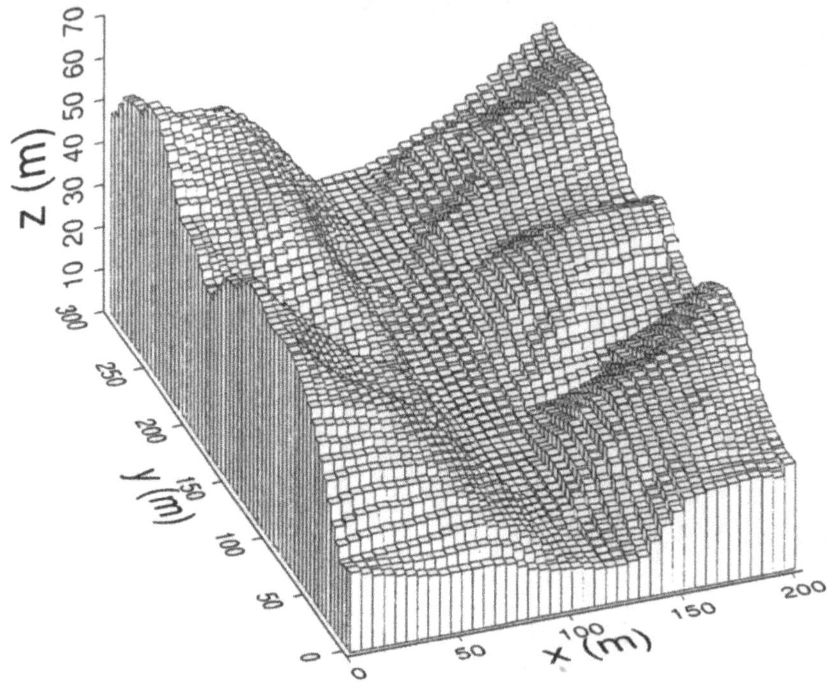

Fig. 14.1. Coordinates and topography of the OFR. The x axis ($x = 0, 1, 2, \ldots, 39$) is almost parallel to the west → east axis, and the y axis ($y = 0, 1, 2, \ldots, 59$) to south → north. The direction of the continuous vertical axis z is defined as lower → higher

quadrat based on topographic data on the OFR plot. Figure 14.1 represents a three-dimensional view of the topography and coordinates of the OFR simulator. As mentioned above, an individual-based model simulates each tree (as represented by a single trunk) located in a quadrat to grow and die independently while interacting with other trees located in the same and the neighboring quadrats.

The census data includes all trees whose diameter at breast height (DBH) is equal to or larger than 5 cm within the 6-ha plot, measured in 1987, 1989, 1991, and 1993 (see Chapters 5 and 6; Masaki et al. 1992; Nakashizuka et al. 1992). Although more than 50 tree species are found in the OFR plot, we selected only 23 species for modeling because it was impossible to determine demographic parameters for those species with only a few individuals in the plot. We established two criteria for inclusion: (1) either more than 60 individuals (trunks) were found or the species was a so-called canopy species; and (2) data on trunk height was available. Only individuals satisfying both criteria were selected. Thus, the included tree species are shown in Table 14.1. Their total density was almost 90% that of the whole community.

Table 14.1. Species selected for modeling and their density and total basal area in 1987

Abbreviation		Species	Density (/ 6 ha)	BA (m^2/ 6 ha)
1	ACA	*Acer amoenum*	497	57.6
2	ACM	*Acer mono var. marmoratum f. dissectum*	177	54.2
3	ACR	*Acer rufinerve*	91	20.7
4	ACS	*Acer sieboldianum*	72	8.0
5	ACT	*Acer tenuifolium*	79	7.3
6	BNJ	*Benthamidia japonica*	75	6.5
7	BTG	*Betula grossa*	40	36.0
8	CLB	*Clethra barbinervis*	165	7.9
9	CRC	*Carpinus cordata*	535	41.7
10	CRJ	*Carpinus japonica*	68	16.0
11	CRL	*Carpinus laxiflora*	544	81.8
12	CRT	*Carpinus tschonoskii*	97	36.5
13	CSC	*Castanea crenata*	62	80.3
14	FGC	*Fagus crenata*	115	160.3
15	FGJ	*Fagus japonica*	695	394.1
16	FRL	*Fraxinus lanuginosa f. serrata*	74	4.8
17	MLM	*Meliosma myriantha*	97	5.7
18	OSJ	*Ostrya japonica*	71	50.1
19	PRV	*Prunus verecunda*	87	46.5
20	QRC	*Quercus crispula*	51	64.5
21	QRS	*Quercus serrata*	318	507.1
22	STO	*Styrax obassia*	571	47.4
23	SWC	*Swida controversa*	149	71.2
24	ZZZ	all others	514	110.5
25	Total	5244	1916.9	

BA, total basal area.

14.3 Growth: Height and DBH

The growth model of a trunk has two components in the individual-based model: height and diameter growth. In this model, tree height growth is assumed to be deterministically subordinated to its DBH. Here we use Ogawa's generalized allometric equation,

$$\frac{1}{H_{i,t}} = \frac{1}{\alpha D_{i,t}^{\ h}} + \frac{1}{H_{\max}} \qquad (1)$$

in which the height of the i-th individual (trunk) in year t or $H_{i,t}$ (m) is represented as a deterministic function of its DBH or $D_{i,t}$ (cm). The values of a, h, and $H_{\max}$ were estimated by applying a nonlinear fitting to the DBH–height relationship data. The observed $H_{i,t}$–$D_{i,t}$ relationships and the estimated curves for each species are shown

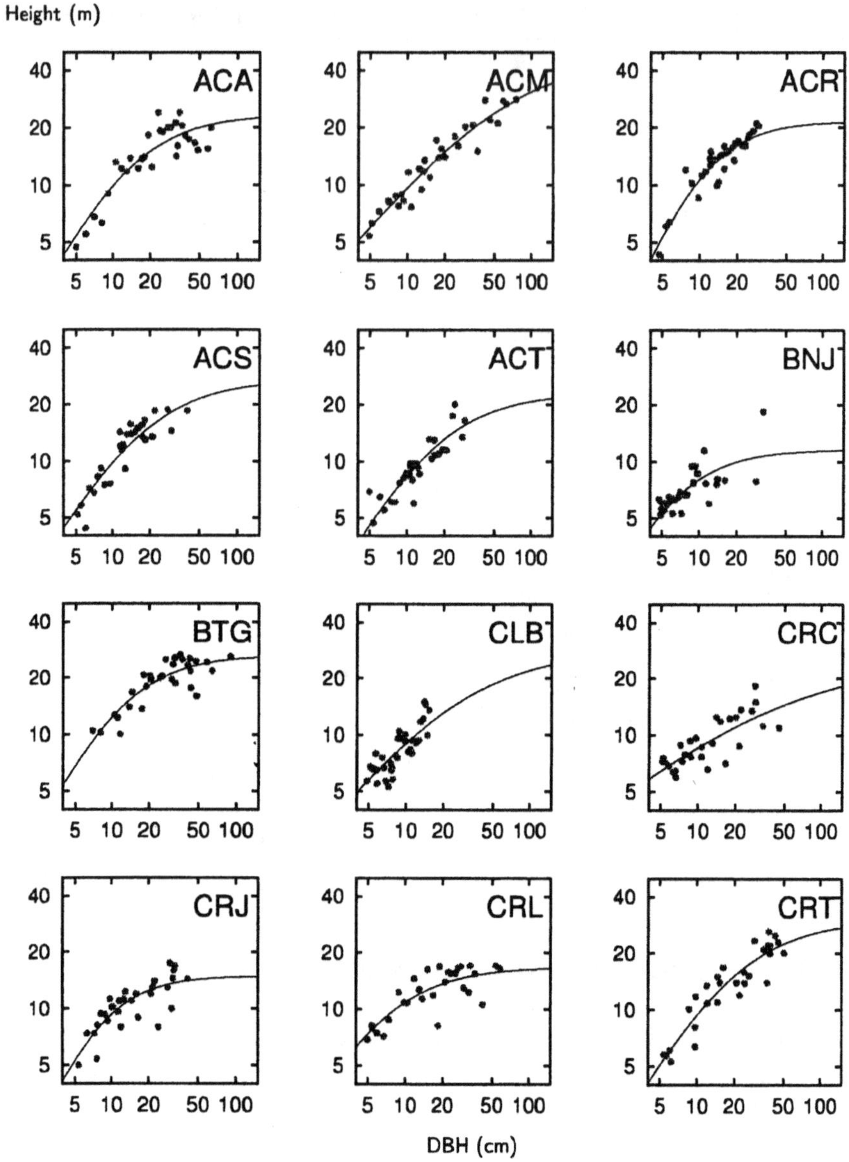

Fig. 14.2. Relationship between tree height and DBH (trunk diameter at breast height) observed in 1987, 1989, and 1991. The *solid lines* indicate the estimated allometric curves. Refer to Table 14.1 for definitions of the abbreviations

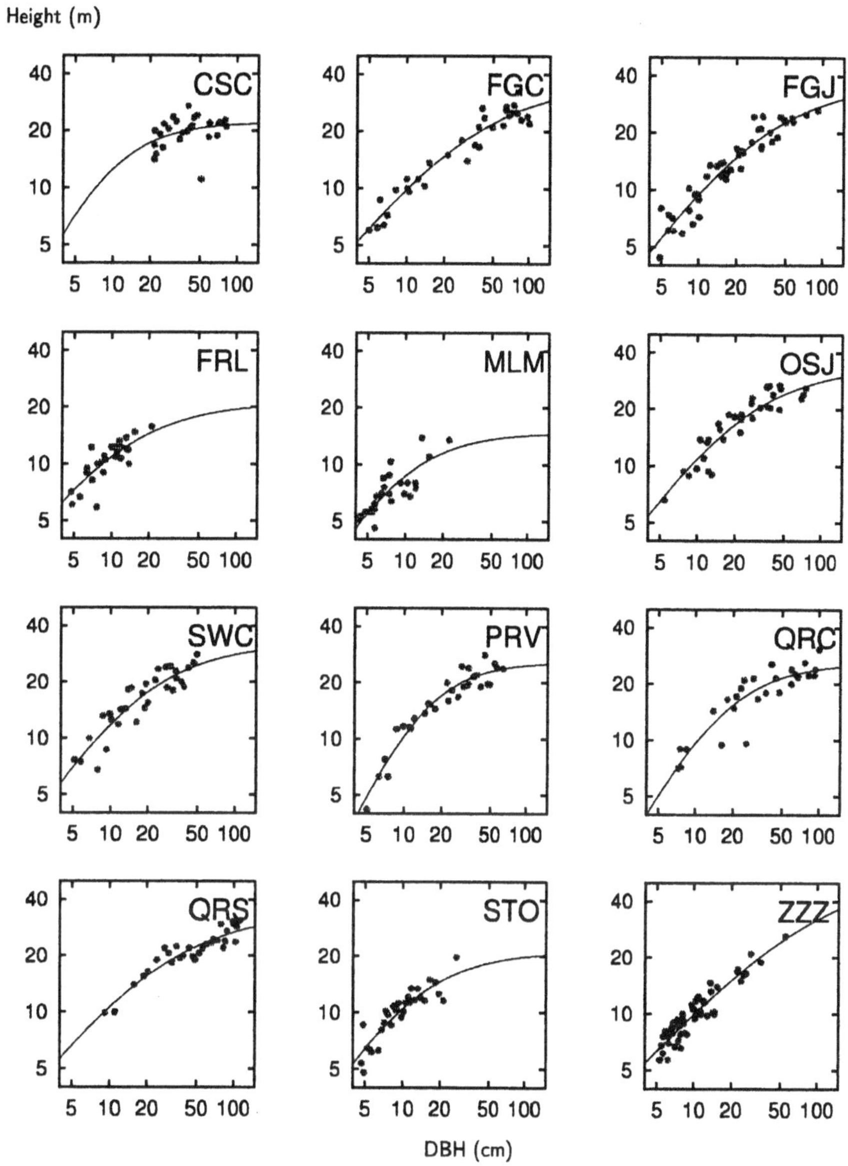

Fig. 14.2. *Continued*

in Fig. 14.2. In the following, the term "height" of a tree is equal to the sum of the trunk height $H_{i,t}$ and the altitude of the quadrat where the tree is located.

Because the observed DBH growth rates (over several years) showed large variation and sometimes lacked a discernible trend, as shown in Fig. 14.3, we generated the DBH growth ras' using a stochastic model, which was based on the tree species and size and on its interactions with neighboring trees.

In this section, we explain how to construct the submodel of growth for each species. Individual size $D_{i,t}$ is the DBH for the i-th trunk. The growth rate, $\mathbf{G}_{i,t}$, during the census interval Δt is defined as

$$\mathbf{D}_{i,t+\Delta t} = D_{i,t} + \mathbf{G}_{i,t} \tag{2}$$

where both $\mathbf{D}_{i,t+\Delta t}$ and $\mathbf{G}_{i,t}$ are random variables. We assume that $\mathbf{G}_{i,t}$ has an exponential distribution whose functional form depends on the attributes of the i-th tree. The probability that $\mathbf{G}_{i,t}$ is in the range from z to z + dz is,

$$\mathbf{Pr}\{z \leq \mathbf{G}_{i,t} < z + \mathrm{d}z\} = \lambda_{i,t} \exp(-\lambda_{i,t}\, z)\, \mathrm{d}z \tag{3}$$

where the parameter $\lambda_{i,t}$ is equal to the inverse of the expectation of $\bar{G}_{i,t}$ or $\lambda_{i,t} = 1/\bar{G}_{i,t}$.

We assigned the functional form of $\bar{G}_{i,t}$ as if it were a variation of the Gompertz growth model, which is modified by the intensity of one-directional competition,

$$\bar{G}_{i,t} = D_{i,t} \exp\{-(g_c + g_d \ln D_{i,t} + g_s\, S_{i,t})\} \tag{4}$$

where $S_{i,t}$ is the competition intensity for the i-th trunk (defined later). The newly introduced parameters are initial growth rate g_c, size dependency g_d, and intensity of one-directional competition g_s.

The term "one directional" expresses a mode of competition between individuals in which the larger individual can reduce the growth and survivability of the smaller one. We defined the index of competition intensity, $S_{i,t}$, for each individual, and its functional forms were chosen by performing a large number of trial-and-error calculations to fit them to the tree growth data,

$$S_{i,t} = \sum_{\forall j} \left(\frac{D_{j,t} \cos\phi}{R_{i,j}} \right)^{\beta} \tag{5}$$

In the above equation, j is the index for all individual trunks whose height is taller than that of the i-th trunk and ϕ is the angle between the top of j-th trunk and the special point (explained later) on the celestial sphere defined as

$$\phi = \cos^{-1} \frac{-\Delta y \cos\theta + \Delta z \sin\theta}{R_{i,j}} \tag{6}$$

where $R_{i,j}$ is the Euclidean distance between the tops of i-th and the j-th trunk, Δy is the distance in the y direction, and Δz is the difference in height.

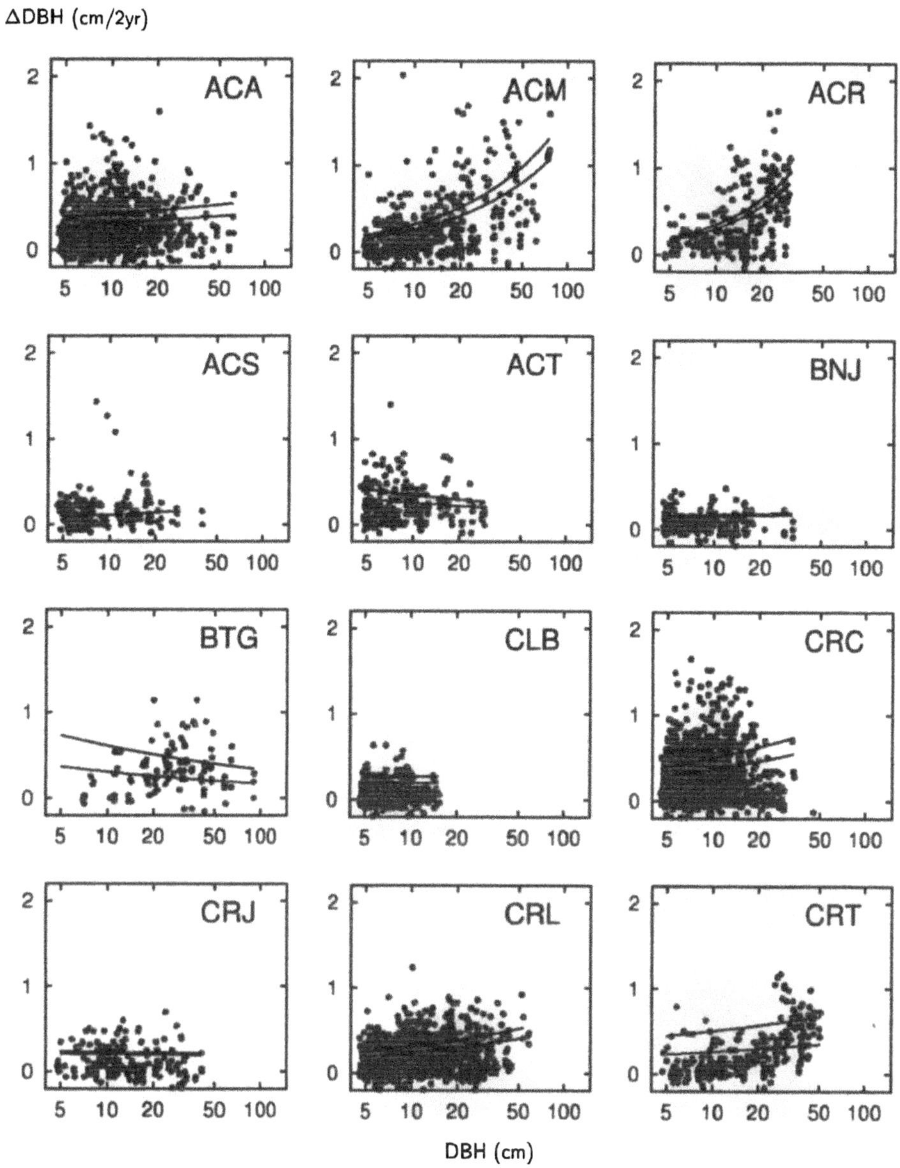

Fig. 14.3. Relationship between DBH (in 1987, 1989, and 1991) and the DBH growth rate (ΔDBH). Dots represents all observations for each tree species during the period from 1987 to 1993. The *bold solid lines* are estimated relationships according to our growth model under conditions without any shading (i.e., $S_{i,t} = 0$ for all individual trunks), while the *thin lines* are estimates assuming a fixed value for the index of shading ($S_{i,t} = 0.5$ for all individual trunks). Refer to Table 14.1 for definitions of the abbreviations

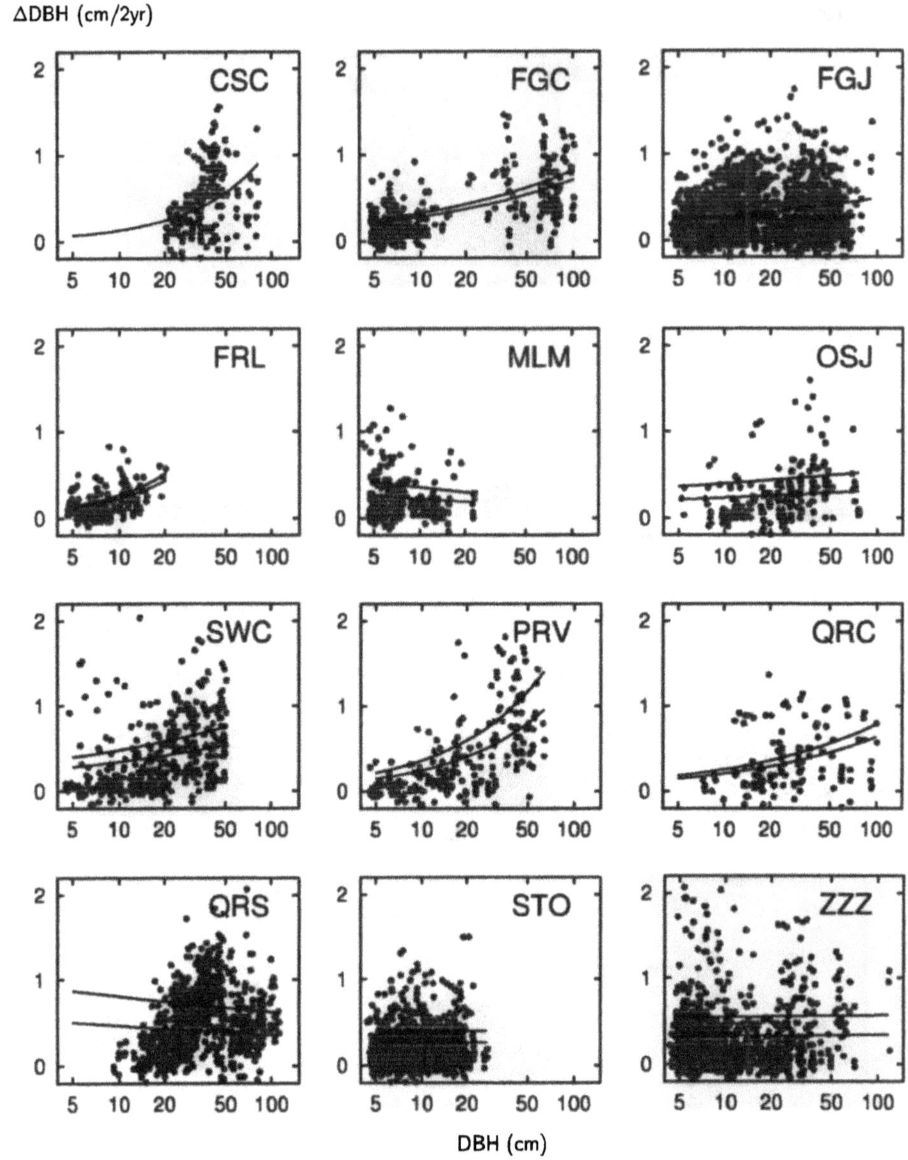

Fig. 14.3. *Continued*

Here we have two constants that are the same for all species, θ and β. The former, θ, is defined as the angle on y–z plane between the top of a trunk and the direction that most affects its growth rate on the celestial sphere. Because the y axis is almost parallel to the south–north axis, $\theta = 0$ is defined as due south, $\theta = \pi/2$ as the zenith, and $\theta = \pi$ as due north and parallel to the horizon. We introduced θ because we predicted that a shading tree located in the direction of around the zenith might have more importance than one in the direction of the horizon. The other parameter, β, is the "degree of focusing" to the point. The values for these two constants will be calculated later from growth data such that they make a statistical maximum, as explained in the following section.

14.4 Parameterization by the Maximum Likelihood Method

We used the maximum likelihood method to evaluate these DBH growth parameters. The likelihood equation, assuming that the probabilistic distribution of the DBH growth rate is exponential, becomes,

$$\begin{aligned}\ln f_G = & \sum_{\forall\tau}\sum_{\forall i}\left\{g_c + g_d \ln D_{i,\tau} + g_s S_{i,\tau}\right\} \\ & - \sum_{\forall\tau}\sum_{\forall i}\frac{G_{i,\tau}}{x_{i,\tau}}\exp\left\{g_c + g_d \ln D_{i,\tau} + g_s S_{i,\tau}\right\} \\ & + (\text{terms without any parameters})\end{aligned} \qquad (7)$$

in which the observed growth rate $G_{i,\tau}$ is defined as

$$G_{i,\tau} = D_{i,\,\tau+\Delta t} - D_{i,\,\tau} \qquad (8)$$

and the year of census, τ, which includes 1987, 1989, and 1991 but not 1993 (because $G_{i,1993}$ could not be calculated).

We can obtain the maximum likelihood estimator by setting all partial differentials of $\ln f_G$ for all growth parameters to zero, that is,

$$\frac{\partial \ln f_G}{\partial g_c} = 0, \frac{\partial \ln f_G}{\partial g_d} = 0 \text{ and } \frac{\partial \ln f_G}{\partial g_s} = 0. \qquad (9)$$

We solved the simultaneous equation numerically by using the Newton-Raphson method. The estimated growth rates are shown in Fig. 14.3 together with the observed growth rates for individual trunks.

The OFR census data might bring a small inconsistency to our growth submodel. As shown in Fig. 14.3, the observed DBH distributions of many species are clearly truncated rather than gradually terminated. This implies that the observed growth

Table 14.2. Estimated parameters for each tree species based on OFR census data (1987–1993)

Species[a]	G_c (/ 2 years)	g_d	g_s × 0.01	D_{max} (cm)	a	h	H_{max} (m)	r_c (/ 2 years)	
ACA	0.17	0.72	0.50	62.1	0.88	1.29	23.4	10.10	ACA
ACM	0.04	0.22	0.00	76.6	1.82	0.82	50.1	3.38	ACM
ACR	0.07	0.15	1.50	30.6	0.55	1.60	21.5	1.55	ACR
ACS	0.10	0.79	0.02	27.8	1.04	1.17	27.2	1.71	ACS
ACT	0.46	1.15	1.29	29.3	0.80	1.22	22.9	5.05	ACT
BNJ	0.13	0.87	0.89	32.7	0.96	1.44	11.6	2.04	BNJ
BTG	0.42	1.00	1.44	90.4	1.06	1.34	26.5	0.63	BTG
CLB	0.22	0.91	1.98	15.4	1.76	0.88	28.2	5.05	CLB
CRC	0.14	0.57	0.46	33.7	3.35	0.59	25.5	10.10	CRC
CRJ	0.19	1.00	0.00	41.5	0.62	1.60	15.0	0.98	CRJ
CRL	0.09	0.55	0.64	53.9	1.81	1.23	16.8	6.73	CRL
CRT	0.10	0.52	1.50	50.9	0.93	1.17	29.9	2.91	CRT
CSC	0.02	0.14	0.00	82.4	1.00	1.46	22.1	0.30	CSC
FGC	0.11	0.55	0.50	102.0	1.75	0.89	35.8	3.37	FGC
FGJ	0.12	0.70	0.00	92.8	1.40	0.96	36.7	10.10	FGJ
FRL	0.07	0.15	2.67	20.4	2.07	1.04	21.1	1.88	FRL
MLM	0.44	1.18	1.71	23.2	1.11	1.28	14.8	6.74	MLM
OSJ	0.11	0.60	1.41	75.7	1.58	1.01	33.9	0.91	OSJ
PRV	0.03	0.01	0.83	64.8	0.49	1.55	25.7	1.72	PRV
QRC	0.07	0.47	0.89	100.9	0.76	1.31	25.9	0.66	QRC
QRS	0.35	0.81	1.88	112.5	1.94	0.90	34.0	1.46	QRS
STO	0.18	0.78	1.11	27.4	1.40	1.17	21.2	6.73	STO
SWC	0.24	0.69	1.26	50.90	1.57	1.08	31.9	2.54	SWC
ZZZ	0.32	0.80	2.38	121.2	2.12	0.75	61.1	10.10	ZZZ

G_c, initial growth rate of DBH (trunk diameter at breast height) growth, which is equal to exp $(-g_c)$; g_d, DBH- dependent coefficient of DBH growth; g_s, shading intensity; D_{max}, maximum DBH; a and h, coefficients for DBH–height allometry; H_{max}, maximum tree height; r_c, expected recruitment per 2 years.

[a]Refer to Table 14.1 for definitions of abbreviations.

rate is still positive at the observed maximum size, so the estimated mode of DBH growth would be unbounded.

In order to avoid this problem, we assumed that the growth of trunk diameter was limited not only by the above size-dependent term but also by the maximum DBH, which is peculiar to each species. Thus, $\overline{G}_{i,t}$ is set to zero when $D_{i,t}$ is greater than the maximum DBH, or D_{max} in the growth submodel. As D_{max} we chose the maximum DBH observed through the whole census period for each species.

We calculated the maximized $\ln f_G$ for each of several sets of the two common parameters, θ and β, mentioned in the previous section. By comparing the maximized likelihoods, we discovered a set of values for these parameters for which $\ln f_G$ was largest. At that point, θ and β were estimated as 0.556π and 2.0, respectively.

All estimated parameters of growth (diameter and height) for each species are shown in Table 14.2.

14.5 Mortality: Breakage and Withering

The mode of tree mortality (that is, trunk mortality) in the OFR plot is understood to consist mainly of two factors (Nakashizuka et al. 1992): disturbance, which depends on the DBH; and withering by shading. In our model, we also separate mortality into two components reflecting these two factors.

Before modeling, we found that the number of dead individual trunks was too small (about 200 over the whole census period) to determine an estimator for each species. Hence, a single mortality model for all species was constructed.

The first component of our mortality model is trunk breakage. We chose the functional form for the probability of breakage per two years as

$$B_{i,t} = m_d \frac{D_{i,t}}{D_{\max}} \tag{10}$$

where m_d is the parameter to be estimated. This functional form was adopted because it provided the highest maximum likelihood (defined later) among several trial functions, such as $D_{i,t}$, $D_{i,t}^2$, and $(D_{i,t}/D_{\max})^2$.

We again used the $D_{\max}$ likelihood method to evaluate this parameter. The likelihood equation for breakage thus becomes

$$\ln f_B = \sum_{\forall\tau}\sum_{\forall i'} \ln B_{i',\tau} + \sum_{\forall\tau}\sum_{\forall i''} \ln\left(1 - B_{i'',\tau}\right) + \left(\text{terms without any parameter}\right) \tag{11}$$

where i' is the index for a broken trunk at $\tau + \Delta t$ and i'' is the index for a survivor. Setting

$$\frac{\partial f_B}{\partial m_B} = 0 \tag{12}$$

and solving it numerically using the Newton-Raphson method, we obtained the maximum likelihood estimator. By applying the maximum likelihood method to the OFR census data, m_d was evaluated as 1.40×10^{-2}.

The other component of mortality is withering caused by shading by taller trunks. Its modeling and parameterization were performed in the same way. The functional form for individual trunk elimination by withering (excluding breakage) was assumed to be

$$W_{i,t} = m_c + m_s S_{i,t} \tag{13}$$

where m_c is the constant rate of random mortality and m_s represents the shade dependency. The likelihood equation is as follows:

$$\ln f_W = \sum_{\forall\tau}\sum_{\forall i'} \ln W_{i',\tau} + \sum_{\forall\tau}\sum_{\forall i''} \ln(1 - W_{i'',\tau}) + (\text{terms without any parameter}) \tag{14}$$

and the parameters m_c and m_s were estimated as 4.50×10^{-3} and 1.79×10^{-2}, respectively.

14.6 Recruitment: Estimated by the "Inverse Growth" Method

The recruitment rate is usually defined as the number of trees that in the current census are larger than or equal to the minimum DBH or $D_{\min}$ but that were smaller in the previous census. Unfortunately, we found again that the number of samples available for estimating the recruitment rate as defined above was too small. Five of the 23 species selected had no recruitment during the entire census period, and the number of trees recruited was not large enough for most other species.

Hence, we developed a method to estimate the recruitment rate similar to that used by Kohyama and Takada (1998) to resolve this difficulty. They inferred the recruitment rate by applying submodels for growth and mortality developed in advance for the relatively smaller individuals in the population.

Because the method of Kohyama and Takada (1998) is for a density model for size structure, we needed to adapt it to our individual-based model. We did so by using an "inverse growth" method applied to each tree species as follows:

1. We introduced an approximated equation to define "inverse growth,"

$$D_{k,t-\Delta t} \approx D_{k,t} - \bar{G}_{k,t} \tag{15}$$

where k is the index of the individual trunk selected for parameterization.

2. The criteria for individual screening to parameterize the recruitment rate are: (a) Select all individuals to be regarded as recently recruited by applying the above "inverse growth" equation. We chose all individuals whose estimated recruitment year was between 1985 and 1991. (b) If the number of samples is less than the target number (set at 20) after procedure (a), add individuals from earlier recruitment years, beginning with 1984, until the target is reached.

3. Set the last year each k-th individual is observed in the census to τ_k.

4. For each selected sample, apply the "inverse growth" equation recursively until $D_{k,t-\Delta t}$ becomes smaller than the minimum DBH or $D_{\min}$. Determine the estimated recruitment year T_k such that $D_{k,Ti-\Delta t} < D_{\min} = D_{k,Ti}$.

5. After detecting T_k for all samples, calculate the average recruitment rate, which is specific for a tree species, using the following equation,

$$r_c = \left(\max_{\forall k} \tau_k - \min_{\forall k} T_k \right)^{-1} \sum_{\forall k} \prod_{t=T_k}^{\tau_k} \left(1 - B_{k,t} - W_{k,t}\right)^{-1} \qquad (16)$$

in which the average, weighted by the mortality rate, is used.

The recruitment rate for all selected species was estimated by using the "inverse growth" method.

The recruitment rate values are shown in Table 14.2. Figure 14.4 shows the distribution of T_i and the average recruitment rate r_c.

The remaining problem for modeling recruitment concerns spatial structure. We found that for all selected species it was difficult to detect any correlation between recruitment and possible mother trees providing seeds because of the sample number limitation. To resolve the problem in our forest simulator, we assumed that a species can recruit only in the quadrats in which the species was present during the census period from 1987 to 1993. A more realistic submodel for recruitment could be developed in the future by using data on seed dispersal (see Chapter 10) and seed bank dynamics (see Chapters 11 and 12).

14.7 Results of the Simulation

As shown in Table 14.3, the results of the simulation over a 400-year period suggest that the tree species would change drastically with the present canopy trees being replaced by those with much higher recruitment rates. Because of this replacement, the estimated basal area (BA) in the forest decreased slightly during the simulation.

The forest configuration of the OFR in 1987 was used as the initial state for all runs of the individual-based simulator, and the simulation period was 400 years. Using the parameter value set calibrated by the field measurements (baseline set), the simulation results predicted that the BA within the plot would decrease to 26.6 m^2/ha after 400 years from the initial value of 32.2 m^2/ha. This decrease is caused by tree species replacement: the density of some of the current canopy species such as *Quercus serrata* (QRS, see Table 14.1) and *Castanea crenata* (CSC) decreased because of their small recruitment rate, and they were replaced by trees with larger recruitment rates such as *Acer amoenum* (ACA), *Carpinus cordata* (CRC), and *Styrax obassia* (STO). Because the maximum DBH of the latter species was smaller than those of the former, the BA in the forest was reduced.

It is interesting that the above result is the complete opposite of the result predicted in Chapter 6. Although the stand-based model (see Chapter 6) predicts that the BA in the OFR will increase in the future because of recovery from large-scale disturbances and because the forest will be more mature, the present individual-based model predicts that the forest will becomes "thin" as larger species are replaced by smaller tree species. Expressed by terms used in Chapter 6, IC-species (shade-intol-

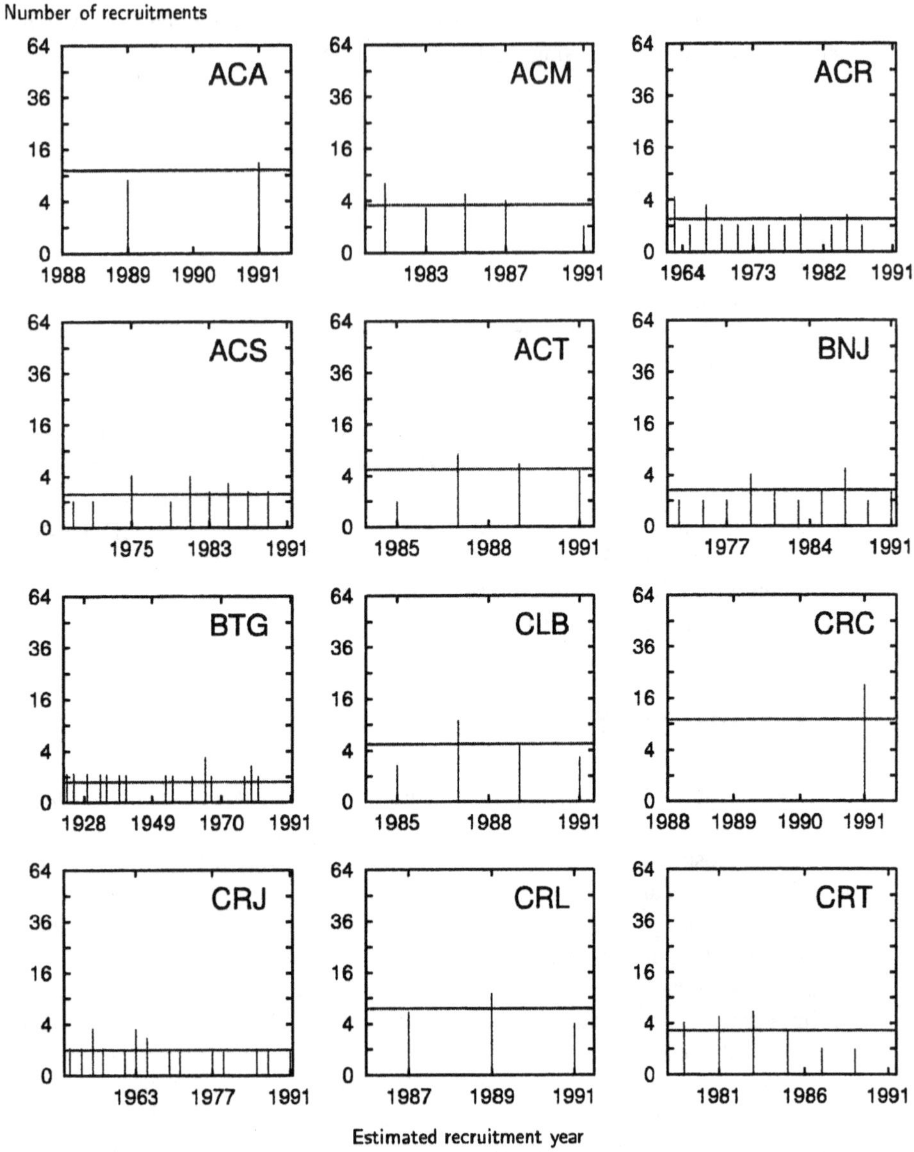

Fig. 14.4. Estimated recruitment year and number of recruitments. The *horizontal axis* indicates the year estimated by "inverse growth" for existing trees, and the *vertical axis* shows the estimated number of recruitments for each year. The *thin horizontal* lines indicate the average recruitment rate for each species. Refer to Table 14.1 for definitions of the abbreviations

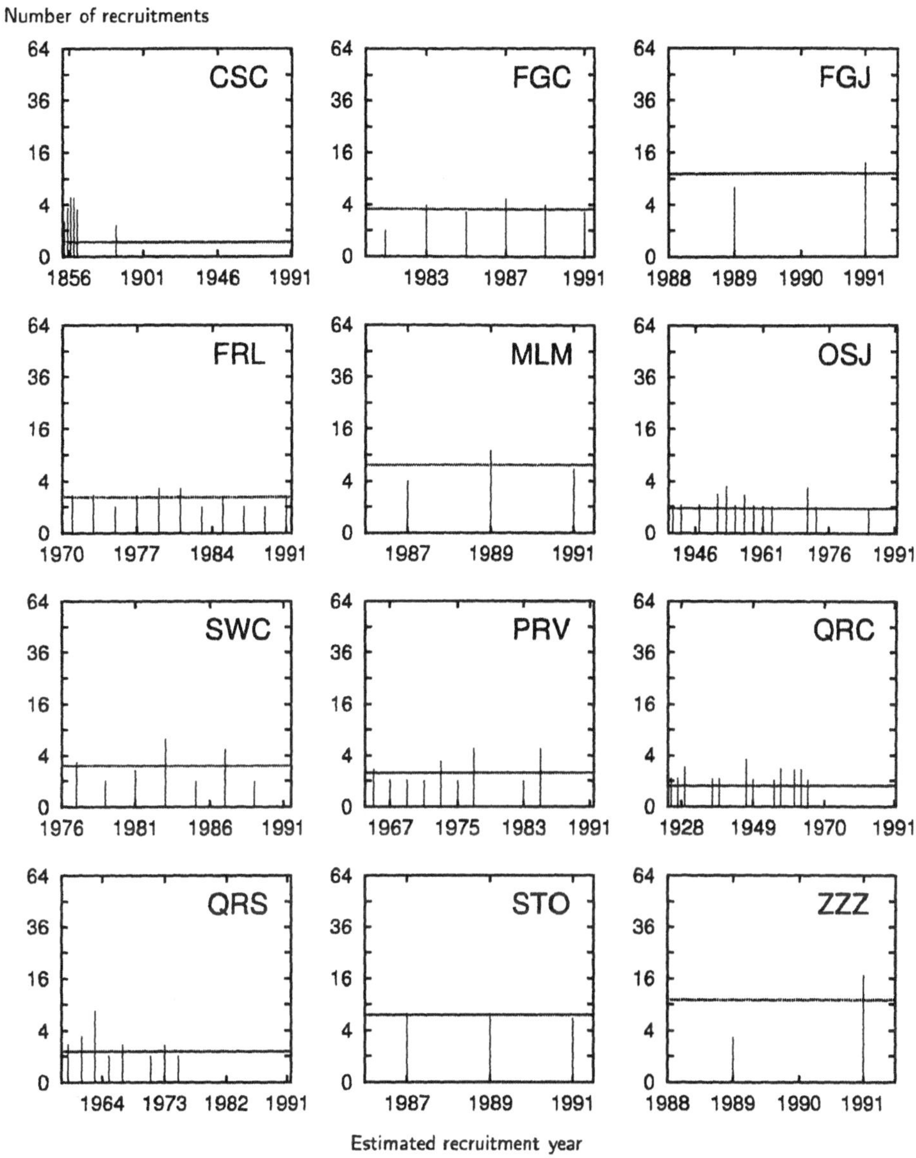

Fig. 14.4. *Continued*

Table 14.3. Predictions by a 400-year computer simulation using the OFR individual-based model

Species[a]	Δ Density(/ 6 ha)			Δ BA(m²/ 6 ha)		
ACA	381	↗	649	4.9	↗	26.0
ACM	146	↘	112	4.5	↗	4.8
ACR	69	↘	37	1.6	↘	1.1
ACS	57	↗	66	0.6	↗	0.7
ACT	60	↗	152	0.6	↗	3.3
BNJ	58	↗	88	0.7	↗	1.2
BTG	38	↘	17	3.2	↘	1.1
CLB	128	↗	204	0.7	↗	2.0
CRC	442	↗	809	3.6	↗	21.3
CRJ	44	↘	28	1.2	↘	0.6
CRL	399	↘	369	6.7	↗	11.0
CRT	75	↘	58	3.0	↗	3.5
CSC	50	↘	5	6.9	↘	0.0
FGC	88	↗	131	14.5	↘	10.8
FGJ	575	↘	308	35.5	↘	13.0
FRL	62	↗	68	0.5	↗	1.2
MLM	75	↗	205	0.5	↗	3.0
OSJ	54	↘	18	4.2	↘	0.5
PRV	74	↘	36	3.8	↘	2.5
QRC	46	↘	13	6.2	↘	0.7
QRS	231	↘	22	39.2	↘	2.0
STO	430	↘	407	3.6	↗	8.4
SWC	124	↘	70	5.9	↘	5.1
ZZZ[b]	390	→	390	9.2	→	9.2

Δ, predicted change from 1987 to 2387; ↗, increase; ↘, decrease; BA, total basal area.

[a]Refer to Table 14.1 for abbreviation definitions.

[b]Assumed to not change during the simulation.

erant, canopy trees such as QRS and CSC) will be succeeded by TC- (shade-tolerant, canopy, such as ACA and CRC) and TU-species (shade-tolerant, understory, such as STO).

The difference in results is caused by differences in the modeling method. The former density model defined by a transition matrix disregards species-specific characteristics (e.g., the maximum size of DBH) for simplification. On the other hand, the behavior of the individual-based model, which has been implemented from 24 tree species, might be much more sensitive to subsequent censuses, because the model includes many parameters estimated from limited measurements. Considering these advantages and disadvantages, the best way of modeling forest dynamics is still an open problem.

To improve this model, we should plug more submodels into the simulator, because many data collected in the OFR plot have not yet been used. For example, we could develop a more realistic recruitment submodel by using seed dispersal data (see Chapter 10), or create a dynamic model for the seedling and sapling stages (i.e., individuals under DBH 5.0 cm) from the corresponding data set (see Chapter 12). A more rational simulator for forest dynamics could be used by ecologists as a virtual experimental field to verify relationships between ecologically interesting details (for example, differences in demographic parameters, responses to episodic disturbance, or gene flow by pollination) and whole-forest dynamics taking into account spatial structure. Some important mechanisms determining the forest community (e.g., the coexistence of tree species having quite different life histories, see Chapter 13) could be tested and examined directly and statistically by doing several hundred replicated trials generated from a less-simplified forest simulator.

References

Botkin DB, Janak JF, Walis JR (1972) Rationale limitations and assumptions of a northeastern forest growth simulator. IBM J Res Dev 16:101–116

Kohyama T, Takada T (1998) Recruitment rates in forest plots: G_f estimates using growth rates and size distributions. J Ecol 86:633–639

Kubo T, Ida H (1998) Sustainability of an isolated beech-dwarf bamboo stand: analysis of forest dynamics with individual based model. Ecol Model 111:223–235

Masaki T, Suzuki W, Niiyama K, Iida S, Tanaka H, Nakashizuka T (1992) Community structure of a species-rich temperate forest, Ogawa Forest Reserve, central Japan. Vegetatio 98:97–111

Nakashizuka T, Iida S, Tanaka H, Shibata M, Abe S, Masakit, Niiyama K (1992) Community dynamics of Ogawa Forest Reserve, a species-rich deciduous forest, central Japan. Vegetatio 103:105–112

Pacala SW, Canham CD, Silander JAJ (1993) Forest models defined by field measurements: the design of a northeastern forest simulator. Can J For Res 23:1980–1988

Pacala SW, Canham CD, Silander JAJ, Kobe RK, Ribbens E (1996) Forest models defined by field measurements: estimation error analysis and dynamics. Ecol Monogr 66:1–43

Shugart HH (1984) A theory of forest dynamics. Springer, New York

Takada T, Nakashizuka T (1996) Density-dependent demography in Japanese temperate broad-leaved forest. Vegetatio 124:211–221

Part 5

Eco-physiological Studies on Tree Growth

15 Differential Analyses of the Effects of the Light Environment on Development of Deciduous Trees: Basic Studies for Tree Growth Modeling

Ichiro Terashima, Kyoko Kimura, Kosei Sone, Ko Noguchi, Atsushi Ishida, Akira Uemura, and Yoosuke Matsumoto

15.1 Introduction

One of the tasks of the ecophysiology team of the Ogawa Forest Reserve (OFR) project was to provide physiological bases for modeling the growth of individual trees, for incorporation into the individual-based model (IBM) of forest dynamics (see Chapter 14).

Some tree functions can be estimated with considerable accuracy. For example, if the photosynthetic properties of the leaves with respect to the photosynthetic photon flux density (PPFD) gradient are known, photosynthesis of a tree can be estimated by theoretical methods such as that of Monsi and Saeki (1953). Moreover, if one can estimate the abundance of the nonphotosynthetic organs of a tree, one can further estimate net production of a tree. The pipe model theory (Shinozaki et al. 1964a, b), which describes quantitative relationships between the biomass of leaves and that of the nonphotosynthetic organs, has been used for this purpose. These approaches enable us to estimate net primary production of individual trees or forest stands. However, to model the forest dynamics, we need to know many other characteristics of the forest in addition to its net production.

In particular, the behavior of the individual tree should be precisely modeled for the IBM. To describe the long-term behavior of a tree, we must understand tree growth in response to various environmental factors. In some pioneering models of tree growth, it is usually assumed that branches are autonomous or independent in terms of the carbon economy (Sprugel et al. 1991). For example, Takenaka (1994) used a model to calculate light interception by branches and determined whether or not they produced sibling branches based on their light interception. Applying a simple branching rule, he succeeded in mimicking tree growth. An essential point of these successful models is that tree growth is expressed in an iterative or differential manner.

The iterative or differential approach is useful for analyses of tree growth. For example, it is possible to separate the effect on growth and the physiological characteristics of the tree of a particular environmental factor, such as light in the current year, from those of previous years. This allows changes in the light environment during tree

Ecological Studies, Vol. 158
Nakashizuka, Matsumoto (eds.) Diversity and Interaction in a Temperate Forest Community: Ogawa Forest Reserve of Japan

growth to be properly treated. It may also be possible to express the yearly growth of nonphotosynthetic organs by a differential form of the pipe model theory. For these purposes, deciduous tree species, which dominate the OFR, provide useful systems. In this chapter, we describe our studies seeking the fundamental rules of development of deciduous trees. First, we analyze mechanisms that determine the properties of current-year shoots in *Fagus japonica*. Second, growth of nonphotosynthetic organs in *Acer mono* var. *marmoratum* f. *dissectum* and *A. rufinerve* are analyzed. All these species are important components of the OFR (see Chapter 6).

15.2 Effects of Current-Year and Previous-Year Light Environments on Properties of the Leaf and Gross Morphology of the Current-Year Shoot

15.2.1 Predetermination of Shoot Properties

A tree can be recognized as an assemblage of modular branches. Harper (1977) first introduced this idea when he applied techniques of population biology to the analysis of clonal plants and trees. Because branches are countable, "Eltonian" population biology Elton (1927) of branches is feasible. However, this approach cannot be applied easily to functional aspects, because, even within a single tree, the environment and properties of the branches are heterogeneous. For example, there is a gradient of the photosynthetically active photon-flux density (PPFD) in the tree canopy. Shoot morphology also differs according to the height of a tree. For species that develop long and short shoots, long shoots are abundant in the upper, exposed parts of the tree, while short shoots dominate in shaded parts (Hallé et al. 1978). Leaves are also different; for example, there is a difference between sun and shade leaves (Björkman 1981). In deciduous tree species having flush-type phenology, all the leaves for the next growth season are prepared in the winter buds (Kozlowski and Clausen 1966). In *Fagus sylvatica*, even the number of mesophyll cell layers, which differs between sun and shade leaves, is determined in the previous autumn in the bud (Eschrich et al. 1989). Koike et al. (1997) found that the PPFD in the previous year has a distinct effect on leaf anatomy and photosynthesis in broad-leaved deciduous species with flush-type phenology such as *Zelkova serrata* and *Quercus serrata*.

Many deciduous trees, including species of *Fagus* and *Acer*, have distinct long and short shoots. Long shoots are characterized by long internodes and large lateral buds, whereas short shoots have short internodes and few lateral buds (Hallé et al. 1978). Maruyama (1983) found that there is a strong relationship between leaf number (or bud size) and shoot type. For the flush-type species, therefore, the shoot type may be determined in the previous year, given that the leaf number is determined in the previous year.

These observations imply that, in tree species with flush-type shoot phenology, leaf characteristics and shoot gross morphology are greatly influenced by the light environment in the previous year. However, the current-year light environment may also affect shoot and leaf properties. Otherwise, species with flush-type phenology could not respond to abrupt changes in the light environment brought about by, for example, gap formation, which is one of the main topics of investigation of the OFR project (see Chapter 12).

We have examined the effects of current-year and previous-year light environments on gross morphology of the shoots and leaf characteristics of *Fagus japonica*, a common species in cool-temperate areas, including the OFR, on the Pacific side of Honshu, the main island of Japan. *F. japonica* trees show typical flush-type shoot phenology and develop long and short shoots. To separate the effects of the current-year PPFD from those of previous years, we artificially shaded a part of the tree canopy and examined the effects of shading on shoot characteristics. We also conducted growth experiments with the seedlings in controlled PPFD regimes.

15.2.2 Effects on Leaf Properties

A 15-m tower enclosing a *Fagus japonica* tree of about 14 m height was constructed in a deciduous forest at Nakoso, Fukushima, Japan (36° 58' N, 140° 36' E; 700 m above sea level, 5 km north of OFR). This tree had several leaf tiers. Except for the uppermost tier, where shoots grew obliquely, the shoots were horizontal and, therefore, the leaves were displayed evenly in the tier. We chose five tiers of 13.9, 11.5, 9.5, 8, and 2 m height and called them tiers 1, 2, 3, 4, and 5, respectively.

To examine the effects of changes in the yearly light environment, we shaded about 200 shoots in the uppermost tier 1 with a shading frame (2 × 2 × 2 m) from 27 April 1995. The PPFD inside the frame relative to that in a fully exposed location above the forest canopy (R-PPFD) was 5%.

To separate the effects of the current-year PPFD from those of PPFDs of previous years more clearly, we also conducted transfer experiments using seedlings. Uniform seedlings collected on the floor of the OFR in April 1995 were grown in the shade frames (4% or 1% R-PPFD). In early May of 1996, some of these seedlings were transferred to the frame of 20% R-PPFD. The R-PPFD on the forest floor measured on 11 July 1995 was about 4%.

Changes in daily PPFD (mol quanta m^{-2} day^{-1}) of each tier measured just above the tier relative to that of the first tier from bud break to full leaf greening are shown in Fig. 15.1. The uppermost tier (tier 1) was fully exposed. R-PPFDs of the lower tiers were considerably attenuated by many branches and by the trunk of the tree itself and surrounding deciduous trees. This effect was considerable even immediately after bud break, which occurred around 8 May 1996. These measurements revealed that the PPFD incident on young leaves in the lower tiers was fairly low from the beginning.

Properties of the leaves in the five tiers are summarized in Table 15.1. There was a typical sun/shade gradient in leaf properties. Leaves from the upper tiers were thicker and had much nitrogen on a leaf-area basis (N/area) compared with those from the lower tiers. The chlorophyll to nitrogen (Chl/N) ratio increased with the decrease in the height.

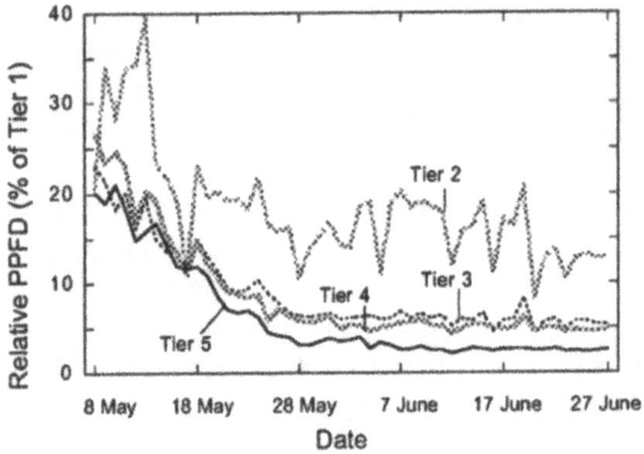

Fig. 15.1. Changes in daily photosynthetically active photon-flux density (PPFD) (mol m^{-2} day^{-1}) just above respective tiers relative to that of tier 1. The PPFDs were monitored with quantum sensors. Daily PPFD was calculated for each calendar day. The daily PPFD of tier 1 was assumed to be 100%

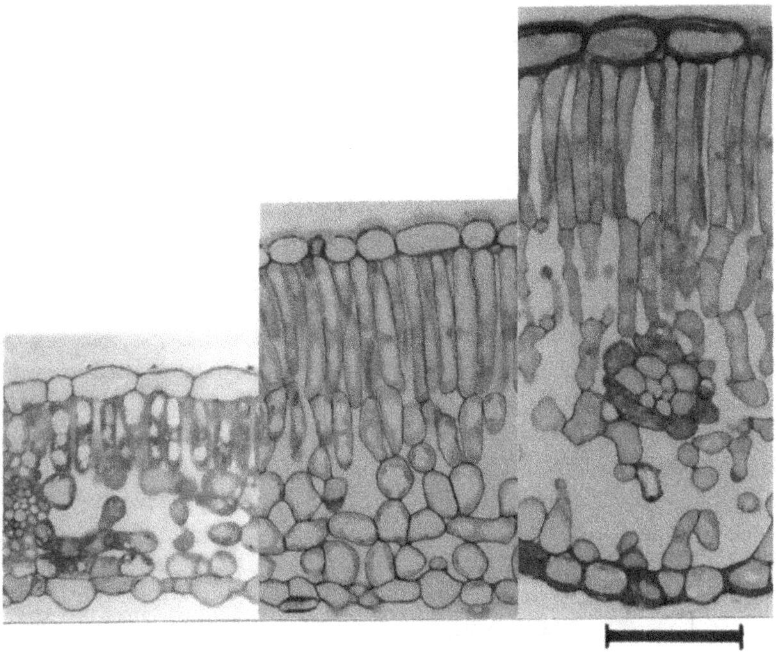

Fig. 15.2. Microphotographs of transverse sections of shade (*left*) and sun (*middle*) leaves of *F. japonica* and a sun leaf of *F. crenata* (*right*). *Bar*, 50 μm

The maximum photosynthetic rate on a leaf-area basis is strongly correlated with N/area (Evans 1989). When N nutrition is sufficient, the maximum photosynthetic rate is also strongly correlated with leaf thickness (Björkman 1981; Koike 1988). As expected from the data shown in Table 15.1 and Fig. 15.2, in situ measurements on the same *F. japonica* tree showed that the maximum rates of net photosynthesis were around 7 and 2 μmol CO_2 m^{-2} s^{-1} for the most exposed and the most shaded leaves, respectively (Uemura et al. 2000).

Table 15.1. Properties of the leaves in five leaf tiers of a *F. japonica* tree

Variables	Tier 1 100%[a]	Tier 2 15%[a]	Tier 3 6%[a]	Tier 4 5%[a]	Tier 5 3%[a]	Shaded 5%[a]
Leaf mass / area (g m^{-2})	96.2 ± 23.3	62.9 ± 9.32	46.6 ± 19.1	38.4 ± 11.2	34.9 ± 8.4	52.0 ± 7.0
Leaf thickness (μm)	224.3 ± 34	186 ± 10	142 ± 29	136 ± 30	131 ± 37	–
Chl / area (mmol m^{-2})	0.70 ± 0.10	0.71 ± 0.10	0.66 ± 0.05	0.64 ± 0.05	0.59 ± 0.06	0.73 ± 0.06
N / area (mmol m^{-2})	167 ± 10.4	122 ± 11.6	99.0 ± 1.0	85.4 ± 3.5	79.2 ± 7.4	110 ± 6.0
Chl / N (mmol mol^{-1})	3.8 ± 0.53	5.4 ± 0.83	6.1 ± 0.36	6.9 ± 0.44	6.8 ± 0.56	6.2 ± 0.34

All data are presented as mean ± SD ($n = 10$).

[a]Percentages denote mean PPFD relative to Tier 1 between 31 May and 27 June 1996.

Table 15.2. Effects of changes in R-PPFD during the 1995 and 1996 growing seasons on leaf properties in 1996

	R-PPFD in 1995/R-PPFD in 1996			*P*-value of the difference	
Variables	4% / 20%	1% / 20%	1% / 1%	4% / 20% vs. 1% / 20%	1% / 20% vs. 1% / 1%
Leaf mass / area (g m^{-2})	33.1 ± 1.8	34.4 ± 2.1	20.7 ± 1.6	NS	< 0.001
N / area (mmol N m^{-2})	64 ± 8	74 ± 9	71 ± 9	NS	NS
Chl / N (mmol mol^{-1})	2.0 ± 0.19	2.9 ± 0.65	6.2 ± 0.34	< 0.05	< 0.001

All data are presented as mean ± SD (n = 4).

Student's *t*-test was used to detect significant differences; NS, not significant.

Effects of the changes in PPFD between the two successive growth seasons on leaf properties of the seedlings are shown in Table 15.2. These results clearly showed that the leaf properties were dependent on the current-year PPFD environment, but not on the previous-year PPFD at all. In this experiment also, the leaf dry mass per area (LMA) depended on the growth R-PPFD. The Chl/N ratio decreased with increasing R-PPFD, indicating that nitrogen was preferentially allocated to light-harvesting components in shaded leaves. However, N/area of the seedlings was not significantly different among R-PPFDs. This could reflect the low nitrogen nutrient level employed in the present transfer experiments. It is known that the dependence of N/area on PPFD is weak when nutrient availability is low (Terashima and Evans 1988; for a theoretical consideration, see Hikosaka and Terashima 1995; Terashima and Hikosaka 1995).

All of these data in Tables 15.1 and 15.2 indicate that the leaves of *F. japonica* trees responded well to the current-year light environment (and nitrogen availability) both anatomically and physiologically.

We had postulated that the properties of leaves would be strongly affected by the previous-year PPFD for two reasons. First, the number of cell layers in the palisade tissue of *F. sylvatica* is determined when the embryonic leaves develop in the winter bud (Eschrich et al. 1989). Second, it is widely thought that the PPFD before bud break is fairly high in the deciduous forest (Parker 1995), implying that the future shade leaves would receive high PPFD for a while after bud break. However, the palisade tissue of *F. japonica* leaves typically had only one cell layer irrespective of the light environment (Fig. 15.2). Moreover, differences in the R-PPFD between the upper and lower tiers were marked even immediately after the bud break (Fig. 15.1). The difference in R-PPFD between the tiers may be large enough to account for the sun/shade differentiation of leaf properties during expansion of the leaves.

Clearly, determination processes of leaf properties in *F. japonica* are different from those described for *F. sylvatica* (Eschrich et al. 1989) and for *Fagus crenata* (Uemura et al. 2000). When expanding young leaves in the sunny parts are shaded, *F. crenata* develops costful sun-type leaves while *F. japonica* develops shade leaves (Uemura et al. 2000). This could explain why "retrogression" survival of *F. japonica* is better than that of *F. crenata* when the trees are shaded (see Chapter 13). In *F. crenata*, the cost of sun-type leaves in the shade would not be paid back by photosynthesis. Thus, "retrogression" survival of *F. crenata* trees would be difficult.

The present study proved that the method used for separating the effects of the current-year light environment from those of the previous-year light environments on leaf properties is very efficient. The method should be applicable to many other species.

15.2.3 Effects on the Gross Morphology of the Shoot

Properties of the gross morphology of the current-year shoots in the five tiers of the tree are summarized in Table 15.3. The number of leaves (LNo), cumulative leaf area (CLA), and length (SL) of the current-year shoots were all greater in the upper tiers than in the lower tiers.

The dependence of SL on LNo is shown in Fig. 15.3A. Shoots with four or fewer leaves were very short and seldom developed large lateral buds. Thus, these shoots can be referred to as short shoots according to the definition of Hallé et al. (1978). Since shoots with five or more leaves tended to be long and to have large lateral buds, these shoots were considered long shoots (Hallé et al. 1978). Although the mechanism underlying the change in internode length between a LNo of four and five is unknown, the drastic change is responsible for the clear dichotomy between short and long shoots. There was a strong linear relationship between CLA and LNo (Fig.15.3B; for details see Kimura et al. 1998).

A part of tier 1 was shaded during the growing season in 1995. Results of the shading experiment on gross shoot morphology are summarized in Table 15.4. For the shoots developed in 1995, LNo values, counted by the leaf scars, were very similar; the difference was significant only marginally. SL was significantly greater in the exposed tier 1 than in the shaded one. This indicates that shoot length was affected by the current-year light environment as well. Probably, photosynthate produced by the young leaves contributed to the shoot growth to some extent. Such effects would be more marked in long shoots than in short shoots.

We can also compare the data between 1995 (the first year of shading) and 1996 (the second year; see Table 15.4). LNo values of the exposed tier 1 were similar in both years. However the SL in 1996 was much greater than that in 1995. This is probably because the year 1995 was a mast year. Thus, there would have been com-

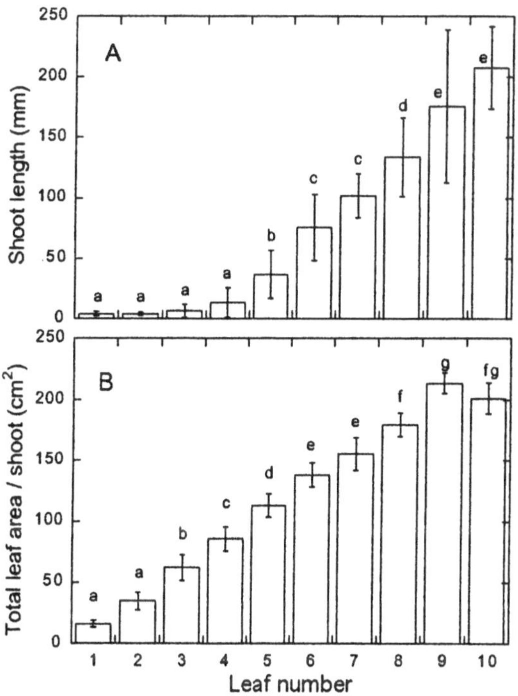

Fig. 15.3. Dependence of shoot length (**A**) and cumulative leaf area per shoot (**B**) on leaf number in *F. japonica*. Different letters indicate statistically significant differences (Sheffe's *F*-test, P < 0.01)

Table 15.3. Properties of the shoots in five leaf tiers

Variables	Tier 1 100%[a]	Tier 2 15%[a]	Tier 3 6%[a]	Tier 4 5%[a]	Tier 5 3%[a]	Shaded 5%[a]
Leaf number	7.62 ± 1.63	4.61 ± 1.02	4.02 ± 1.60	3.86 ± 0.97	3.40 ± 1.81	3.94 ± 0.89
Shoot length (mm)	128.8 ± 58.2	32.5 ± 8.5	29.1 ± 26.2	13.1 ± 15.5	28.1 ± 47.5[b]	13.4 ± 10.4
Cumulated leaf area (cm^2 $shoot^{-1}$)	165 ± 32.5	99.0 ± 27.7	81.6 ± 38.8	79.2 ± 2.5	64.1 ± 34.0	88.0 ± 12

All data are presented as mean ± SD ($n = 45 - 53$).

[a]Percentages denote mean PPFD relative to Tier 1 between 31 May and 27 June 1996.

[b]The lowest tier 5 contained some very long shoots directly emerged from the trunk.

Table 15.4. Effects of shading of Tier 1 on leaf number and shoot length (Kimura et al. 1998)

	Variables	Shaded ($n = 50$)	Exposed ($n = 45$)	P
1995	Leaf number	6.92 ± 1.21	7.40 ± 1.19	< 0.05
	Shoot length (mm)	31.8 ± 11.4	47.8 ± 20.1	< 0.001
1996	Leaf number	3.94 ± 0.89	7.62 ± 1.63	< 0.001
	Shoot length (mm)	13.4 ± 10.4	128.8 ± 58.2	< 0.001

petition for photosynthates between reproductive organs and shoots. This again indicates that shoot growth was partly dependent on current-year production.

The contribution of current-year photosynthates to shoot growth has not been reported previously for species showing flush-type shoot growth. However, it has been reported in species with continuous-type shoot growth. For example, Kozlowski and Clausen (1966) shaded early leaves of two *Betula* species and found that growth of the shoot and later leaves was suppressed.

Transfer experiments with seedlings indicated that the gross morphology of the current-year shoots was determined by the light environment of the previous year but not by that of the current year. In other words, the effects of the current-year PPFD on SL were not observed at all in the transfer experiment (data not shown; for details, see Kimura et al. 1998). This could be attributed to the fact that the LNo of the current-year shoots of the plants used in the transfer experiments was mostly less than five and the shoots were all short shoots.

As in many flush-type species, LNo is determined in winter buds in *F. japonica*. The embryonic leaves were several millimeters long in the winter bud. It is, therefore, likely that most cell divisions were complete by the beginning of winter and the number of cells are determined independently of the current-year photosynthetic production. If LNo and cell number are both influenced by the production of the previous year, and if cell number per leaf is not markedly different, the strong relationship between LNo and CLA (Fig. 15.3B) can be understood.

All these results with respect to gross morphology of the current-year shoot (Tables 15.3 and 15.4) indicate that the PPFD of the previous year has a great influence on the determination of gross shoot morphology. These results allow us to speculate that the number of leaves is determined by the abundance of reserve materials such as carbohydrates and nitrogenous compounds in the parental shoots. Such materials in the shoots also contribute to shoot growth in the next year, at least in the early phase. The reserve could be sufficient for the growth of short shoots. However, the growth of long shoots would partly depend on current-year photosynthates. Further experiments are needed to understand the phenomenon quantitatively, and this experimental system for differentiating the effects of the previous-year PPFD from those of the current-year PPFD is useful for such studies.

15.2.4 Shoot Autonomy

The artificial shading of a part of the uppermost tier decreased the PPFD of the shaded part to about 5% that of the initial, fully exposed condition. The shaded condition was similar to the PPFD level of tier 4, and the properties of shoots newly produced in the shade frame were similar to those in tier 4 (Tables 15.1 and 15.3). This indicates that, apparently, properties of shoots and leaves are mainly determined by the PPFD, irrespective of the positions of the shoots within a tree. This also indicates that one can study shoot dynamics by studying the light environment.

However, even under similar PPFD conditions, the behaviors of shoots of the mature tree and the seedling are different. For an extreme example, seedlings can survive at an R-PPFD of 2%, while there are no branches under such dark parts of the tree. Factors responsible for the differences in behavior between mature trees and seedlings should be addressed. One possibility is that the weight ratio of nonphotosynthetic organs to photosynthetic organs is markedly different between the mature tree and the seedling. Thus, the weight of nonphotosynthetic organs supported by a unit area of leaf is different. To understand tree growth based on the notion of "shoot autonomy," we need to know how growth of nonphotosynthetic organs is regulated.

15.3 Differential Analyses of the Pipe Model Relationship

The pipe model theory, which was originally proposed by Shinozaki et al. (1964a, b), has been extensively used for examining the architecture of the nonphotosynthetic organs of a tree. This theory claims that, for a tree or a forest canopy, the cumulated dry mass of the leaves above height z is proportional to the mass of nonphotosynthetic organs between heights z and z + Δz. Although the statement is very simple, many studies have supported this theory. However, its physiological bases are not clear. In particular, the light factor, which exerts crucial effects on photosynthesis and transpiration, is not directly considered. We have paid particular attention to the light factor in this study and analyzed the growth of individual branches.

Three trees of species *Acer mono* var. *marmoratum* f. *dissectum*, 1–2 m in height, were selected in the OFR. This is a canopy species with rather stable recruitment (see Chapter 3). Hemispherical photographs were taken just above the current-year shoots in October 1997. By analyzing the photographs with a computer (Hemiphoto software, programmed by H. ter Steege), we obtained the site factor for every current-year shoot. We also used six individuals of *Acer rufinerve* Sieb. et Zucc., 1–3 m in height, in the Ashu Experimental Forest of Kyoto University, Kyoto, Japan. This study site is located at 35° 20' N, 135° 45' E and is 700 m above sea level. This area is dominated by *Fagus crenata* Blume and *Quercus crispula* Blume. There also are many *Carpinus* and *Acer* species. For *A. rufinerve*, the R-PPFD for each of the current-year shoots was measured with quantum sensors in September 1998. R-PPFD determined with quantum sensors corresponded well to the site factor (data not shown). *A. rufinerve* is a subcanopy species that occurs in gaps in the OFR (see Chapter 3).

In these *Acer* species, the growth of the stem or trunk occurs well after the bud breaks (Komiyama et al. 1987). Thus, the growth of nonphotosynthetic organs can be mostly attributed to the current-year photosynthetic production.

After the measurement of the light environment, leaves were collected separately from each current-year shoot. Leaf area and dry mass were measured to obtain cumulative leaf mass (CLM) and cumulative leaf area (CLA) for each current-year shoot. By multiplying CLA by R-PPFD (or site factor), we obtained cumulative light interception (CLI). The branch was cut at a right angle to the axis, and the greatest diameters (D) of the annual ring of the current year and that of the previous year (D′) were measured. As an index of the cross-sectional area of the xylem, D^2 was used. For the index of the yearly increase in sectional area, $D^2 - D'^2$ was used.

As already mentioned, the original pipe model theory describes relationships between the CLM above height z of a tree or a forest canopy and the dry mass of the nonphotosynthetic organs that exist between heights z and z + Δz. In this study, we used D^2 instead of mass of nonphotosynthetic organs between heights z and z + Δz. We used the cumulative leaf area or weight for the shoots upstream of the point under consideration, in accordance with a suggestion by Chiba (1991). In other words, the physiological connection, rather than the height per se, is appreciated. This modification is important when physiological analyses are conducted.

We applied the theory to the analysis of the architecture of a single tree. There is a strong relationship between CLM and D^2 for various positions on the tree (Fig. 15.4). This clearly indicates that the pipe model theory is relevant for analyzing the architecture of nonphotosynthetic organs including young branches.

Shinozaki et al. (1964a, b) claimed that, for trees of various sizes, there is a unique and strong relationship between total leaf mass of the tree and the square of the trunk diameter just below the lowest living branch (Da^2). The present analyses also revealed a strong relationship between the CLM of the tree and Da^2. The relationship between CLA and Da^2 was also very strong, while the relationship between CLI and Da^2 was weakest (Fig. 15.5A–C).

On the other hand, the strongest correlation was found between CLI and $Da^2 - Da'^2$ (Fig. 15.5D–F). The correlations between CLM and $Da^2 - Da'^2$ and CLA and $Da^2 - Da'^2$ were much weaker. These results indicate that the current-year wood growth is influenced by the current-year light interception. Although further analyses are needed, we have thus succeeded in differentiating the growth of nonphotosynthetic organs.

We are planning to extend this study using various techniques such as a transfer experiment and artificial changes in the light environment. We can also use stable isotopes such as $^{13}CO_2$ as a tracer. Experimental approaches of this kind are very important for the quantitative understanding of growth responses of a tree to the environment. Through such studies, we can more precisely predict the growth pattern of nonphotosynthetic organs in relation to photosynthetic organs.

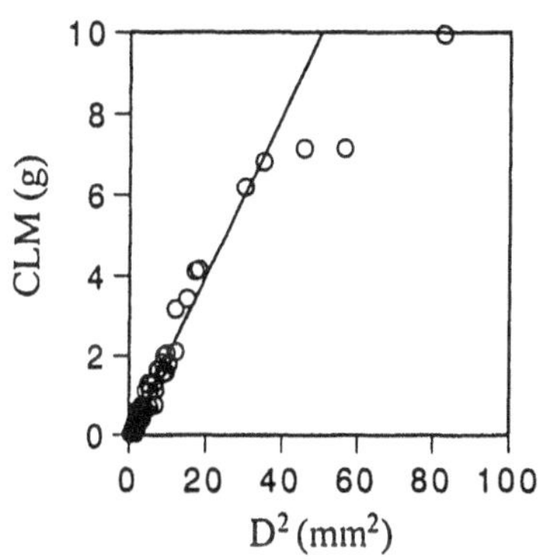

Fig. 15.4. Relationships between upwardly cumulative leaf mass at various positions and cross-sectional area at the respective positions in an *Acer mono* var. *marmoratum* f. *dissectum* tree (1.4 m in height)

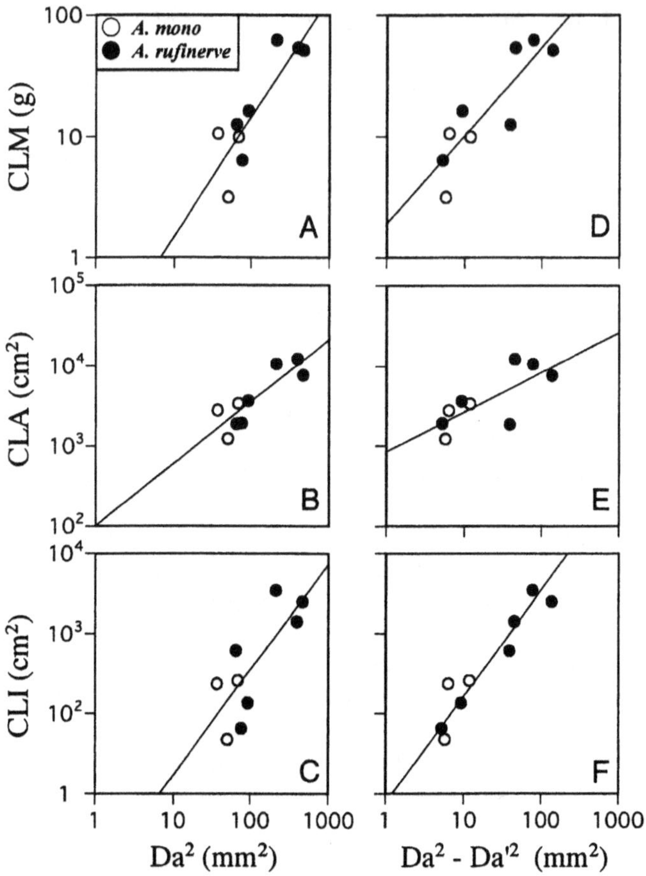

Fig. 15.5. Relationships between leaf parameters and basal area just below the lowest living branch (**A**, **B**, and **C**) and leaf parameters and current-year growth in the basal area just below the lowest living branch (**D**, **E**, and **F**). Leaf parameters are cumulative leaf mass (*CLM*), cumulative leaf area (*CLA*), and cumulative light interception (*CLI*). Pearson's correlation coefficients for these relationships are as follows: **A** $r = 0.847$ ($P = 0.0023$); **B** $r = 0.811$ ($P = 0.0057$); **C** $r = 0.705$ ($P = 0.0317$); **D** $r = 0.794$ ($P = 0.0081$); **E** $r = 0.615$ ($P = 0.079$); and **F** $r = 0.850$ ($P = 0.0021$). *Open circles*, *Acer mono* var. *marmoratum* f. *dissectum*; *solid circles*, *A. rufinerve*

15.4 Concluding Remarks

In this chapter, we have shown that deciduous trees provide good systems for studying mechanisms of tree growth, acclimation of shoot properties, and differentiation of sun and shade leaves. Studies based on these models would be very useful for the mechanistic prediction of forest dynamics. Although the results of the present studies have not been incorporated into the current IBM described in Chapter 14, incorporation of data analyzed by the present methods for the many other species in that model will strengthen the next-generation model.

Acknowledgments

The studies described in this chapter were supported by a grant (B-52.3.2) from the Environmental Agency, Japan. We thank Drs. T. Nakashizuka, H. Tanaka, and M. Shibata for discussion, and Drs. A. Takenaka and S. Funayama-Noguchi for useful comments.

References

Björkman O (1981) Responses to different quantum flux densities. In: Lange OL, Nobel PS, Osmond CB, Ziegler H (eds) Physiological plant ecology. I. Responses to the physiological environment. Springer, New York, pp 57–107

Chiba Y (1991) Plant form based on the pipe model theory II. Quantitative analysis on ramification in morphology. Ecol Res 6:21–28

Elton CS (1927) Animal Ecology. Sidgwick & Jackson, London

Eschrich WR, Burchardt R, Essiamah S (1989) The induction of sun and shade leaves of the European beech (*Fagus sylvatica* L.); Anatomical studies. Trees 3:1–10

Evans JR (1989) Photosynthesis and nitrogen relationships in leaves of C_3 plants. Oecologia 78:9–19

Hallé F, Oldeman RAA, Tomlinson PB (1978) Tropical trees and forests. Springer, Berlin

Harper JL (1977) Population biology of plants. Academic, London

Hikosaka K, Terashima I (1995) A model of the acclimation of photosynthesis in the leaves of C_3 plants to sun and shade with respect to nitrogen use. Plant Cell Environ 18:605–618

Kimura K, Ishida A, Uemura A, Matsumoto Y, Terashima I (1998) Effects of current-year and previous-year PPFDs on shoot gross morphology and leaf properties in *Fagus japonica*. Tree Physiol 18:459–466

Koike T (1988) Leaf structure and photosynthetic performance as related to the forest succession of deciduous broad-leaved trees. Plant Species Biol 3:77–87

Koike T, Miyashita N, Toda H (1997) Effects of shading on leaf structural characteristics in successional deciduous broad-leaved tree seedlings and their silvicultural meaning. For Resources Environ 35:9–25

Komiyama A, Inoue S, Ishikawa T (1987) Characteristics of the seasonal diameter growth of twenty-five species of deciduous broad-leaved trees. J Jpn For Sci 69:379–385

Kozlowski TT, Clausen JJ (1966) Shoot growth characteristics of heterophyllous woody plants. Can J Bot 44:827–843

Maruyama K (1983) Shoot characteristics as a function of the bud length on Japanese beech trees. J Jpn For Sci 65:43–51

Monsi M, Saeki T (1953) ‹ber den Lichtfaktor in den Pflanzengesellschaften und seine Bedeutung f‚r die Stoffproduktion. Jpn J Bot 14:22–52

Parker GG (1995) Structure and microclimate of forest canopies. In: Lowmann MD, Nadkarm NM (eds) Forest canopy. Academic, San Diego, pp 73–106

Shinozaki K, Yoda K, Hozumi K, Kira T (1964a) A quantitative analysis of plant form—the pipe model theory. I. Basic analysis. Jpn J Ecol 14:97–105

Shinozaki K, Yoda K, Hozumi K, Kira T (1964b) A quantitative analysis of plant form—the pipe model theory. II. Further evidence of the theory and its application in forest ecology. Jpn J Ecol 14:133–139

Sprugel DG, Hinckley TM, Schaap W (1991) The theory and practice of shoot autonomy. Annu Rev Ecol Syst 22:309–334

Takenaka A (1994) A simulation model of tree architecture development based on growth response to local light environment. J Plant Res 107:321–330

Terashima I, Evans JR (1988) Effects of light and nitrogen nutrition on the organization of the photosynthetic apparatus in spinach. Plant Cell Physiol 29:143–155

Terashima I, Hikosaka K (1995) Comparative ecophysiology of leaf and canopy photosynthesis. Plant Cell Environ 18:1111–1128

Uemura I, Ishida A, Nakano T, Terashima I, Tanabe H, Matsumoto Y (2000) Acclimation of leaf characteristics of *Fagus* species to previous-year and current-year solar irradiances. Tree Physiol 20:945–951

16 Gas Exchange Characteristics of Major Tree Species in Ogawa Forest Reserve

YOOSUKE MATSUMOTO and YUTAKA MARUYAMA

16.1 Introduction

Plants grow and develop through photosynthetic production. Photosynthesis is affected by external and internal factors such as sunlight, temperature, and other climatic conditions, nutrition status, and the morphological and anatomical features of the leaves. Since Boysen–Jensen's original study (1932, etc.), researchers have sought to identify both, qualitatively and quantitatively, the relationship between various factors and the photosynthetic rate of numerous plant species, and there have been many reports on the results (Tokari 1977; Tazaki 1978; Hatano and Sasaki 1987; Furukawa 1991; Larcher 1995; Koike 1988, 1996; etc.). Nevertheless, the target species in such reports have mostly been herbaceous plants, with the focus on agricultural crops, and there have been relatively few studies on woody plants. According to the literature survey conducted by Linder (1979, 1981), for example, since 1891, only 130 species worldwide have been the subject of studies on photosynthesis (and the only trees studied were coniferous). This is in spite of the fact that in Japan alone there are 500 species of woody plants.

Although the photosynthesis of woody plants in Japan has been studied widely since the work of Kuroiwa (1960) and Negisi (1966), the number of species which have been considered is still limited. This is especially true for the deciduous broad-leaved tree species which are distributed in the Ogawa Forest Reserve (OFR) (Koike 1988; Han et al. 1999; Saito and Kakubari 1999; Uemura et al. 2000, etc.). The procedures use by these authors were applied to the 56 tree species (see Chapter 6) in OFR, whose photosynthetic properties were mostly unknown.

In this chapter, we present the gas-exchange characteristics of individual leaves based on a study conducted to identify the photosynthetic rate, transpiration rate, and water-use efficiency of representative broad-leaved trees in Japan (Matsumoto et al. 1999), including the tree species in OFR. In addition, we show the relationship between the gas-exchange characteristics and the morphology of leaves, sunlight, and temperature, which vary depending on the leaf's position in the crown, for four species of the *Carpinus* genus. These ecophysiological characteristics will contribute to an understanding of the structure and composition of OFR, and will be appli-

Ecological Studies, Vol. 158
Nakashizuka, Matsumoto (eds.) Diversity and Interaction in a Temperate Forest Community: Ogawa Forest Reserve of Japan

Table 16.1. Name and distribution properties of tree species

Species name[a]	Family name[a]	Distribution in OFR[b]	Type[c]	Forest zone[d]	Moisture regime[e]	Shade tolerance[f]
Acer mono var *marmoratum* f. *dissectum*	Aceraceae	+	DBL	Cool	S.Wet–M	Low–M
Betula grossa	Betulaceae	+	DBL	Cool	M	Low
Betula maximowicziana	Betulaceae		DBL	Cool	S.Wet–M	Low
Betula platyphylla var. *japonica*	Betulaceae		DBL	Cool	M–S.Dry	Low
Cercidiphyllum japonicum	Cercidiphyllaceae		DBL	Cool	Wet	Low
Fagus japonica	Fagaceae	+	DBL	Cool	Wide	M–High
Prunus sargentii	Rosaceae		DBL	Cool	M–S.Dry	Low
Quercus crispula	Fagaceae	+	DBL	Cool	M–S.Dry	Low
Stewartia pseudo-camellia	Theaceae		DBL	Cool	S.Dry	M
Acer cissifolium	Aceraceae	+	DBL	Cool–Warm	S.Wet	Low
Benthamidia japonica	Cornaceae	+	DBL	Cool–Warm	Wide	M
Carpinus japonica	Betulaceae	+	DBL	Cool–Warm	Wide	M
Carpinus laxiflora	Betulaceae	+	DBL	Cool–Warm	M–S.Dry	Low–M
Carpinus tschonoskii	Betulaceae	+	DBL	Cool–Warm	M	Low–M
Clethra barvinervis	Clethraceae	+	DBL	Cool–Warm	S.Dry–Dry	M–High
Elaeagnus umbellata	Elaeagnaceae		DBL	Cool–Warm	S.Dry–Dry	Low
Hamamelis japonica	Hamamelidaceae	+	DBL	Cool–Warm	S.Dry	M
Ilex macropoda	Aquifoliaceae	+	DBL	Cool–Warm	S.Dry–Dry	M
Lindera obtusiloba	Lauraceae		DBL	Cool–Warm	M	Low
Magnolia praecocissima	Magnoliaceae		DBL	Cool–Warm	S.Wet–M	Low–M
Quercus acutissima	Fagaceae		DBL	Cool–Warm	M–S.Dry	Low
Sorbus alnifolia	Rosaceae	+	DBL	Cool–Warm	S.Dry	M
Swida controversa	Cornaceae	+	DBL	Cool–Warm	M	Low
Zelkova serrata	Ulmaceae	+	DBL	Cool–Warm	S.Wet–M	M

Castanopsis sieboldii	Fagaceae	EgBL	Warm	M	Low
Cinnamomum japonicum	Lauraceae	EgBL	Warm	M	High
Ilex chinensis	Aquifoliaceae	EgBL	Warm	M	Low
Ilex pedunculosa	Aquifoliaceae	EgBL	Warm	S.Dry–Dry	M
Quercus gilva	Fagaceae	EgBL	Warm	M	M
Quercus glauca	Fagaceae	EgBL	Warm	M–S.Dry	M–High
Quercus myrsinaefolia	Fagaceae	EgBL	Warm	M	Low
Aphananthe aspera	Ulmaceae	DBL	Warm–SubT	M–S.Dry	Low
Cinnamomum camphora	Lauraceae	EgBL	Warm–SubT	M	Low
Ehretia dicksonii	Boraginaceae	DBL	Warm–SubT	M	Low
Ilex rotunda	Aquifoliaceae	EgBL	Warm–SubT	S.Wet–M	M
Lithocarpus edulis	Fagaceae	EgBL	Warm–SubT	M–S.Dry	Low
Litsea coreana	Lauraceae	EgBL	Warm–SubT	M	High
Michelia compressa	Magnoliaceae	EgBL	Warm–SubT	M	M–High
Ternstroemia gymnanthera	Theaceae	EgBL	Warm–SubT	M–S.Dry	M
Tilia miqueliana	Tiliaceae	DBL	Warm–SubT	S.Wet–M	Low–M
Ulmus parvifolia	Ulmaceae	DBL	Warm–SubT	S.Wet	M

[a]The scientific names are in accordance with Satake et al. (1989a, 1989b).

[b]+, species distributed in OFR.

[c]DBL, deciduous broad-leaved trees; EgBL, evergreen broad-leaved trees.

[d,e,f]The descriptions of the distribution areas, moisture site locations, and shade tolerance are based on the descriptions in Yanagisawa (1981), Forest Development Technological Institute (1985), and Tanimoto (1990).

[d]Cool, species distributed in the cool-temperate zone; Warm, species distributed in the warm-temperate zone; SubT, species distributed in the subtropical zone.

[e]Wide, distributed over wide areas; Wet, distributed in wet sites; S.Wet, distributed in semi-wet sites; M, distributed in adequately moist sites; S.Dry, distributed in semi-dry sites; Dry, distributed in dry sites.

[f]High, high shade tolerance; M, medium shade tolerance; Low, low shade tolerance.

cable to the conservation and management of other forest ecosystems with a similar floristic composition.

16.2 Maximum Gas Exchange Rates

To investigate the gas exchange rates, we used the tree species in the arboretum on the campus of the Forestry and Forest Products Research Institute (located at 140°08′E, 36°00′N, and 20–25 m above sea level) (Table 16.1), i.e., a total of 41 tree species, including 9 cool–temperate tree species, 15 cool–temperate to warm–temperate tree species, 7 warm–temperate tree species, and 10 warm–temperate to subtropical tree species. Of these, 15 species are found in OFR.

Trees growing well with a height of 10 m or more were selected for the measurements. Fully expanded, intact, and exposed mature leaves, about 6 m above ground level on the south side of each individual tree, were carefully chosen. Ten leaves or more were measured for each tree species. The photosynthetic rate of the leaves was measured using a portable photosynthesis measurement system (SPB-H2 and SPB-H3, Shimazu (Kyoto, Japan)–ADC (Hoddesdon, UK)), and the transpiration rate and stomatal conductance were measured with a steady-state porometer (Li-1600, Li-Cor, Lincoln, USA).

The measurements were carried out on June 26–28, 1991, July 24–26, 1991, June 4–5, 1992, and July 21–27, 1992. In order to obtain the light-saturated photosynthetic rate, only leaves which were receiving direct sunlight were used for the measurements. To minimize the effects of stomatal closure due to high vapor pressure deficit (VPD) at midday, the measurements were conducted from about 8:00 a.m. to 11:00 a.m., when the humidity of the ambient air was relatively high.

16.2.1 Maximum Net Photosynthetic Rate

In this section, the maximum net photosynthetic rate (Pn_{max}), defined as the mean of the top three Pn values, is compared among species.

The Pn_{max} of all the species studied ranged from 9 to 24 μmol CO_2 m^{-2} s^{-1} (Fig. 16.1), which is within the same range as that of leaves of deciduous trees in the sun (10–15(25) μmol CO_2 m^{-2} s^{-1}) reported by Larcher (1995), and of deciduous broad-leaved trees (4–27 μmol CO_2 m^{-2} s^{-1}) reported by Ceulemans and Saugier (1991), but is relatively high compared with that of deciduous broad-leaved trees in Hokkaido (4–13 μmol CO_2 m^{-2} s^{-1}) reported by Koike (1988).

The Pn_{max} of evergreen broad-leaved trees was 9–20 μmol CO_2 m^{-2} s^{-1}, which is higher than that of trees from the warm–temperate to subtropical zones (6–12 (20) μmol CO_2 m^{-2} s^{-1}) reported by Larcher (1995), but falls within the range of that of evergreen broad-leaved trees (3–23 μmol CO_2 m^{-2} s^{-1}) reported by Ceulemans and Saugier (1991). Among the species studied here, the Pn_{max} of deciduous broad-leaved trees was relatively high compared with that of evergreen broad-leaved trees, which is consistent with the results summarized by Larcher (1995).

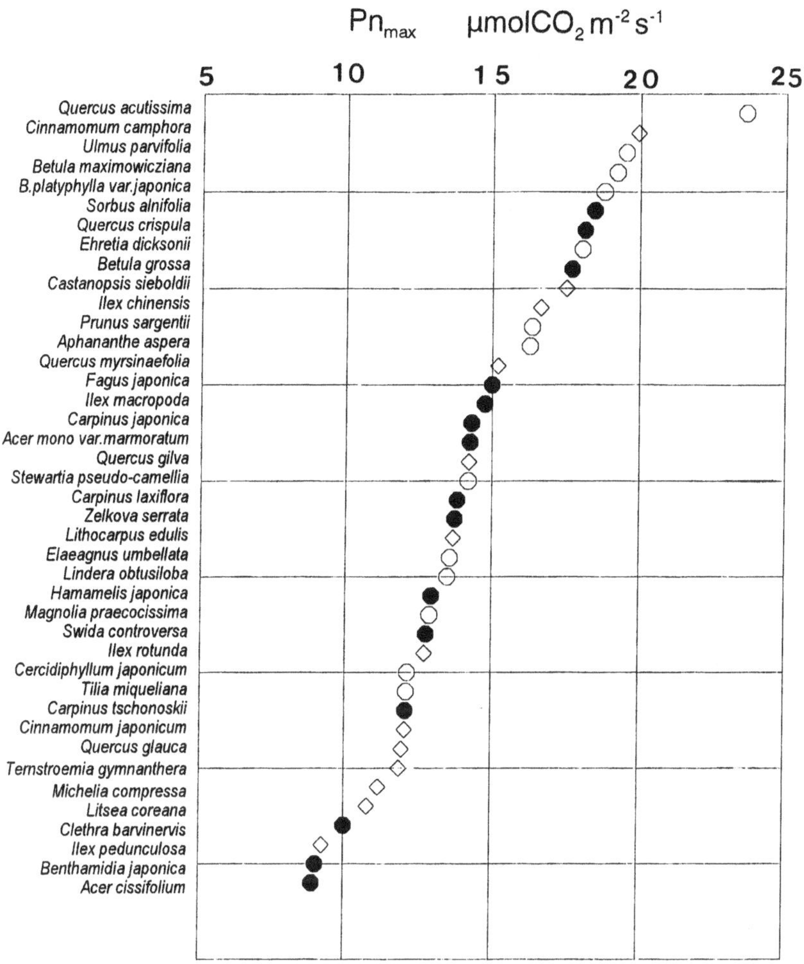

Fig. 16.1. Maximum net photosynthetic rates (Pn_{max}). *Solid circles*, deciduous broad-leaved trees growing in OFR; *open circles*, deciduous broad-leaved trees not growing in OFR; *open squares*, evergreen broad-leaved trees

Quercus acutissima and Betulaceae, which are typical light-demanding pioneer species in temperate forests in Japan, showed a relatively high Pn_{max} compared with the other species. Among the dominant tree species in OFR, large-size canopy trees (maximum height and diameter are more than 20 m and 100 cm, respectively) such as *Quercus crispula*, *Betula grossa*, *Fagus japonica*, and *Acer mono* var. *marmoratum* f. *dissectum* had a relatively high Pn_{max}. *Sorbus alnifolia*, *Ilex macropoda,* and *Carpinus japonica,* which are medium-size canopy trees in OFR (their height reaches 20 m, but their diameter rarely exceeds 50 cm), also had a high Pn_{max}.

On the other hand, subcanopy tree species in OFR (maximum tree height approximately 10–15 m, and diameter rarely more than 30 cm) with a comparatively high shade tolerance, such as *Acer cissifolium*, *Benthamidia japonica*, and *Clethra barvinervis,* had a relatively low Pn_{max}. However, we did not find a clear difference in Pn_{max} between the shade-tolerant and shade-intolerant canopy species that are categorized in Chapter 6.

16.2.2 Maximum Vapor Diffusion Conductance

The transpiration rate (Tr) is proportional to the atmospheric vapor pressure deficit (VPD) and the vapor diffusion conductance (Gw):

$$Tr = VPD \cdot Gw \qquad (1)$$

Although VPD may influence Gw, Tr is strongly affected by Gw. Thus, we consider

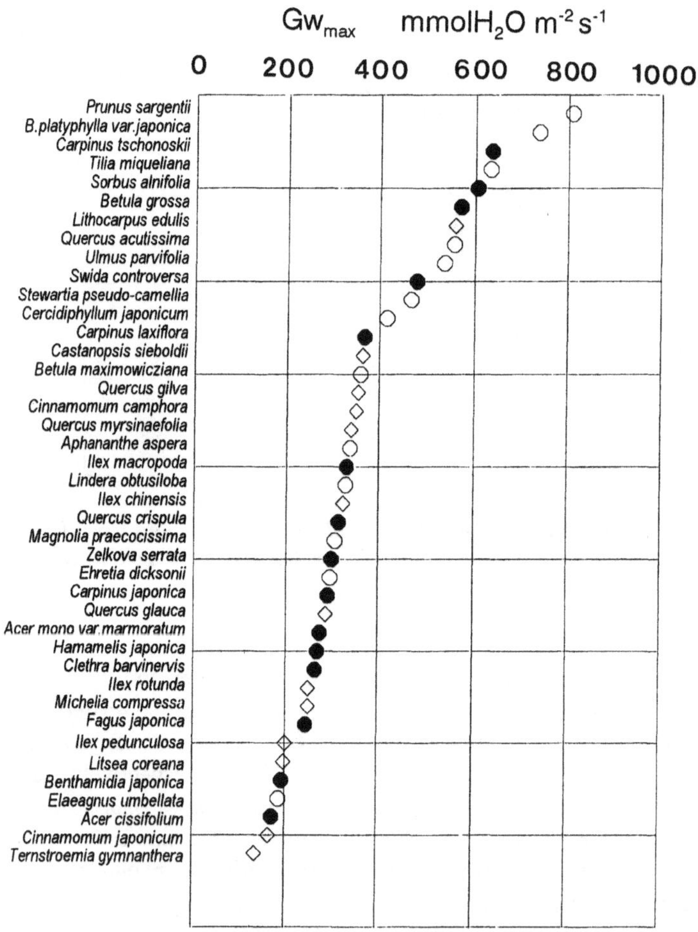

Fig. 16.2. Maximum vapor diffusion conductance (Gw_{max}). Symbols as in Fig. 16.1

Gw to be a rough index of the moisture demand of the species (Matsumoto et al. 1992).

In this section, Gw_{max} (the maximum vapor diffusion conductance) is defined as the mean of the top three Gw values for each species.

We found that Gw_{max} ranged from 170 to 810 mmol H_2O m^{-2} s^{-1} and from 130 to 560 mmol H_2O m^{-2} s^{-1} for deciduous broad-leaved trees and evergreen broad-leaved trees, respectively (Fig. 16.2). The values were relatively high in the shade-intolerant canopy species such as *Prunus sargentii* and *Betula platyphylla* var. *japonica*, and were relatively low in the shade-tolerant and/or the late successional species such as *Ternstroemia gymnanthera*, *Cinnamomum japonicum*, *Acer cissifolium*, *Elaeagnus umbellata*, *Benthamidia japonica*, and *Litsea coreana*.

Among the tree species found in OFR, Gw_{max} was relatively high in *Carpinus tschonoskii*, *Sorbus alnifolia*, *Betula grossa*, and *Swida controversa*, which are categorized in Chapter 6 as shade-intolerant canopy species, and relatively low in *Acer cissifolium* and *Benthamidia japonica*.

16.2.3 Water-Use Efficiency

The water-use efficiency (WUE) of productivity is defined as the ratio of the dry matter production and water consumption of plants during a given period. The ratio of Pn and Tr is also expressed as the water-use efficiency of photosynthesis (Larcher 1995). Ehleringer et al. (1993) defined the relationship between Pn and Gw as the intrinsic water-use efficiency of photosynthesis (IWUE, where IWUE = Pn/Gw).

In this section, the intrinsic water-use efficiency (IWUE), defined as the ratio of Pn_{max} and Gw_{max} (Pn_{max}/Gw_{max}), is compared among species.

The IWUE of the 41 temperate broad-leaved tree species studied here ranged from 19 to 88 × 10^{-6} mol CO_2 per mol H_2O (Fig. 16.3). The IWUE of OFR tree species ranged from 20 to 60 × 10^{-6} mol CO_2 per mol H_2O. Among the OFR tree species, shade-tolerant canopy species such as *Fagus japonica*, *Quercus crispula*, *Acer mono* var. *marmoratum* f. *dissectum*, and *Acer cissifolium* (Chapter 6) showed a relatively high IWUE. *Quercus crispula* and *Fagus japonica* also had a high Pn_{max}, suggesting that they are likely to have superior gas-exchange characteristics.

Shade-intolerant canopy species such as *Carpinus tschonoskii*, *Swida controversa*, *Sorbus alnifolia*, and *Betula grossa* (Chapter 6) showed a relatively low IWUE, suggesting that they are water-consuming species.

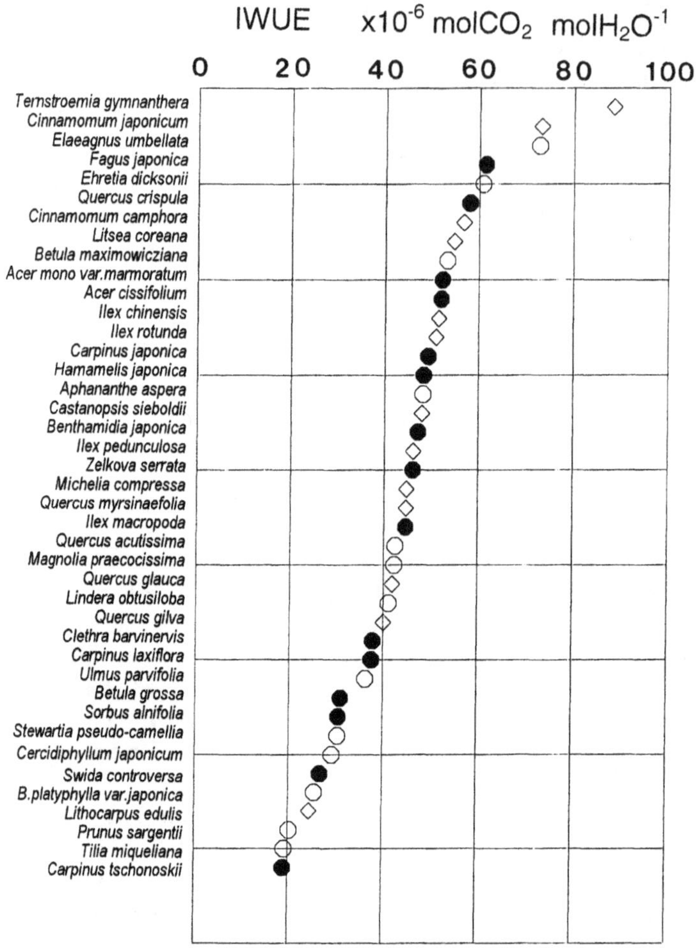

Fig. 16.3. Intrinsic water use efficiency (IWUE). The ratio between the maximum net photosynthetic rate (Pn_{max}) and the maximum vapor diffusion conductance (Gw_{max}) shown in Figs 16.1 and 16.2. Symbols as in Fig. 16.1

16.3 Photosynthetic Characteristics of *Carpinus*

Carpinus laxiflora and *C. cordata* are shade-tolerant canopy species, and *C. cordata* is generally distributed in valley bottoms. *C. japonica* and *C. tschonoskii* are shade-intolerant canopy species, and *C. tschonoskii* is commonly distributed in young stands. These four species have slightly different habitats, as mentioned above, although they are in the same genus (Chapter 6). This section presents the effects of leaf morphology (specific leaf area), light intensity, and temperature on photosynthesis in four species belonging to the *Carpinus* genus.

Leaves in the sun and leaves in the shade were collected from mature trees (with a tree height of 8–15 m) in a secondary forest near OFR. Their net photosynthetic rate and transpiration rate were measured using a portable photosynthesis and transpiration measurement system (SPB-H2, Shimazu (Kyoto, Japan)–ADC (Hoddesdon, UK)) under controlled environmental conditions. The cut end of the petiole of each sample leaf was put in a small vial filled with air-free water in order to maintain the supply of water to the leaf during the measurements. To determine the light–photosynthesis relationship, the air temperature was kept at 23°C ± 1°C, and the photosynthetically active photon flux densities (PPFD) were changed from 0 to 1000 μmol m^{-2} s^{-1}. To the determine temperature–photosynthesis relationship, the PPFD was kept at 500 ± 50 μmol m^{-2} s^{-1} and the air temperature was changed from 15 to 35°C. To avoid stomatal closure due to high VPD, the relative humidity of the air was kept above 70%.

16.3.1 Specific Leaf Area and Photosynthetic Rate

The specific leaf area (SLA, leaf area/leaf dry weight) increases as the shade increases during growth (Tanaka et al. 1994). The net photosynthetic rate (Pn) of leaves in the sun is generally higher than that of leaves in the shade for most tree species. In particular, the SLA is small and the Pn is high for leaves in the sun, while the SLA is large and the Pn is low for leaves in the shade. Figure 16.4 shows the relationship between SLA and Pn for the four *Carpinus* species. In all these species, Pn decreased as SLA increased. As the SLA increased, the Pn decreased linearly in the case of *C. japonica* and *C. cordata*, but leveled off in the case of *C. tschonoskii* and *C. laxiflora*. This suggests that *C. tschonoskii* and *C. laxiflora* have adapted to shad-

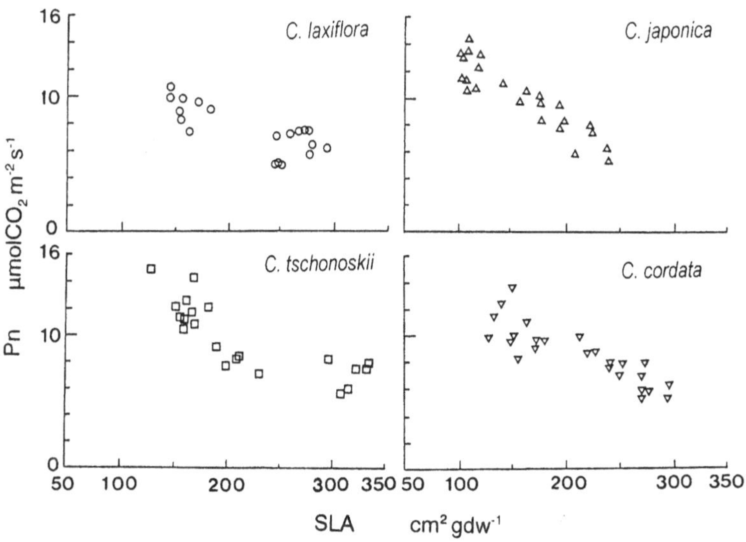

Fig. 16.4. Specific leaf area (SLA) and net photosynthetic rate (Pn) of the four *Carpinus* species

ing by developing thinner leaves under limited light conditions, but without too much loss of photosynthetic efficiency.

16.3.2 Light–Photosynthesis Relationship

Figure 16.5 shows the light–photosynthesis curves in relation to the maximum Pn of the temperature–photosynthesis curves (see Fig. 16.6). Except for *C. japonica*, the PPFD of light saturation of the three *Carpinus* species was about 500 μmol m^{-2} s^{-1} for leaves in the sun and about 350 μmol m^{-2} s^{-1} for leaves in the shade. On the other hand, the PPFD of light saturation of *C. japonica* was about 650 μmol m^{-2} s^{-1} for leaves in the sun and about 400 μmol m^{-2} s^{-1} for leaves in the shade, which is larger than that of the other *Carpinus* species. This result may partly explain the response of the *Carpinus* species to light conditions; *C. japonica* had the fewest seedlings under closed canopies (Chapter 13).

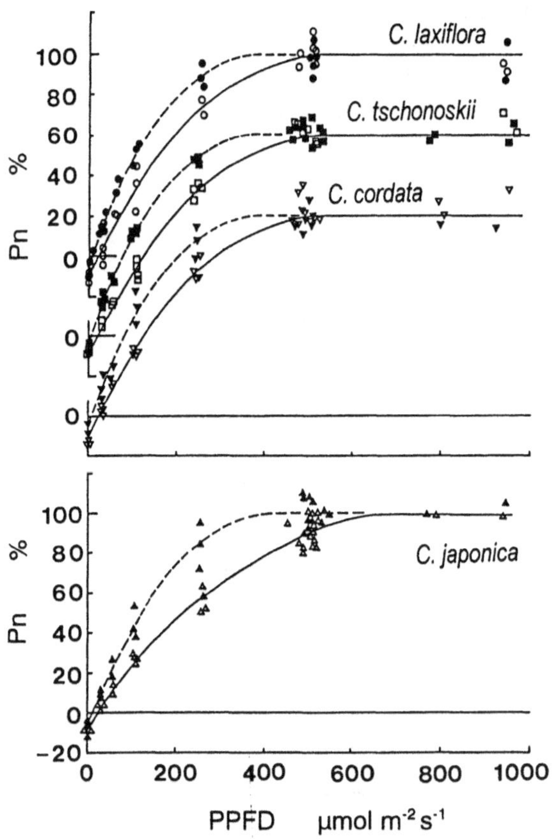

Fig. 16.5. Light–photosynthesis relationship of the four *Carpinus* species. *Solid lines with open symbols*, leaves in the sun; *broken lines with closed symbols*, leaves in the shade

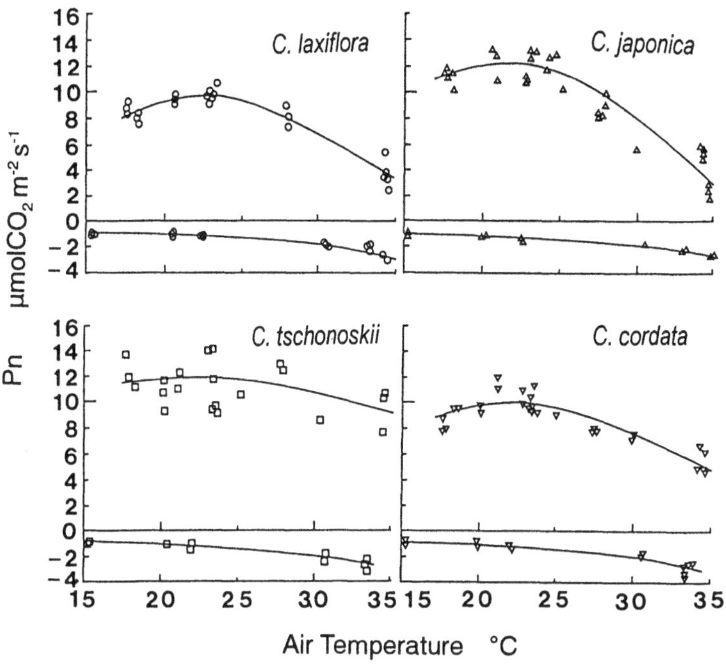

Fig. 16.6. Temperature–photosynthesis relationship and dark respiration curve of leaves in the sun from the four *Carpinus* species

16.3.3 Temperature–Photosynthesis Relationship

For the tree species in this study, the optimum temperature for photosynthesis is roughly equivalent to the average daytime temperature of the habitat. According to the temperature–photosynthesis curves shown in Fig. 16.6, the optimum temperature for photosynthesis was approximately 23°C for all species. This approximately coincides with the monthly average temperature in OFR in July, which is about 21°C (see Chapter 2), and the daytime temperature is usually several degrees higher than this.

The warmth index (WI) of the distribution area of *C. tschonoskii* and *C. laxiflora* ranges from 70 to 100 °C month, while that of *C. cordata* and *C. japonica* ranges from 55 to 100 °C month (Kira 1949; Shidei 1974). The northern limit of the horizontal distribution of *C. tschonoskii* and *C. japonica* is almost at OFR, while the southern limit is Kyushu Island. This shows that their normal distribution range is in warmer regions than those of *C. cordata* and *C. laxiflora*, which are distributed from Hokkaido to Kyushu (JFTA 1964, 1968), and that *C. tschonoskii* normally grows in a warmer environment than the other three species. Figure 16.6 show that the decline of Pn at high temperatures is smaller in *C. tschonoskii* than in the other three species.

These photosynthetic temperature responses are related to the distribution of *Carpinus* species in Japan and the species richness of *Carpinus* in OFR.

16.4 Conclusions

This chapter has compared the photosynthetic properties and gas-exchange characteristics of the major broad-leaved tree species, including those found in OFR. The tree species diversity in OFR can partly be explained by the differences in their gas-exchange responses to the environmental regimes. Further quantitative analysis, including a study of root systems and soil conditions, should be undertaken in the future.

References

Boysen-Jensen P (1932) Die Stoffproduktion der Pflanzen. Gustav Fischer, Jena

Ceulemans JR, Saugier B (1991) Photosynthesis. In: Raghavendra AS (ed) Physiology of trees. Wiley, New York, pp 21–50

Ehleringer JR, Hall AE, Farquhar GD (eds) (1993) Stable isotopes and plant carbon–water relations. Academic, San Diego

Forest Development Technological Institute (1985) Information on useful broad leaf trees: nursing and planting methods (in Japanese). Forest Development Technological Institute, Tokyo

Furukawa A (1991) Photosynthesis in leaves and environmental factors. Jpn J Ecol 41:279–297

Han Q, Yamaguchi E, Odaka N, Kakubari Y (1999) Photosynthetic induction responses to variable light under field conditions in three species grown in the gap and understory of a *Fagus crenata* forest. Tree Physiol 19:625–634

Hatano K, Sasaki S (eds) (1987) Growth and environment of trees (in Japanese). Yohken-do, Tokyo

JFTA (Japan Forest Technical Association) (ed) (1964) Illustrated important forest trees of Japan (in Japanese and English). Chikyu, Tokyo

JFTA (Japan Forest Technical Association) (ed) (1968) Illustrated important forest trees of Japan, vol2 (in Japanese and English). Chikyu, Tokyo

Kira T (1949) Forest zones in Japan (Ringyo-Kaisetu series No. 17) (in Japanese). Japan Forest Technical Association, Tokyo

Koike T (1988) Leaf structure and photosynthetic performance as related to the forest succession of deciduous broad-leaved trees. Plant Species Biol 3:77–87

Koike T (1996) Leaf morphology and anatomy affecting the net photosynthetic rate of 33 deciduous broadleaved tree species. For Res Environ 34:25–32

Kuroiwa S (1960) Ecological and physiological studies on the vegetation of Mt. Shimagare. IV. Some physiological functions concerning matter production in young *Abies* trees. Bot Mag Tokyo 73:133–141

Larcher W (1995) Physiological plant ecology, 3rd ed. Springer, Berlin

Linder S (1979) Photosynthesis and respiration in conifers. A classified reference list 1891–1977. Studia Forestalia Suecica 149, Uppsala

Linder S (1981) Photosynthesis and respiration in conifers. A classified reference list, Supplement 1. Studia Forestalia Suecica 161, Uppsala

Matsumoto Y, Maruyama Y, Morikawa Y (1992) Some aspects of water relations on large *Cryptomeria japonica* D. Don trees and climatic changes on the Kanto plains in Japan in relation to forest decline (in Japanese with English summary). Jpn J For Environ 34:2–13

Matsumoto Y, Tanaka T, Kosuge S, Tanbara T, Uemura A, Shigenaga H, Ishida A, Okuda S, Maruyama Y, Morikawa Y (1999) Maximum gas exchange rates in current sun leaves of 41 deciduous and evergreen broad-leaved tree species in Japan (in Japanese with English summary). Jpn J For Environ 41:113–121

Negisi K (1966) Photosynthesis, respiration and growth in 1-year-old seedlings of *Pinus densiflora*, *Cryptomeria japonica* and *Chamaecyparis obtusa*. Bull Tokyo Univ For 62:1–115

Saito H, Kakubari Y (1999) Spatial and seasonal variations in photosynthetic properties within a beech (*Fagus crenata* Blume) crown. J For Res 4:27–34

Satake Y, Hara H, Watari S, Tominari T (eds) (1989a) Wild flowers of Japan. Woody plants I (in Japanese). Heibonsha, Tokyo

Satake Y, Hara H, Watari S, Tominari T (eds) (1989b) Wild flowers of Japan. Woody plants II (in Japanese). Heibonsha, Tokyo

Shidei T (1974) Climate and the distribution of vegetation zones. In: Numata M (ed) The flora and vegetation of Japan. Kodansha, Tokyo; Elsevier, Amsterdam; American Elsevier, New York, pp 20–27

Tanaka T, Matsumoto Y, Shigenaga H, Uemura A (1994) Specific leaf area, photosynthetic capacity, and chlorophyll content of current-year leaves in under-storied *Chamaecyparis obtusa* Endl. in a multi-storied forest (in Japanese with English summary). Jpn J For Environ 36:22–30

Tanimoto T (1990) Ecology for broad-leaved forest management (in Japanese). Sobun, Tokyo

Tazaki T (ed) (1978) Kankyo-syokubutu-gaku (in Japanese). Asakura, Tokyo

Tokari Y (ed) (1977) Photosynthesis and dry matter production of crops (in Japanese). Yohken-do, Tokyo

Uemura A, Ishida A, Nakano T, Terashima I, Tanabe H, Matsumoto Y (2000) Acclimation of leaf characteristics of *Fagus* species to previous-year and current-year solar irradiances. Tree Physiol 20:945–951

Yanagisawa F (1981) Management of deciduous broad leaf forest. 1. Natural forest (in Japanese). In: Forestry Agency (Supervised) Broad leaf forest and its management. Chikyusha, Tokyo, pp 117–173

17 Habitat-Related Responses to Water Stress and Flooding in Deciduous Tree Species

Yong-Mok Park and Yasushi Morikawa

17.1 Introduction

Every plant has its own unique habitat. Even within a given plant community, local differences in the distribution of species are commonly observed. To inhabit a given place, plants must adapt to the prevailing environmental. Among environmental factors, water is the most important in relation to plant distribution because plants need water in extremely great quantities for transpiration. Thus, water is absolutely necessary for plant survival. In any given local area, however, the quantity of available water does not stay within the optimal range but varies with time, diurnally and seasonally. Therefore, sometimes an excess or deficiency of water is imposed on plants. Consequently, their ability to cope with this excess or deficiency of water is an important factor determining the local distribution of plants (Park 1989, 1990).

In the temperate deciduous forests of northern East Asia such as those in Japan and Korea it has been generally considered that the moisture environment is suitable for supporting forests because of the high annual precipitation and mild weather conditions during the plant growing season. However, the precipitation pattern, in which most precipitation occurs in the rainy season, from mid-June to mid-July, preceded and followed by days without rainfall, makes the environment stressful for plants. Thus, forest plants in this region are likely to be subjected to severe water stress before and after the rainy season. In fact, Tazaki (1960) found that the amount of evapotranspiration exceeding the amount of precipitation in late May, 1959, brought about severe water stress in plants and killed pine seedlings growing on sand dunes. Moreover, in early August, just after the end of rainy period, the prolonged rainless days have been shown to impose severe water stress on plants. During periods of 8–20 rainless days during August, seedlings of plants were killed on sand dunes and in a volcanic sand dune area (Maruta 1976; Park 1989; Yura 1988), resulting in differences in local distribution among closely related plant species (Park 1989).

In addition to the precipitation pattern, in temperate deciduous forests, the foliage phenology also affects the moisture regime of a particular forest. For instance, in a deciduous forest in central Korea, leaf development from winter buds starts in late March, and most leaf growth finishes in early June, irrespective of leaf phenology

Ecological Studies, Vol. 158
Nakashizuka, Matsumoto (eds.) Diversity and Interaction in a Temperate Forest Community: Ogawa Forest Reserve of Japan

patterns. In springtime, before the canopy is completely closed, an influx of direct incident light to the forest floor increases the air temperature and lowers the humidity in the forest, resulting in an increase in the vapor pressure deficit on the forest floor (D.W. Lee, personal communication). This condition makes the moisture regime in the forest stressful with regard to the heat budget. For example, clear-cutting or thinning by removal of a tree enhances the light intensity and increases the temperature in a forest stand and, in turn, causes the trees to have a more reduced midday xylem pressure potential (Matsumoto 1984; Morikawa et al. 1986).

Carpinus cordata and *Carpinus laxiflora*, which are shade-tolerant species with a seedling bank strategy of regeneration, are representative deciduous tree species, being distributed throughout northern East Asia, including Korea, China, and Japan (Kim 1992; Masaki et al. 1992; Shibata and Nakashizuka 1995; Shibata et al. 1998). However, they show different features in their local distribution. In particular, in Ogawa Forest Reserve (OFR) (described in detail in previous chapters), a temperate deciduous forest of Japan, the two species show contrastive distribution patterns depending on topographical characteristics. The distribution of *C. cordata* is restricted to locations near streams, while *C. laxiflora* is broadly distributed from ridge crest to mountain slopes (Masaki et al. 1992; Park 1996; Shibata and Nakashizuka 1995). A difference in moisture regime is expected from the different topographical characteristics of the streamside and ridge-crest sites where the two species are found. Thus, this segregation of the two species with regard to habitat urged us to examine their responses to the moisture regimes (excess and deficiency) expected to result from their habitat specificity. Our expectation is that a comparison of water relations in *C. cordata* and *C. laxiflora* may give us the key for solving the problem of their segregation by habitat in this temperate deciduous forest. We have experimentally examined responses to water stress and flooding with field-grown saplings of these two species showing habitat segregation. This information on the habitat-related responses of species to water stress and flooding will be useful for understanding habitat specificity of species in a temperate forest, and hence for the analysis of species composition within the community, and it may also give us a tool for efficient management of natural vegetation.

17.2 The Characteristics of Water Relations in Temperate Deciduous Tree Species

Fig. 17.1 shows diurnal changes in the microclimate on three days after the rainy period at two places where *C. cordata* and *C. laxiflora* are distributed in a temperate deciduous forest. The trends in the changes in the vapor pressure deficit reflect the microclimate and water availability (Figs. 17.1, 2). Despite the proximity of the two places, the difference in the moisture regime resulting from the difference in topography created a different evaporational demand. The moisture regime on the ridge crest, where *C. laxiflora* is distributed, is more stressful than that along the stream, where *C. cordata* grows (Fig. 17.1). The changes in soil water content in the two habitats are shown in Fig. 17.2. During the days after rainfall, soil water content

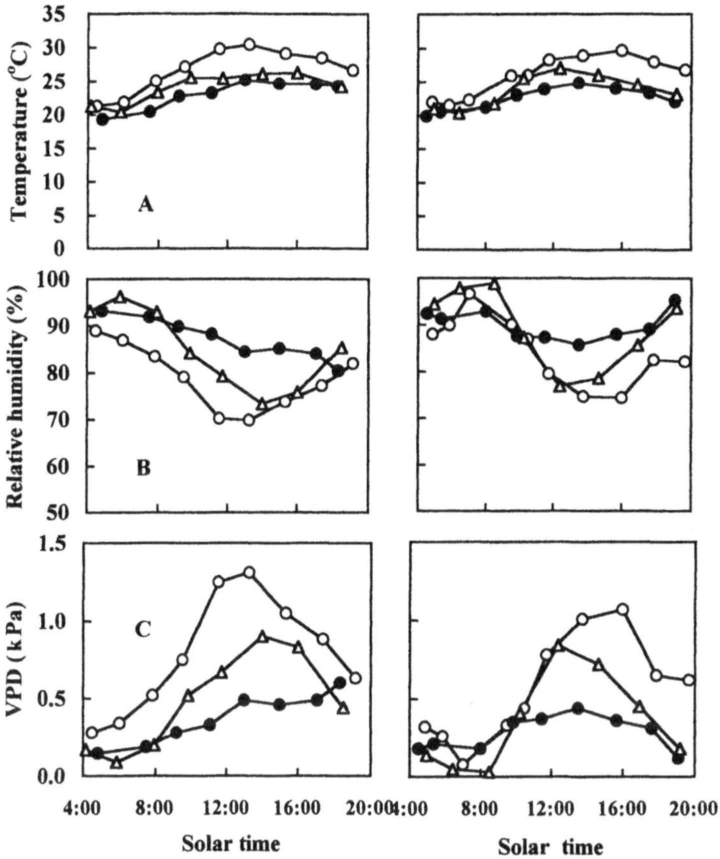

Fig. 17.1. Diurnal changes in air temperature (**A**), relative humidity (**B**), and vapor pressure deficit (VPD) (**C**) in the habitats of *C. laxiflora* (*left*) and *C. cordata* (*right*) on Mt. Keomo, South Korea. Jul. 30 (*open circles*), Aug. 19 (*solid circles*) and Aug. 30 (*open triangles*), 1991 (from Park 1996)

decreased at both sites, but the more rapid decrease occurred in the habitat where *C. laxiflora* is distributed. On 30 July, after 13 rainless days, soil water content 15 cm below the soil surface was below the field capacity (Fig. 17.2B). From these results it is evident that the moisture regime in the habitat of *C. laxiflora* is more stressful.

To evaluate the resistive characteristics of these species to water stress, a comparative study was carried out. In September 1989, 2- to 3-year-old field-grown saplings of *C. laxiflora* and *C. cordata* were taken from the OFR, transplanted into plastic pots, and grown in a greenhouse at 25°C under ambient light intensity and humid conditions for 6 months. These saplings were further acclimated for 2 weeks in a growth chamber at 25°C and 70% humidity at 500 μmole m^{-2} s^{-1} light (14 L:10 D). To determine the photosynthetic rate, stomatal conductance, and water status of the plants, a portable photosynthetic system (LCA-2, ADC, Hertfordshire, UK), steady state porometer (LI-1600, Li-Cor, Lincoln, NE, USA), and pressure chamber (PWSC,

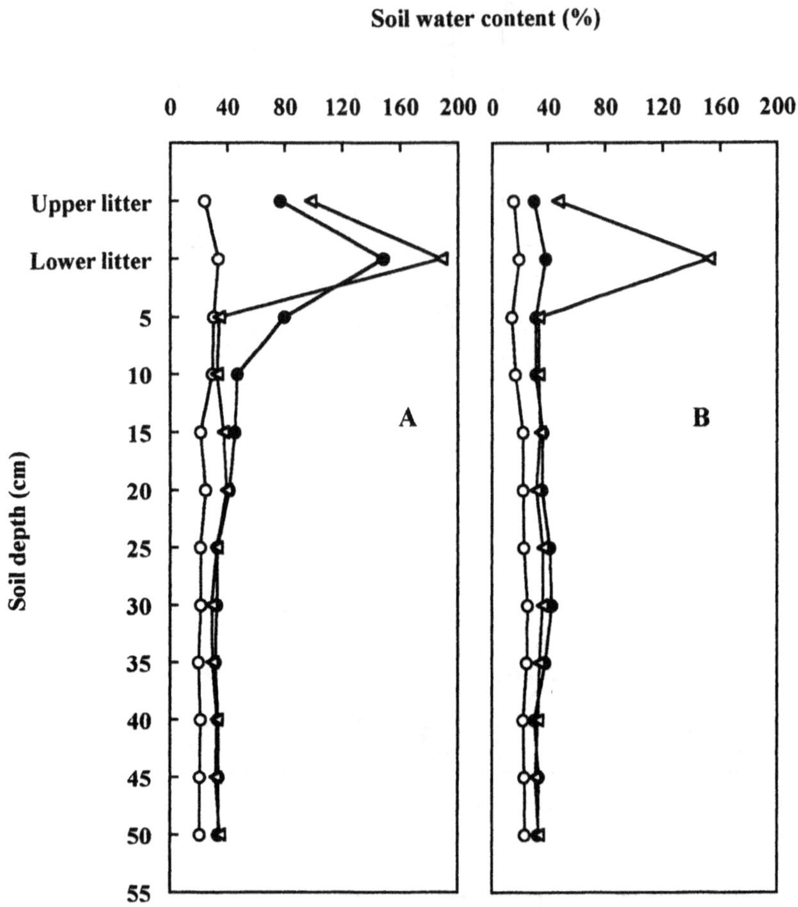

Fig. 17.2. Changes in soil water content at the *C. cordata* site (**A**) and *C. laxiflora* site (**B**) on Mt. Keomo. Jul. 30 (*open circles*), Aug. 19 (*solid circles*) and Aug. 30 (*open triangles*), 1991. Each point represents the mean of triplicate measurements (from Park 1996)

Soil Moisture Equipment Corp, Santa Barbara, CA, USA) were used, respectively. Then stomatal conductance, photosynthetic rate and the water relations of the two species were compared.

Changes in leaf water potential and osmotic potential of the two species with the progressive soil drying are shown in Fig. 17.3. Under well-watered conditions, leaf water potential changed similarly in both species and maintained higher levels compared with that under water-stressed conditions. Water stress caused by withholding water had no effect on leaf water potential, which remained comparable to that of well-watered plants, until 3 days after the water stress treatment, irrespective of the species. Thereafter, leaf water potential in both species decreased rapidly, but became lower in *C. laxiflora* than in *C. cordata*, so that a difference of about –0.5 MPa in water potential between the two species was observed at the end of the experiment.

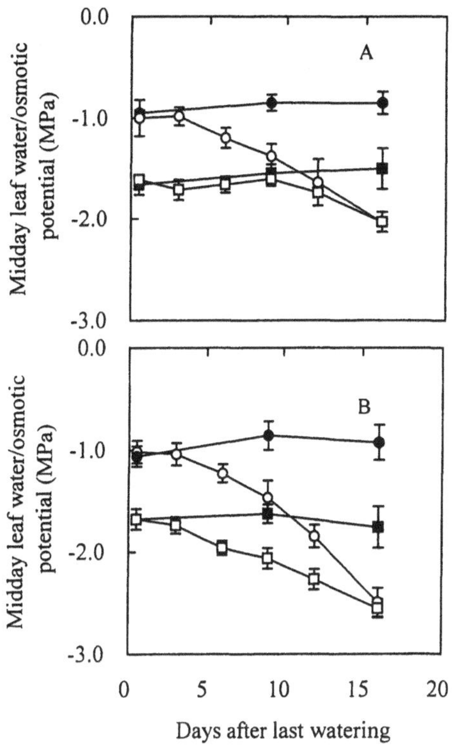

Fig. 17.3. Changes in midday leaf water potential (*circles*) and midday osmotic potential (*squares*) of *C. cordata* (**A**) and *C. laxiflora* (**B**). *Solid* and *open* symbols denote well-watered and water-stressed plants, respectively. *Bars* represent standard deviation of the mean of three to five replications

In spite of the remarkable decrease in leaf water potential under water-stressed conditions, *C. laxiflora* maintained higher turgor pressure than *C. cordata* until the 12th day after water was withheld. The maintenance of high turgor in *C. laxiflora* must be closely related to the osmotic potential, which decreased synchronically with the reduction in leaf water potential in this species during water stress. The osmotic adjustment in *C. laxiflora* contributed to the maintenance of turgor pressure comparable to control plants until the 9th day after withholding water (Fig. 17.3). On the other hand, *C. cordata* did not show a parallel decrease in osmotic potential with water potential reduction. As a result, the turgor potential in *C. cordata* was rapidly reduced to under 50% of the initial turgor pressure on the 6th day after withholding water (Fig. 17.3). The difference in the ability to osmotically adjust under water stress between the two species is well consistent with the data diurnally measured in the field (Park 1996). *C. laxiflora* maintained higher turgor pressure than *C. cordata* in the field, in spite of having a more negative leaf water potential during diurnal changes. Among the responses of plants to water stress, the most sensitive one is the cessation of growth resulting from the decrease in turgor pressure (Boyer 1970; Hsiao 1973). Thus, many plants have developed the ability to adjust osmotic pressure to maintain turgor pressure under water-stressed conditions (Osonubi and Davies 1981; Turner et al. 1987). The maintenance of turgor pressure by osmotic adjustment plays an important role in extending the period of water uptake and car-

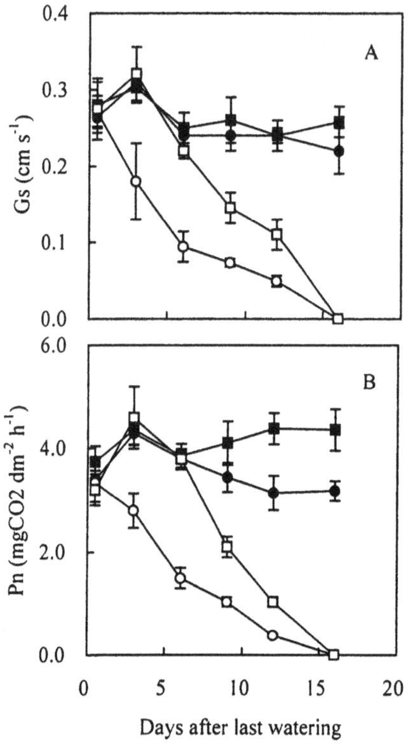

Fig. 17.4. Changes in stomatal conductance (*Gs*) (**A**) and net photosynthesis (*Pn*) (**B**) of *C. cordata* (*circles*) and *C. laxiflora* (*squares*) under well-watered (*solid symbols*) and water-stressed (*open symbols*) conditions. *Bars* represent standard deviation of the mean of three to five replications

bon gain in plants under declining conditions of water availability (Osmond et al. 1987; Lo Gullo and Salleo 1988). In addition to osmotic adjustment, the bulk elastic modulus, which represents the elasticity of the cell wall, is associated with the maintenance of turgor pressure as leaf water content changes (Lo Gullo and Salleo 1988). In this experiment, however, there were no significant differences in the elastic modulus between the two species, although its value was slightly higher in *C. laxiflora* (data not shown). Thus, the effect, if any, of the elastic modulus on changes in turgor pressure in these two species is small. These characteristics are closely related to the strategy of the plants in the field (Lo Gullo and Salleo 1988).

Stomatal conductance and the photosynthetic rate as water stress progresses with soil drying are shown in Fig. 17.4. Under well-watered conditions, both species showed higher stomatal conductance than under water-stressed conditions. On the other hand, under water-stressed conditions, there was a difference in stomatal conductance between the two species (Fig. 17.4). As water stress increased, stomatal conductance responded more sensitively in *C. cordata* than in *C. laxiflora*. The stomatal conductance in *C. cordata* was reduced to 40% of the initial level by the 3rd day after watering was stopped, and steeply decreased until the 6th day, to about 30% of the initial level (Fig. 17.4A). Variations in the photosynthetic rate in the two species mainly depended on the change in stomatal conductance, irrespective of the water conditions. Under water-stressed conditions, the photosynthetic rate in *C. laxiflora* was higher than that in *C. cordata* throughout the experiment, except at the very end. The photosynthetic rate of *C. laxiflora* was almost identical to its photosynthetic rate under well-watered conditions until the 6th day of withholding water, whereas *C. cordata* maintained a high photosynthetic rate only until the 3rd day after withholding water. Thereafter, a decrease in photosynthesis corresponding to the reduction in stomatal conductance was observed in both species (Fig. 17.4). In both species, the pattern of photosynthesis was nearly parallel to that of stomatal conductance for the entire period of the experiment. Thus, it seems that the difference in their ability to maintain turgor pressure must be responsible for most of the differences in stomatal conductance and photosynthetic rate between the two species under water-stressed conditions. The decrease in stomatal conductance of *C. cordata* on the 3rd day after withholding water may be regulated by another factor as well. It is generally accepted that stomatal closure is mainly responsible for decreases in the photosynthetic rate, and that stomatal conductance is regulated to maximize water use efficiency (Farquhar et al. 1980). Stomatal conductance is affected by ambient vapor pressure deficit, transpiration, leaf turgor potential, and soil water conditions (Gollan et al. 1985; Meinzer et al. 1997; Turner et al. 1984). Thus, the rapid reduction of stomatal conductance in *C. cordata* on day 3 after watering stopped must have resulted from efficient regulation of the stomata (Fig. 17.4A). A rapid stomatal closure under water stress is one mechanism to avoid water stress found in relatively sensitive plants (Picon et al. 1996). In addition, the concept that abscisic acid (ABA) originated in the root as a result of soil drying mediates stomatal closure is largely accepted for herbaceous plants (Zhang and Davies 1990) and even for tree species (Loewenstein and Pallardy 1998), although there are some problems yet to be resolved (Fuchs and Livingston 1996; Saliendra et al. 1995). The difference in stomatal conductance between these two species is well matched with the data measured in the field, where the stomatal conductance of *C. laxiflora* was highly maintained as leaf water potential severely decreased at midday (Park 1996). The difference in response to water stress between the two species must originate from their different strategies for coping with water stress.

Consequently, it is concluded that *C. cordata* is more sensitive to water stress than *C. laxiflora*, which pursues a tolerance strategy. In Japan, the seedlings of tree species emerged from mid-June to mid-July during the long rainy period first face water stress during the following rainless period with high solar radiation. Thus, the difference in resistance to water stress during this period must determine whether the emerged seedlings can establish themselves in a given location. In particular, water stress would be more severe in seedlings growing on ridge crests or mountain slopes, which are likely to dry out. As a result, a small difference in drought resistance between *C. cordata* and *C. laxiflora* seedlings may result in a great difference in the mortality of their seedlings and in turn result in the difference in seedling establishment. Considering that these species have a seedling bank strategy as their reproduction mechanism (Shibata and Nakashizuka 1995), it is highly reasonable that the superiority of its resistive characteristics to water stress makes *C. laxiflora* able to inhabit ridge crests or mountain slopes in the cool temperate forests of Japan and Korea.

17.3 The Response to Flooding in a Temperate Deciduous Forest

To compare the resistive characteristics to flooding of *C. laxiflora* and *C. cordata*, an artificial flooding experiment was carried out with field-grown saplings. The saplings used in this experiment were prepared as for the drought experiment. Two- to 3-year-old saplings of *C. cordata* and *C. laxiflora* were taken from the OFR and transplanted into specially manufactured polyvinylchloride pots (7 cm in diameter and 80 cm deep) in September 1989. They were grown in the greenhouse at a temperature of 25°C under ambient air conditions for 6 months. After that, these saplings were further acclimated in the growth chamber at 25°C air temperature and 70% humidity at 500 μmole m^{-2} s^{-1} light (14 L:10 D) for 2 weeks. The flooding treatment was conducted by submerging the polyvinylchloride pots into two big plastic tanks (100 cm in height and 70 cm in diameter) filled with tap water until the end of experiment. Control plants transplanted into the polyvinylchloride pots were grown under the same conditions without flooding treatment. Then the water relations, stomatal conductance and photosynthetic rate were measured.

The response of stomatal conductance to soil flooding is shown in Fig. 17.5. The flooded *C. cordata* plants maintained high stomatal conductance comparable to that

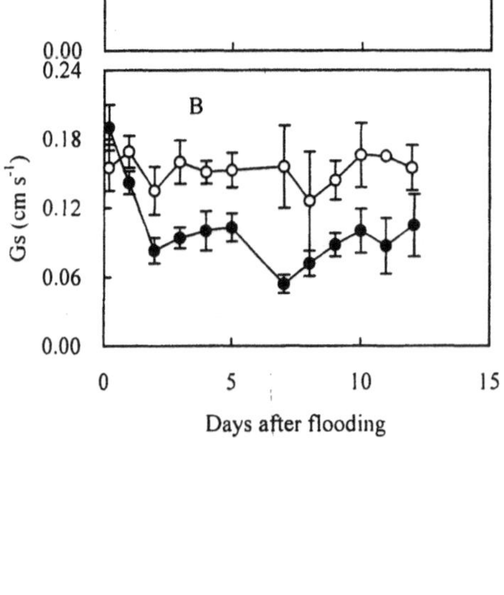

Fig. 17.5. Changes in stomatal conductance in *C. cordata* (**A**) and *C. laxiflora* (**B**) under flooded (*solid symbols*) and nonflooded (*open symbols*) conditions. *Bars* represent standard deviation of the mean of three to five replications (from Park 1997)

of nonflooded plants, and it was maintained throughout the experiment. Moreover, under flooded conditions, a stomatal conductance slightly higher than that under nonflooded conditions was maintained from the start to the 8th day after flooding.

On the other hand, the stomatal conductance of *C. laxiflora* under flooded conditions significantly declined to a level about 50% that in nonflooded plants beginning the 2nd day after the treatment (Fig. 17.6B). Furthermore, the reduced stomatal conductance was maintained without any recovery even at the end of the experiment. Considering that stomatal closure has been observed on the 4th day after flooding treatment even in flood- sensitive plants such as *Quercus nigra*, *Q. alba*, and *Fraxinus pennsylvanica* (Gravatt and Kirby 1998), *C. laxiflora*, which decreased stomatal conductance strikingly, must be a plant very sensitive to flooding.

Flooding affects not only stomatal conductance and photosynthesis, but even local plant distribution in the field. Among the effects of flooding on plant physiology, much attention has been focused on the change in stomatal conductance, which controls carbon gain. In recent years, many studies have been centered on the direct cause of stomatal closure by flooding in a variety of species (Else et al. 1996; Fuchs and Livingston 1996; Jackson et al. 1988; Pezeshki and Chambers 1985, 1986; Zhang and Davies 1987). In particular, several studies have attempted to characterize a messenger originated from oxygen-deficient roots, which is considered to be a trigger causing stomata to close during soil flooding (Else et al. 1996; Jackson et al. 1988; Sojka 1992; Zhang and Davies 1987). On the basis of these studies, for the most part it is accepted that stomatal closure by flooding is associated with increases in the concentration of ABA in the leaves, which is transported through xylem sap flow from the roots (Jackson et al. 1988; Zhang and Davies 1987). However, on the

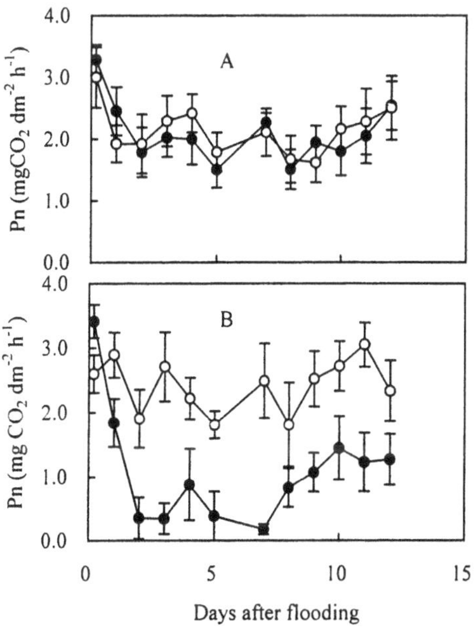

Fig. 17.6. Changes in net photosynthesis (*Pn*) in *C. cordata* (**A**) and *C. laxiflora* (**B**) under flooded (*solid symbols*) and nonflooded (*open symbols*) conditions. *Bars* represent standard deviation of the mean of three to five replications (from Park 1997)

basis of their research on the substance inducing rapid stomatal closure from flooding, Else et al. (1996) suggested that the substance promoting stomatal closure in flooded plants is not ABA. Although there is some disagreement on the trigger substance affecting stomatal conductance in flooded plants, it is obvious that flooding affects stomatal conductance differently in different species as well as at different growth stages (Gravatt and Kirby 1998; Smith and Huslig 1990; Sojka 1992).

The difference in stomatal behavior between the two *Carpinus* species after flooding treatment led to differences in the photosynthetic responses (Fig. 17.6). Flooded *C. laxiflora* plants showed a severe retardation in the photosynthetic rate from the 2nd day after the flooding treatment, to about 20% the level in the controls. This lowered photosynthetic rate continued for 6 days; thereafter, a slight recovery in photosynthesis was observed. This recovery, however, was still lower than the level of the photosynthetic rate in nonflooded plants. On the other hand, *C. cordata* under the flooded conditions maintained a high photosynthetic rate comparable to that under nonflooded conditions throughout the experiment (Fig. 17.6A). The patterns of the photosynthetic rates in both species well matched with those of stomatal conductance, with little variation irrespective of flooding treatment. Thus, the reduction of photosynthesis in *C. laxiflora* under flooded conditions must depend on the stomatal conductance. A similar phenomenon of flooding-induced closure of stomata being responsible for a decrease in photosynthesis has been shown in various plant species (Gravatt and Kirby 1998). However, under flooded conditions, the degree to which stomatal conductance contribute to the change in the photosynthetic rate differs considerably among species (Regehr et al. 1975; Pezeshki and Chambers 1985; Zaerr 1983). Thus, more experiments are needed to clarify the relationship between stomatal conductance and photosynthesis under flooded conditions.

Flooding limits gas exchange between the roots and atmosphere, leading to anoxia near the rhizosphere (Armstrong et al. 1994), which limits root respiration and, in turn, restricts water absorption. Thus, the restriction of water absorption under flooded conditions often reduces xylem pressure potential in plants (Zaerr 1983). The increased resistance to water movement through the plant during flooding reduces the xylem pressure potential, which is a factor in the closing of the stomata (Pezeshki and Chambers 1985, 1986; Zaerr 1983). *C. cordata* showed no significant difference in predawn leaf water potential between the flooded and nonflooded conditions throughout the experiment (Fig. 17.7). Furthermore, the midday leaf water potential under flooded conditions was rather higher than that under nonflooded conditions. In contrast, in *C. laxiflora* the predawn and midday leaf water potentials under flooded conditions decreased more than those under nonflooded conditions on the 6th day after flooding treatment. An interesting point is that the reduced leaf water potentials in *C. laxiflora* plants recovered even to the level of the control plants on the 12th day after the flooding treatment. This recovery of leaf water potential in *C. laxiflora* might be due to the development of the new adventitious roots that were confirmed at the end of the experiment (data not shown). The production of new fine roots after flooding reduces the water stress enough for plants of some species to reopen stomata (Kozlowski and Pallardy 1979; Regehr et al. 1975; Sojka 1992). *C. laxiflora*, however, could not increase stomatal conductance and the pho-

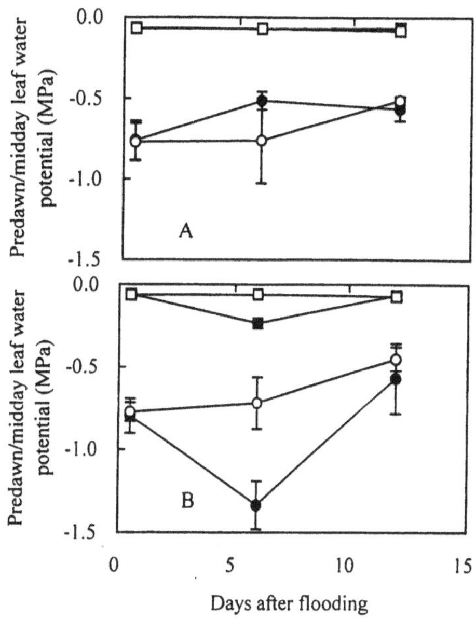

Fig. 17.7. Changes in predawn leaf water potential (*squares*) and midday leaf water potential (*circles*) in *C. cordata* (**A**) and *C. laxiflora* (**B**) under flooded (*solid symbols*) and nonflooded (*open symbols*) conditions. *Bars* represent standard deviation of the mean of three to five replications (from Park 1997)

tosynthetic rate to the level of the control plants in spite of the recovery in leaf water potential at the end of the experiment.

Some species inhabiting lowlands and floodplains have developed a certain degree of adaptation to go along with their specific strategies (Gravatt and Kirby 1998; He et al. 1999; Pezeshki and Cambers 1986;, Voesenek et al. 1988). Thus, the resistance mechanism to flooding is closely associated with their strategy with regard to habitat characteristics. For instance, *Ranunculus repens*, a species inhabiting frequently flooded areas showed a tolerance mechanism with a metabolic adjustment, while *R. sceleratus* showed an avoidance mechanism with the development of intercellular air space and the ability to increase the elongation rate of the petiole under water (He et al. 1999). From this point of view, *C. cordata* must be a flood-tolerant plant, while *C. laxiflora* can be classified as a flood-intolerant plant. It is evident that the seedling stage of plants must be more susceptible to environmental stress than the sapling stage used in this experiment. Taking these conditions into account, the differences in resistive characteristics to flooding may be more obvious in the seedling stage. Thus, the seedling stage would give us a definite answer to the question why *C. cordata* and *C. laxiflora* show habitat segregation in spite of being taxonomically closely related species. Consequently, the results obtained experimentally here are well supported by the allopatric distribution of *C. cordata* and *C. laxiflora* in the temperate deciduous forests of northern East Asia (Kim 1992; Masaki et al. 1992; Shibata and Nakashizuka 1995; Shibata et al. 1998).

17.4 Conclusions

Plants growing in the field often undergo stresses originating from an excess or deficiency of water. Thus, the resistive characteristics to the excess or deficiency of water in the seedling or sapling stages sometimes plays an important role in determining whether a species can grow in a given habitat. In this respect, we have clearly showed that *C. laxiflora* is able to grow on ridge crests and mountain slopes because it is able to resist water stress, while the resistive characteristics of *C. cordata* to flooding make it able to inhabit streamside habitats. The most important conclusion reached in this paper is that the different resistive characteristics to a given moisture regime between *C. laxiflora* and *C. cordata* seem to account for the habitat segregation observed in the temperate forests of Japan.

Acknowledgments

This research was carried out as a part of a Science Technology Agency Fellowship in Japan given to Yong-Mok Park. We thank Drs. Y. Matsumoto and Y. Maruyama for their helpful assistance throughout the experiment and Dr. I. Terashima for the constructive review of the manuscript. Park's special thanks are extended to Drs. T. Inoue, S. Sakurai, and Ms. K. Nakamura for their encouragement during his stay in Forestry and Forest Products Research Institute.

References

Armstrong W, Brandle R, Jackson MB (1994) Mechanisms of flood tolerance in plants. Acta Bot Neerl 43:307–358

Boyer JS (1970) Leaf enlargement and metabolic rates in corn, soybean and sunflower at various leaf water potentials. Plant Physiol 46:233–235

Else MA, Tiekstra AE, Croker SJ, Davies WJ, Jackson MB (1996) Stomatal closure in flooded tomato plants involves abscisic acid and a chemically unidentified anti-transpirant in xylem sap. Plant Physiol 112:239–247

Farquhar GD, Schulze ED Kupper M (1980) Responses to humidity by stomata of *Nicotiana glauca* L. and *Corylus avellana* L. are consistent with the optimization of carbon dioxide uptake with respect to water loss. Aust J Plant Physiol 7:315–327

Fuchs EE, Livingston NJ (1996) Hydraulic control of stomatal conductance in Douglas fir [*Pseudotsuga menziesii* (Mirb.) Franco] and [*Alnus rubra* (Bong)] seedlings. Plant Cell Environ 19:1091–1098

Gollan T, Turner NC, Schulze ED (1985) The responses of stomata and leaf gas exchange to vapour pressure deficits and soil water content. III. In the sclerophyllous woody species *Nerium oleander*. Oecologia 65:356–362

Gravatt DA, Kirby CJ (1998) Patterns of photosynthesis and starch allocation in seedlings of four bottomland hardwood tree species subjected to flooding. Tree Physiol 18:411–417

He JB, Bogemann GM, Van de Steeg HM, Rijnders JGHM, Voesenek LACJ, Blom CWPM (1999) Survival tactics of *Ranunculus* species in river floodplains. Oecologia 118:1–8

Hsiao TC (1973) Plant responses to water stress. Annu Rev Plant Physiol 24:519–570

Jackson, MB, Young, SF, Hall, KC (1988) Are roots a source of abscisic acid for the shoots of flooded pea plants? J Exp Bot 209:1631–1637

Kim JW (1992) Vegetation of northeast Asia, on the syntaxonomy and syngeography of the oak and beech forests. Ph.D. Dissertation. Vienna Univ., Vienna

Kozlowski TT, Pallardy SG (1979) Stomatal responses of *Fraxinus pennsylvanica* seedlings during and after flooding. Physiol Plant 46:155–158

Loewenstein NJ, Pallardy SG (1998) Drought tolerance, xylem sap abscisic acid and stomatal conductance during soil drying: a comparison of canopy trees of three temperate deciduous angiosperms. Tree Physiol 18:431–439

Lo Gullo MA, Salleo S (1988) Different strategies of drought resistance in three Mediterranean sclerophyllous trees growing in the same environmental conditions. New Phytol 108:267–276

Maruta E (1976) Seedling establishment of *Poligonum cuspidatum* on Mt. Fuji, Japan. Jpn J Ecol 26:101–105

Masaki T, Suzuki W, Niiyama K, Tanaka H, Nakashizuka T (1992) Community structure of a species-rich temperate forest, Ogawa Forest Reserve, central Japan. Vegetatio 98:97–111

Matsumoto Y (1984) Photosynthetic production in *Abies veitchii* advance growths growing under different light environmental conditions. III. Diurnal fluctuation in xylem water potential. Bull Tokyo Univ For 73:253–262

Meinzer FC, Hinckley TM, Ceulemans R (1997) Apparent responses of stomata to transpiration and humidity in a hybrid poplar canopy. Plant Cell Environ 20:1301–1308

Morikawa Y, Hattori S, Kiyono Y (1986) Transpiration of a 31-year-old *Chamaecyparis obtusa* stand before and after thinning. Tree Physiol 2:105–114

Osmond CB, Austin MP, Berry JA, Billings WD, Boyer JS, Dacey JWH, Nobel PS, Smith SD, Winner WE (1987) Stress physiology and the distribution of plants. The survival of plants in any ecosystem depends on their physiological reactions to various stresses of the environment. Bioscience 37:38–48

Osonubi O, Davies WJ (1981) Root growth and water relations of oak and birch seedlings. Oecologia 51:343–350

Park YM (1989) Factors limiting the distribution of *Digitaria adscendens* and *Eleusine indica* on coastal sand dunes in Japan. Ecol Res 4:131–144

Park YM (1990) Effects of drought on two grass species with different distribution around coastal sand-dunes. Funct Ecol 4:735–741

Park YM (1996) Diurnal changes of tissue water relations in two allopatric tree species (in Korean with English summary). Kor J Ecol 19:453–463

Park YM (1997) The effect of soil flooding on photosynthesis and water relations of *Carpinus cordata* and *Carpinus laxiflora* (in Korean with English summary). Kor J Ecol 20:175–179

Pezeshki SR, Chambers JL (1985) Stomatal and photosynthetic response of sweet gum (*Liquidambar styraciflua*) to flooding. Can J For Res 15:371–375

Pezeshki SR, Chambers JL (1986) Variation in flood-induced stomatal and photosynthetic response of three bottom-land tree species. For Sci 32:914–913

Picon C, Guehl JM, Ferhi A (1996) Leaf gas exchange and carbon isotope composition responses to drought in a drought-avoiding (*Pinus pinaster*) and a drought-tolerant (*Quercus petraea*) species under present and elevated atmospheric CO_2 concentrations. Plant Cell Environ 19:182–190

Regehr DL, Bazzaz FA, Bogges WR (1975) Photosynthesis, transpiration, and leaf conductance of *Populus deltoides* in relation to flooding and drought. Photosynthetica 9:52–61

Saliendra NZ, Sperry JS, Comstock J (1995) Influence of leaf water status on stomatal response to humidity, hydraulic conductance, and soil drought in *Betula occidentalis*. Planta 196:357–366

Shibata M, Nakashizuka T (1995) Seed and seedling demography of four co-occurring *Carpinus* species in a temperate deciduous forest. Ecology 76:1099–1108

Shibata M, Tanaka H, Nakashizuka T (1998) Causes and consequences of mast seed production of four co-occurring *Carpinus* species in Japan. Ecology 79:1099–1108

Smith MW, Huslig SM (1990) Influence of flood-preconditioning and drought on leaf gas exchange and plant water relations in seedlings of Pecan. Environ Exp Bot 30:489–495

Sojka RE (1992) Stomatal closure in oxygen-stressed plants. Soil Sci 154:269–279

Tazaki T (1960) On the growth of pine yearlings in coastal dune regions with special reference to their drought resistance. Jpn J Bot 17:239–277

Turner NC, Schulze ED, Gollan T (1984) The responses of stomata and leaf gas exchange to vapor pressure deficits and soil water content. I. Species comparisons at high soil water contents. Oecologia 63:338–342

Turner NC, Stern WR, Evans P (1987) Water relations and osmotic adjustment of leaves and roots of lupines in response to water deficits. Crop Sci 27:977–983

Voesenek LACJ, Blom CWPM, Pouwels RHW (1988) Root and shoot development of *Rumex* species under waterlogged conditions. Can J Bot 67:1865–1869

Yura H (1988) Comparative ecophysiology of *Larix kaempfri* (Lamb.) Carr. and *Abies veitchii* Lindl. 1. Seedling establishment on bare ground on Mt. Fuji. Ecol Res 3:67–73

Zaerr JB (1983) Short-term flooding and net photosynthesis in seedlings of three conifers. For Sci 29:71–78

Zhang J, Davies WJ (1987) ABA in roots and leaves of flooded pea plants. J Exp Bot 38: 1649–1659

Zhang J, Davies WJ (1990) Changes in the concentration of ABA in the xylem sap as a function of changing soil water status can account for changes in leaf conductance and growth. Plant Cell Environ 13:277–285

18 Microenvironments and Growth in Gaps

MORIYOSHI ISHIZUKA, YUKIHITO OCHIAI, and HAJIME UTSUGI

18.1 Introduction

Canopy gaps, created by tree-fall disturbances, greatly alter local microenvironmental conditions and the availability of resources for regenerating tree seedlings (Chazdon and Fetcher 1984; Bazzaz and Wayne 1994). These disturbance-induced local modifications in environmental conditions play a significant role in regulating the population dynamics and coexistence of forest species (see Chapter 7; Pickett 1983; Canham and Marks 1985; Denslow 1987).

In general, gap formation can temporarily increase the availability of light, water, and soil nutrients (Canham and Marks 1985; Bazzaz and Wayne 1994), which is believed to promote the regeneration of tree seedlings. However, only a limited number of studies have been made in assessing spatial and temporal patterns of local micro-environmental conditions in canopy gaps, and the responses of tree seedlings to these patterns.

In 1991, we began artificial gap experiments to investigate experimentally the consequences of small- to medium-scale forest disturbances on the physiological ecology and regeneration processes in a secondary hardwood forest in Nakoso near the Ogawa Forest Reserve (OFR). In this chapter, we describe some of the patterns of light and soil moisture regimes in gaps according to gap size, as well as the responses of regenerating trees to these patterns.

18.2 Study Site and Methods

18.2.1 Site

The experimental site, located about 5 km north of the Ogawa Forest Reserve, is a 0.7-ha secondary deciduous hardwood stand (36° 58' N, 140° 36' E) in Nakoso National Forest. The stand occupies a flat ridge at 650–660 m elevation. According to the Japanese Soil Type Classification System (Forest Soil Division 1975), the soil around the stand is classified as a Moderately Moist Black soil (drier subtype). Canopy

Ecological Studies, Vol. 158
Nakashizuka, Matsumoto (eds.) Diversity and Interaction in a Temperate Forest Community: Ogawa Forest Reserve of Japan

dominants (at 15–18 m height) are *Carpinus laxiflora*, *Quercus serrata*, and *Q. crispula*. The stand originated after clear-cutting in the early 1920s and was treated as a coppice forest until the mid-1940s. A detailed description of the stand structure is given in Ochiai et al. (1994b).

18.2.2 Artificial Gaps and Regeneration Monitoring Transects

Five different-sized gaps, with radii ranging from 3–8 m, were made artificially for this study. In March 1991, two single-tree openings (3 and 3.5 m in radius) and three multitree openings (5.5, 6, and 8 m in radius) were created by cutting trees lower than 0.3 m above the ground (Fig. 18.1). To eliminate the influences of trees remaining in the gaps, all stems > 100 cm tall in each gap were cleared and carefully removed from the gaps. The gaps were located far enough apart so that no gap could influence any of the others.

For monitoring the regeneration processes in each gap (except for the smallest

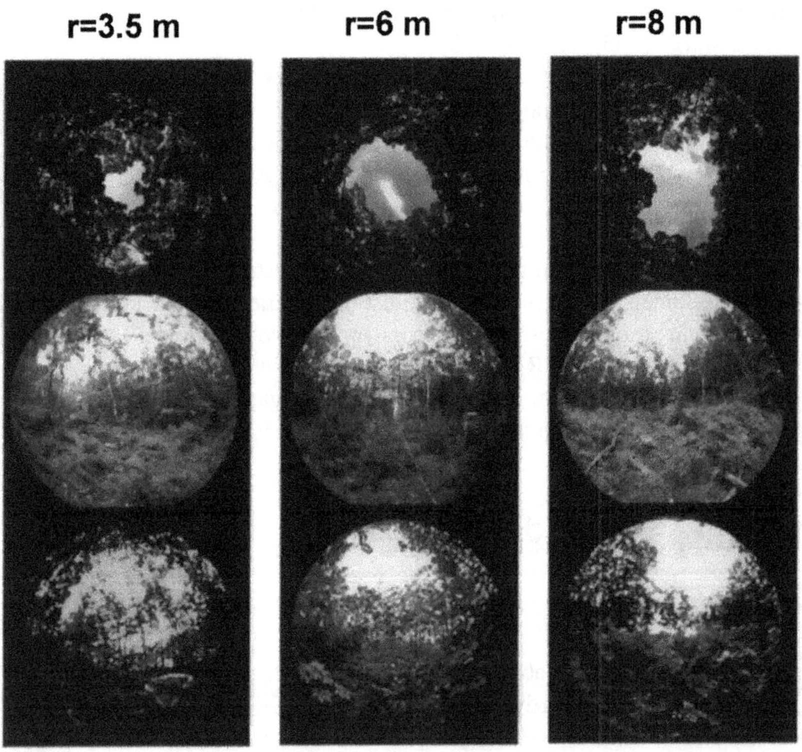

Fig. 18.1. Canopy gaps with radii of 3.5 (*left*), 6 (*center*), and 8 m (*right*), artificially created in the Nakoso Experimental Site in March 1991. Fisheye views of the canopy in July 1991 (*top*) and of the vegetation in May 1992 (*middle*) and July 1997 (*bottom*)

gap), a 2-m-wide strip transect was established running north and south from the gap center into the surrounding canopied area. In the 8-m-radius gap, the transect did not extend past the gap edges. After gap creation, all stems in each transect taller than 50 cm were recorded at the end of the growing season every year for species, origin of stem (advanced growths, sprouts from stumps or roots, seedlings from seeds), location, diameter at 10-cm height, and tree height.

18.2.3 Light and Soil Moisture Availability

For the microclimate factors, we mainly monitored light and soil matric potentials in the gaps and under the closed canopy nearby. Light regimes in the gaps and under the closed canopy were evaluated from the analysis of hemispherical photographs and from measurements of incident photosynthetic photon flux density (PPFD). Hemispherical photographs (vertical exposures at 1.5 m height) were taken monthly in each gap from late April (before leafing out) to late November (after leaf fall) at three points around the gap center from 1992 to 1998. Hemispherical photographs were also taken along the north–south transect at 1- or 2-m intervals during the leafy period in 1993–1995. Data on the incident PPFD, averaged over 10-min periods, were collected at 1.3 m height at the center of each of the 3- and 5.5-m-radius gaps during 1995–1998 using a PPFD sensor (LI-190SB, Li-Cor, Lincoln, NE, USA) connected to a data logger (KADEC-UP Corner System, Sapporo, Japan). The same PPFD sensor-logger system was also installed above (on the top of a scaffold tower) and under the intact canopy. The amounts of PPFD incident on the center (at 1.5 m height) of gaps with radii of 3.5, 6, and 8 m during the growing season were estimated from the hemispherical photographs of the canopy gaps taken monthly, and the data on full-sun (above canopy) PPFD was separated into diffuse and direct components using the empirical formula by Udagawa (1986). A factor of 4.6 was used to convert radiation (W m^{-2}) to PPFD (μmol m^{-2} s^{-1}) (McKree 1981).

Hemispherical photographs were used to compute a gap light index (GLI) that, as proposed by Canham (1988), specifies the percentage of incident PPFD transmitted through a gap to any particular point in the course of a growing season. The GLI is of the form

$$\mathrm{GLI} = [(Tdiffuse\ Pdiffuse) + (Tbeam\ Pbeam)]\cdot 100.0$$

where *Pdiffuse* and *Pbeam* are the relative amounts of the incident seasonal PPFD received at the top of the canopy as either diffuse skylight or direct radiation, respectively, and *Tdiffuse* and *Tbeam* are the relative amounts of diffuse and beam radiation, respectively, that are transmitted through the gap to a point in the understory. We defined the growing season as the period from 1 May to 31 September, and *Pbeam* for this period was set to 0.4, based on a measurement we made in Tsukuba (36° 01' N, 140° 08' E). A simulation approach was also introduced to assess GLI profiles in idealized gaps of various sizes. Following Takenaka (1988) and Canham (1988), a gap was assumed to be an opening of cylindrical shape in a flat canopy of uniform thickness. In addition, we gave a specific leaf area index (LAI) and extinction coefficient (*k*) to the hypothetical canopy. Local parameters, such as latitude, longitude, canopy height, LAI, and *k*, were adjusted to the parameters of the Nakoso Experimental Site. We examined only level sites because the gaps in the study site were on a flat area of a wide ridge.

To compare soil moisture availability between gaps and closed canopy, the soil matric potential (as an analog of soil water potential) was monitored by Ochiai et al. (1994b) from June 1991 to September 1992 with tensiometers installed at the center of the gaps with radii of 3.5 and 6 m and under the closed canopy. In June 1991, a tensiometer (MM.S30I, Nippon Aleph, Yokohama, Japan) was installed at 15- and 30-cm depths at each site, and then another tensiometer (MM-S301L, Nippon Aleph Corp.) was set at 5-cm depth in April 1992. A rainfall gauge with a capacity of 1,256 cm^2 (34-T, Ota Keiki Seisakusho, Tokyo, Japan) was installed at the center of the 6-m-radius gap and under the closed canopy to monitor precipitation during the same period as the soil moisture measurements were being collected. Hourly soil matric potential and rainfall were recorded by data loggers (MDL-1000, Karter Art Landscape Consultant, Tokyo, Japan) connected to the sensors at each site.

18.3 The Effects of Gap Size on Light and Soil Moisture Regimes

18.3.1 Gap Light Regimes in Relation to Gap Size

Although the relationship between light flux and gap size is not linear, the dependence of both diffuse and direct light levels on gap size is the most apparent evidence of light regimes in canopy openings (Nakashizuka 1985; Takenaka 1988; Barton et al. 1989; Brown 1993). Our study in Nakoso also revealed that PPFD penetration levels (percentage penetration of full-sun PPFD) depended on gap size, although the penetration rates in each gap varied slightly according to the substantial yearly weather

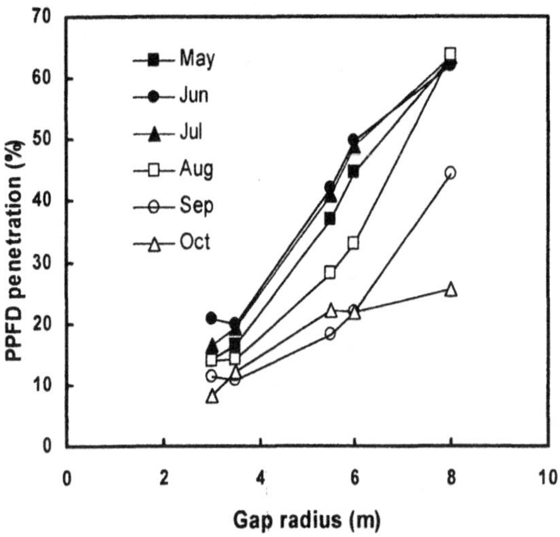

Fig. 18.2. Monthly photosynthetic photon flux density (PPFD) penetration (as a percentage of full-sun PPFD) in the center of five gaps with a range of radii 3–8 m (data from 1995)

variations. The data from 1995 are shown in Fig. 18.2 as an example under normal weather conditions. The gap-size dependence of PPFD penetration in the centers of gaps is remarkable from May to August. After September, however, PPFD penetration in larger gaps showed marked decreases due to the interception of direct beam light. Because the proportion of direct light in the small gaps was lower than 30%, monthly PPFD penetration in the small gaps was relatively constant and was approximately equal to the diffuse light levels. Consequently, single-tree gaps provide PPFD penetrations of not more than 10%–20%, but multiple-tree gaps (two or three trees, at least) can provide PPFD penetrations of more than 35%–50%, especially during the early growing season. This result suggests that gap creation by group cutting of a few adjacent trees is far more effective than single-tree cutting for enhancing understory light levels.

In the northern hemisphere, the northern part of a gap potentially receives much higher direct beam-light penetration or GLI than the southern part (Canham 1988; Takenaka 1988; Canham et al. 1990). We recognized evidence of this in all the gaps, yet the substantial yearly variations in weather conditions changed the amplitude of this pattern. Monthly PPFD penetration patterns along the N–S transect of the 6-m-radius gap are shown in Fig. 18.3. The data from 1995 are shown as an example under normal weather conditions. These patterns are more or less changeable, depending on the weather conditions during each month of the year. It is notable, however, that PPFD distributions along the N–S transect are different among months. PPFD gradients from north to south are more apparent during August to September than those during May to July (Fig. 18.3), which indicates that the potential light gradients from north to south are mainly formed in the late growing season.

Figure 18.4 demonstrates simulated GLI distributions at various heights along the N–S transect for hypothetical gaps with radii of 2.5, 5, 7.5, and 10 m, showing

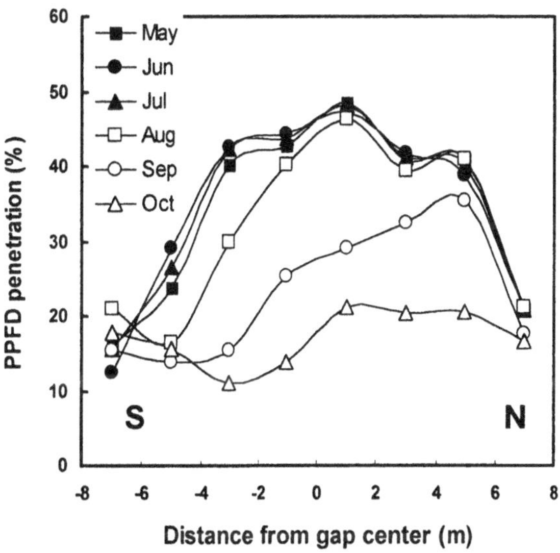

Fig. 18.3. Monthly PPFD penetration (as a percentage of full-sun PPFD) along the north–south transect through the center of the 6-m-radius gap

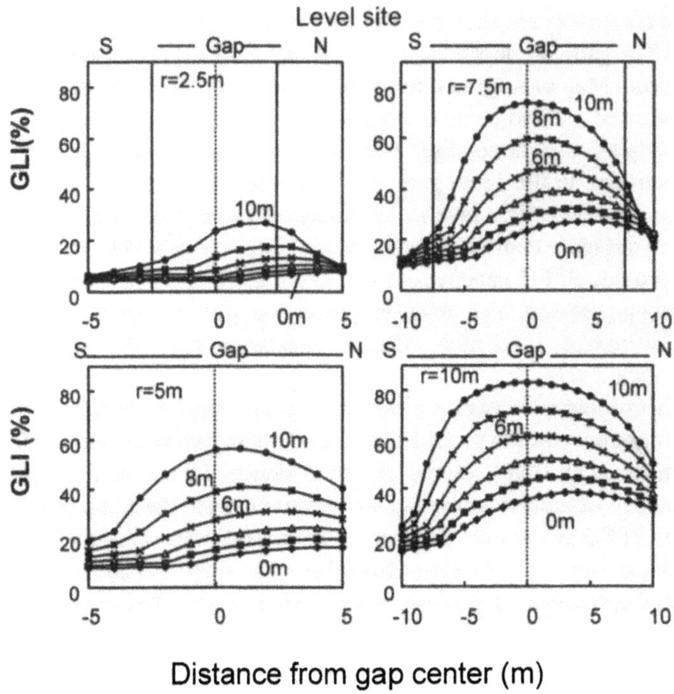

Fig. 18.4. Simulated gap light index (GLI) distribution along a north–south transect through the center of hypothetical gaps with radii of 2.5, 5, 7.5, and 15 m. Model parameters were adjusted to the Nakoso Experimental Site: canopy height, 16 m; LAI, 3.8; *k*, 0.85; location, 36° 58′ N, 140° 36′ E, *Pbeam*, 0.4, growing season, 1 May to 31 September

the predictable mode of GLI in the northern parts of the gaps (Canham et al. 1990). Simulations were run for a canopy height of 16 m. When examined in detail, however, these profiles reveal differences in GLI gradients from north to south with respect to height and gap size. As the height increases, the gradients from north to south increase in small gaps with a radius of less than 5 m; but in larger gaps, modes of GLI shift from the north to the center of the gap. Even in a 5-m-radius gap, the mode of GLI approaches the center of the gap at 6-m height. Consequently, the N–S gradient of light is important evidence in a relatively small gap, such as one with a radius of less than 5 m, and especially for shrub-layer levels (2 m or less).

18.3.2 Soil Moisture Regimes in Gaps and Under the Canopy

Available soil water is usually higher in gaps or in thinned sites than in the intact forest; this is believed to be due to reduced transpiration by canopy trees (Minckler and Woerheide 1965; Becker et al. 1988; Ashton 1992; Ochiai et al. 1994a). A similar observation was made, and a piece of evidence that supports the importance of transpiration's effect on soil moisture conditions was obtained, in the study site at Nakoso (Ochiai et al. 1994b).

Three characteristic periods with different soil moisture conditions were identified during the measurements of soil water potential (matric potential) and precipitation. Here we describe the soil moisture regimes in gaps and under the canopy focusing on these three periods. The three periods are as follows: Gwet, a growing season with frequent rainfall (June–August 1991); Gdry, a growing season with little rainfall (June–September 1992); and Pwet, a pregrowing season with periodic rainfall (April–May 1992).

Soil water conditions were closely related to rainfall patterns. When it rained, the soil became wet immediately, and it dried gradually after a rain (Figs. 18.5, 18.6). Soil water conditions also varied by depth below the soil surface. The deeper soils were for the most part wetter than shallow soils, except during a period of rainfall. During the growing seasons of both wet (Gwet) and dry (Gdry) periods, the drying processes of the soils after a rainfall differed according to gap size. During these two periods, the soils dried faster in less-disturbed sites. In the case of Gwet, the differ-

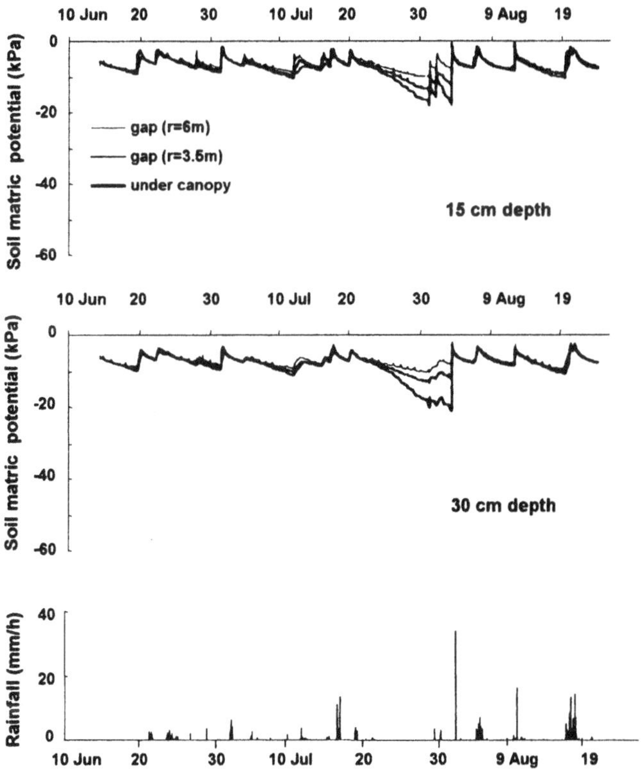

Fig. 18.5. Soil water (matric) potentials in the center of the 3.5- and 6-m-gaps, and under the closed canopy (*above*), and rainfall in the 6-m-radius gap during Gwet, a growing season with frequent rainfalls (June–August 1991). Modified from Ochiai et al. (1994b)

ences in the soil water potentials between the three sites were relatively small, except for the period of no rainfall (Fig. 18.5). When drought conditions lasted longer (Gdry), the soils under the canopy dehydrated very severely (Fig. 18.6). The soil near the surface of the small gap showed a similar course of dehydration. In contrast to these drought conditions, the soils of the 6-m-radius gap maintained relatively moist conditions at each depth.

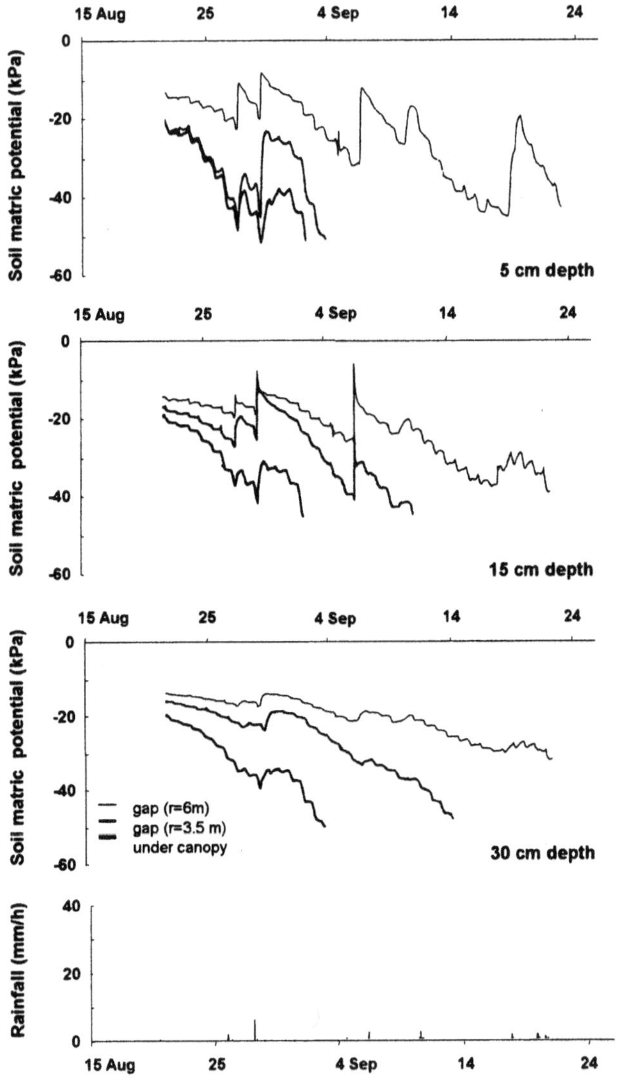

Fig. 18.6. Soil water (matric) potentials in the center of the 3.5- and 6-m-gaps, and under the closed canopy (*above*), and rainfall in the 6-m-radius gap during Gdry, a growing season with little rainfall (June–September 1992). Modified from Ochiai et al. (1994b)

In the pregrowing season (Pwet), however, the differences in soil water matric potentials between the three sites were very small, and the soil at each depth at each site stayed wetter (> −10 kPa) even when there was no rainfall over several days (Fig. 18.7). This means that between gaps and canopied areas, soil moisture conditions differed only during the leafy growing seasons, which supports the theory that gap formation increases available soil water due to reduced transpiration by canopy trees.

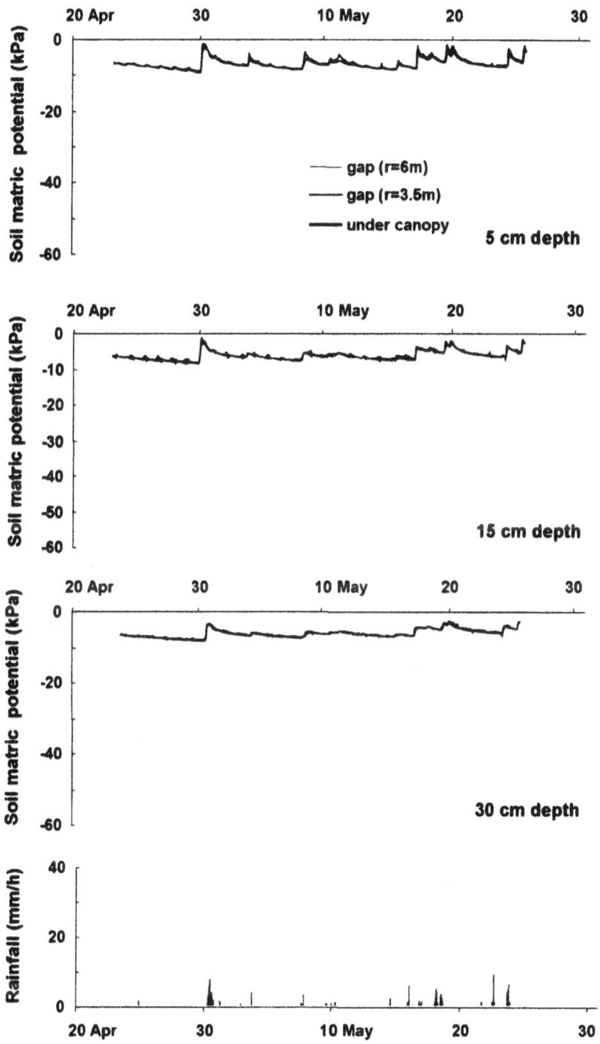

Fig. 18.7. Soil water (matric) potentials in the center of the 3.5- and 6-m gaps, and under the closed canopy (*above*), and rainfall in the 6-m-radius gap during Pwet, a pregrowing season with periodic rainfall (April–May 1992). Modified from Ochiai et al. (1994b)

18.4 Regeneration Response to Microenvironments in Artificial Gaps

18.4.1 Species and Origin of Regenerated Trees

The main components of regenerating trees after gap creation were advanced growths, trees that were present before the area was disturbed, just as occurs in the OFR under the natural disturbance regime (see Chapter 12). Advanced regeneration seems to be a more important strategy to maintain the sapling composition than new regeneration in both natural and artificial gaps, as was also observed in the OFR (see Chapter 12). *Ilex macropoda* was the most abundant in all artificial gaps. *Fraxinus lanuginosa* f. *serrata*, *Acanthopanax sciadophylloides,* and *Rhus trichocarpa* were also commonly distributed in most gaps. *Styrax obassia*, *S. japonica*, *Acer nipponicum*, *Clethra barbinervis,* and *Fagus crenata* were also important but occurred only in one or two gaps.

After a few years, many sprouts grew up from small stumps or roots and, consequently, they dominated in all gaps. Most of these sprouts were seedling sprouts (defined as sprouts coming from stumps less than 5 cm in diameter), which have a potential for superior growth, particularly in an open site (Smith 1985). Because artificial gap creation produced many small stumps, sprouting seemed to play a more important role in gap regeneration in artificial gaps than in natural gaps in the OFR (see Chapter 12). Large stumps of felled canopy trees, such as *Quercus* species, also produced stump sprouts, but they dwindled rapidly.

Seedlings of some species, such as *Q. crispula* and *A. sciadophylloides*, were newly germinated immediately after the disturbance. For example, up to 20–30 new seedlings of these species were observed in a subplot (2×2 m^2) of the transect established in the 6-m-radius gap in 1991–1993 (Tashiro et al. 1994). However, due to the high mortality rate and suppressed growth under the deep shade of sprouts and advanced growths (Tashiro et al. 1994), these seedlings are not expected to play an important role in the early stage of regeneration in these artificial gaps, especially in the small gaps. The proportion of these seedlings in the total basal area of regenerating trees was less than 15%. Some pioneer species, such as *Lespedeza bicolor*, actually regenerated in the largest gap, but we were not able to examine gap-size dependency of this species in this experiment with only four artificial gaps. In the OFR, however, distributional tendencies with respect to the gap-size gradient were recognized in upper-slope gaps (see Chapter 12).

18.4.2 Growth Response to Microenvironments

The height growth processes of regenerating trees following gap creation along the N–S transects established in the 3.5-, 6- and 8-m-radius gaps are presented in Figs. 18.8–18.10. The GLI and the diffuse site factor (Anderson 1964) along the same transect are also shown in the figures. It is apparent that the modes of tree height were formed by growth in the northern part of the gaps, especially in gaps with radii of less than 6 m. The mode of tree height in each gap agrees with the mode of GLI better than it does with that of the diffuse site factor, which implies that the growth of trees responded more strongly to direct light than to diffuse light during the growing season. In the large gap with a radius of 8 m, the GLI was still higher in the northern part than in the southern part, but its peak shifted to near the center of the gap, which was also predicted by the simulation of GLI distribution in idealized gaps (Fig. 18.4). The patterns of GLI and the diffuse site factor were similar in the large gap, but the detailed pattern of tree height appears to agree with GLI better than it does with the diffuse site factor.

Although we did not measure moisture, it has been suggested that the northern sides of gaps may be drier than elsewhere due to higher insolation and, subsequently, to greater growth and transpiration of woody plants (Collins et al. 1985). We noticed no signs of growth stress on regenerating trees in the northern part of the gaps, but an experimental study of seedlings planted in the gaps of the Nakoso study site may suggest insolation-induced growth stress on the growth of seedlings, especially dur-

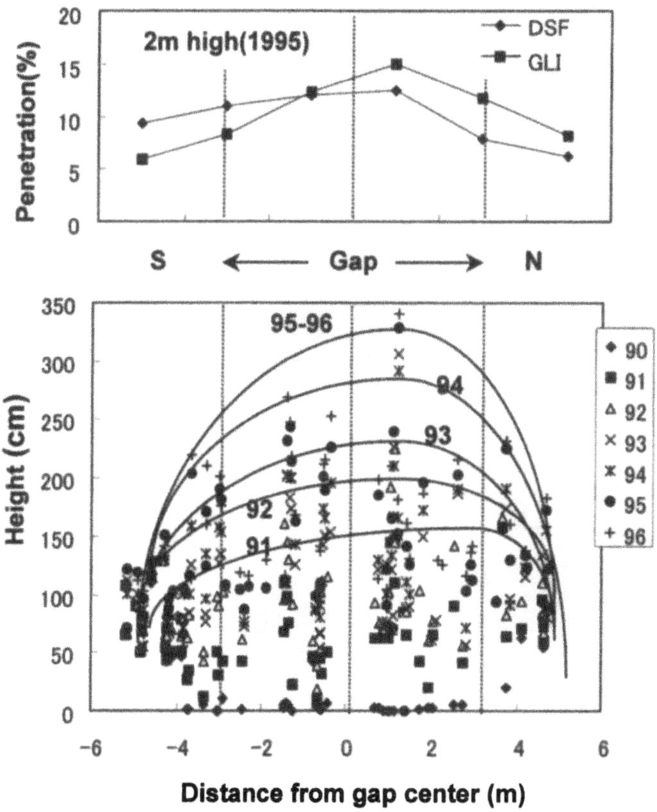

Fig. 18.8. The GLI and the diffuse site factor (DSF) (*above*) and growth of regenerating trees (*below*) along the N–S transect through the center of the 3.5-m-radius gap created in March 1991

ing a dry summer (Kondo and Ishizuka 1997). The study was conducted with four species seedlings, including *Q. crispula* and *A. sciadophylloides*, which were planted in 1993 in the 3- and 5.5-m-radius gaps under two distinct weather regimes: moist conditions in 1995 and dry conditions in 1996. The results showed no significant difference between the correlations of the relative growth rate (RGR) of seedlings to the GLI and to the diffuse site factor. In addition, the correlations between RGR and the light level in the dry year were apparently lower than those in the moist year. These analyses may suggest insolation-induced growth stress on tree seedlings in a dry year.

As described earlier, when drought conditions lasted longer in the growing season, the surface soil in the small gap dried up severely, yet the soils in the 6-m-radius gap remained relatively moist (Fig. 18.6). These results imply that droughts in the growing season can be critical in small gaps, especially for small seedlings with

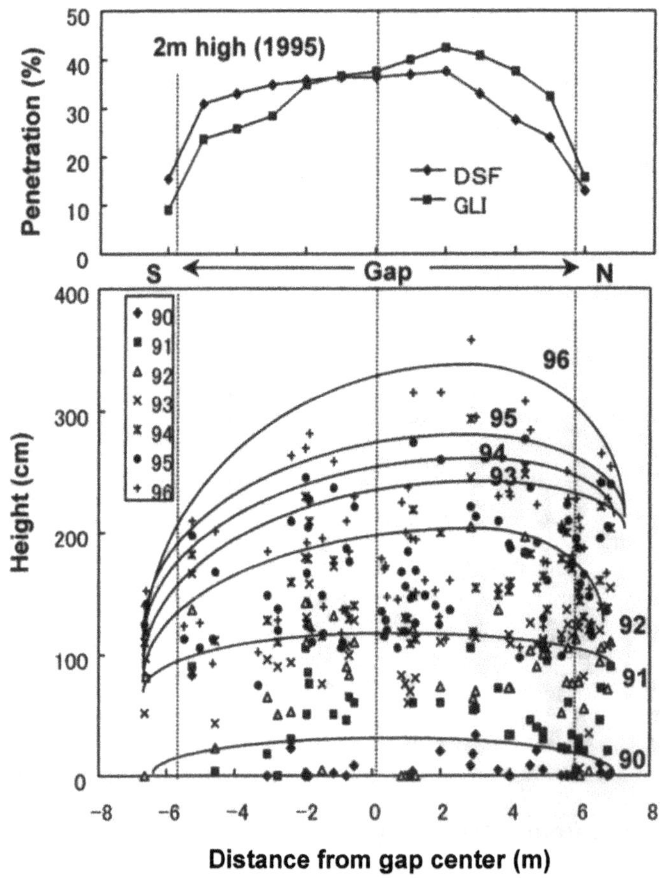

Fig. 18.9. The GLI and DSF (*above*) and growth of regenerating trees (*below*) along the N–S transect through the center of the 6-m-radius gap created in March 1991

shallow root systems. On the other hand, low light levels in smaller gaps are also critical for the growth and survival of trees, especially for small seedlings near the ground. Gap formation generally increases the availability of light, water, temperature, and soil nutrients simultaneously, each factor interacting with the others (Canham and Marks 1985; Bazzaz and Wayne 1994). In addition, the significance of the vertical structure of gaps should be considered, because seedlings or saplings in a small area within a gap can be exposed to different environments, according to their physical size (see Chapter 12). We observed a high mortality rate of newly germinated seedlings under the deep shade of sprouts and advanced growths even in the 6-m-radius gap (Tashiro et al. 1994). Therefore, the whole aspect of growth responses to these interacting factors is highly complicated.

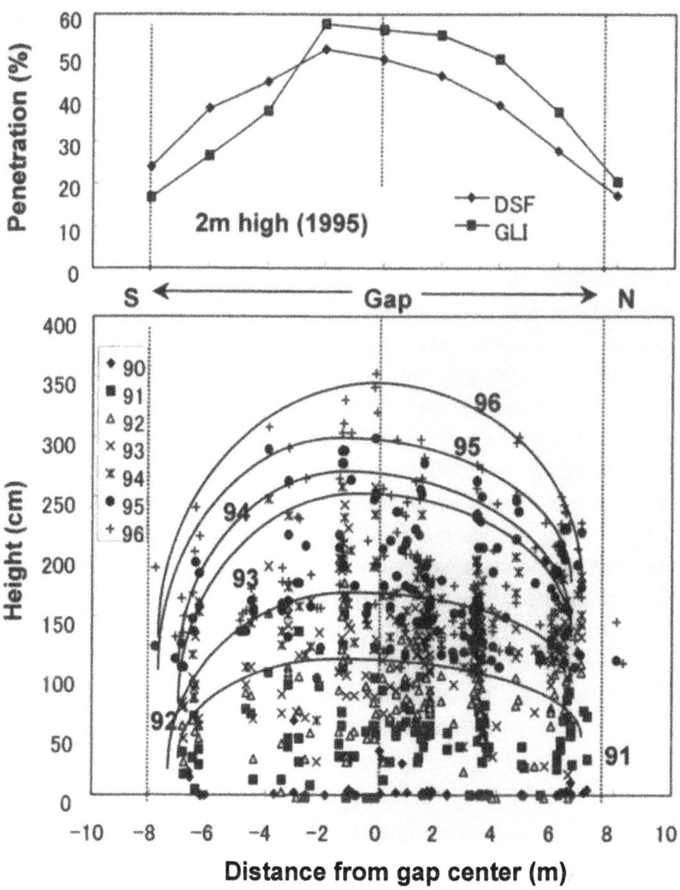

Fig. 18.10. The GLI and DSF (*above*) and growth of regenerating trees (*below*) along the N–S transect through the center of the 8-m-radius gap created in March 1991

18.5 Conclusions

Many environmental factors maintain complex interdependencies. For example PPFD and air and soil temperatures in gaps usually have strong relationships to one another (Bazzaz and Wayne 1994). Moreover, as described above, gap formation increases the availability of light and soil moisture simultaneously. In small gaps, drought, as well as low light levels, during the growing season can be critical for the growth and survival of small seedlings. The presence of environmental complexes suggests that particular demographic or physiological responses of tree seedlings cannot be attributed to particular environmental factors. Our study focused mostly on the gap light environment because the light condition is believed to be the most critical factor for regeneration in gaps; but the growth response is not always dependent on light conditions only (Kondo and Ishizuka 1997). In addition, the physiological response and consequent growth response do not always coincide. For example, the west side of a gap is considered to be more favorable for photosynthesis because of lower temperatures and higher humidity in the morning. Wayne and Bazzaz (1993) found the predicted differences in the diurnal course of photosynthesis on the east and west sides, but no differences in growth. Therefore, clarifying the whole aspect of growth responses to these interacting environmental factors is still challenging.

The preceding studies on seedlings in natural gaps, such as those conducted in the Harvard Forest (see Bazzaz and Wayne 1994), revealed that the within-gap heterogeneity is sufficient to allow differentiation of species responses, which may play an important role in the mechanism of species coexistence and community organization by gap partitioning among species (Sipe and Bazzaz 1994; also see Chapter 12). However, even though we have not yet analyzed the specific response to the heterogeneous light environment in gaps, gap partitioning among tree species does not seem to be advantageous in the artificial gaps created by cuttings, because the newly regenerated trees in the artificial gaps are mostly seedling sprouts. They are often superior to seedlings germinated from seeds because sprouts grow more rapidly as a result of having established root systems to nourish them (Smith 1985). Such root systems also have the advantage of being able to use deep soil water during drought periods, allowing the seedlings to concentrate on receiving higher light penetration.

According to our data, the initial height growth rate of seedling sprouts is nearly twice that of advanced growth. In addition, since seedling sprouts sprout out from the stumps or roots of advanced growth, the species composition of the regenerating community in artificial gaps may be determined by the distribution of advanced growth, at least in the initial stage of regeneration.

References

Anderson MC (1964) Studies of the woodland light climate. I. The photographic computation of light conditions. J Ecol 52:27–41

Ashton PMS (1992) Some measurements of the microclimate with a Sri Lankan tropical rainforest. Agr Forest Meteorol 59:217–235

Barton AM, Fetcher N, Redhead S (1989) The relationship between treefall gap size and light flux in a neotropical rain forest in Costa Rica. J Trop Ecol 5:437–439

Bazzaz FA, Wayne PM (1994) Coping with environmental heterogeneity: The physiological ecology of tree seedling regeneration across the gap-understory continuum. In: Caldwell MM, Pearcy RW (eds) Exploitation of environmental heterogeneity by plants: ecophysiological process above and below ground. Academic, San Diego, pp 349–390

Becker PE, Rabenold E, Indol JR, Smith AP (1988) Water potential gradients for gaps and slopes in a Panamanian tropical moist forest's dry season. J Trop Ecol 4:173–184

Brown N (1993) The implications of climate and gap microclimate for seedling growth conditions in a Bornean lowland rainforest. J Trop Ecol 9:153–168

Canham CD (1988) An index for understory light levels in and around canopy gaps. Ecology 69:1634–1638

Canham CD, Marks PL (1985) The response of woody plants to disturbance: patterns of establishment and growth. In: Pickett STA and White PS (eds) The ecology of natural disturbance and patch dynamics. Academic, Orlando, pp 196–216

Canham CD, Denslow, JS, Platt WJ, Runkle J R, Spies TA, White PS (1990) Light regimes beneath closed canopies and tree-fall gaps in temperate and tropical forests. Can J For Res 20:620–631

Chazdon RL, Fetcher N (1984) Photosynthetic light environments in a lowland tropical rain forest in Costa Rica. J Ecol 72:553–564

Collins BK, Dunne KP, Pickett STA (1985) Responses of forest herbs to canopy gaps. In: Pickett STA, White PS (eds) The ecology of natural disturbance and patch dynamics. Academic, Orlando, pp 217–234

Denslow JS (1987) Tropical rainforest gaps and tree species diversity. Annu Rev Ecol Syst 18:431–451

Forest Soil Division (1975) Classification of forest soil in Japan (Japanese with English summary). Bull Gov For Exp Sta 280:1–28

Kondo H, Ishizuka M (1997) Gap light index and growth of four tree species planted in canopy gaps (in Japanese). Trans Jpn For Soc 108:255–256

McKree KJ (1981) Photosynthetically active radiation. In: Lange OL, Nobel PS, Osmond CB, Ziegler H (eds) Physiological plant ecology I. Responses to the physical environment (Encyclopedia of plant physiology 12A), Springer, Berlin Heidelberg New York, pp 41–55

Minckler LS, Woerheide JD (1965) Reproduction of hardwood: 10 years after cutting as an opening size. J For 63:103–107

Nakashizuka T (1985) Diffused light conditions in canopy gaps in a beech (*Fagus crenata* Blume) forest. Oecologia, 66:472–474

Ochiai Y, Alimanar M, Yusop A R (1994a) Natural distribution and suitable method for plantation of two *Dryobalanops* species in Negara Brunei Darussalam. Bull For & For Prod Res Inst 366:31–56

Ochiai Y, Okuda S, Sato A (1994b) The influence of canopy gap size on soil water conditions in a deciduous broad-leaved secondary forest in Japan. J Jpn For Soc 76:308–314

Pickett STA (1983) Differential adaptation of tropical species to canopy gaps and its role in community dynamics. Trop Ecol 24:68–84

Smith DM (1985) The practice of silviculture. Wiley, New York, pp 468–487

Sipe TW, Bazzaz F A (1994) Gap partitioning among maples (*Acer*) in central New England: shoot architecture and photosynthesis. Ecology 75:2318–2332

Takenaka A (1988) An analysis of solar beam penetration through circular gaps in canopies of uniform thickness. Agr Forest Meteorol 2:307–320

Tashiro T, Ochiai Y, Sakai A, Okuda S, Sato S (1994) Influences of man-made gap to the growth and survival of seedlings in a secondary deciduous broad-leaved forest. Trans Jpn For Soc 105:44–450

Udagawa M (1986) Computing methods for air conditioning by personal computer (in Japanese). Ohmsha, Tokyo, pp 61–63

Wayne PM, Bazzaz F A (1993) Morning vs afternoon sun patches in experimental forest gaps: consequences of temporal incongruency of resources to birch regeneration. Oecologia 94:235–243

Part 6

Genetic Studies of Tree Populations

19 Demographic Genetics of the *Fagus crenata* Population in Ogawa Forest Reserve

KEIKO KITAMURA and SHOICHI KAWANO

19.1 Introduction

In view of population dynamics, gap formation is a major mechanism for regeneration among temperate deciduous climax tree species (Nakashizuka 1987; Yamamoto 1989), which leads to the development of a various sizes of mosaic structure within the forest ecosystem. As a result, long-lived forest tree populations are made up of individuals of various ages with patch structures of various sizes. Under these circumstances, spatio-temporal genetic substructurings and/or localizations in various scales are significant consequence within a local population.

The question of how to maintain genetic diversity within a local population of long-lived woody species consisting of individuals of overlapping generations for several hundred years has long been important for researchers. Spatial autocorrelation analyses are useful tools for detecting small-scale genetic structures (Epperson 1989), and a number of studies on the spatial genetic structures of plant populations have been reported (Epperson and Clegg 1986; Dewey and Heywood 1988; Epperson and Allard 1989; Schoen and Latta 1989; Schnabel and Hamrick 1990; Argyres and Schmitt 1991; Knowles 1991; Perry and Knowles 1991; Schnabel et al. 1991; Xie and Knowles 1991; Geburek and Tripp-Knowles 1994; Williams 1994; Shapcott 1995). However, temporal genetic differentiation within the age class structure and size class structure have not yet been well documented, except for a few pioneering studies (Schaal and Levin 1976; Clegg et al. 1978; Alvarez-Buylla and Garay 1994; Leonardi et al. 1996). The present population status reflects the consequences of past reproductive events or environmental changes in established sites, which must be well manifested within the genetic structure of overlapping generations.

We report the demographic genetic substructuring of *Fagus crenata*, which is one of the main canopy tree species in Ogawa Forest Reserve (OFR). Our results demonstrate extremely localized patterns of genetic structures by means of systematic samplings with thorough analyses of 46 alleles in 11 polymorphic allozyme loci. The subject matter presented here is twofold: first, we report on the fine-scale spatio-temporal genetic substructurings that developed in a hierarchical size class structure because of the environmental heterogeneities of the sites where the trees

Ecological Studies, Vol. 158
Nakashizuka, Matsumoto (eds.) Diversity and Interaction in a Temperate Forest Community: Ogawa Forest Reserve of Japan

were established; and second, we describe the genetic differentiation among patch populations within a local population in light of the metapopulation concept and provide estimations of the genetic neighborhood size and the effective population size. Further information and discussion of this *F. crenata* population appeared in Kitamura et al. (1997a, b), and Kawano and Kitamura (1997).

19.2 Fine-Scale Spatio-Temporal Genetic Structure

Two transects (55 × 10 m and 50 × 10 m) were established in the vicinity of the 6-ha permanent plot in the OFR in 1994 (Fig. 19.1), the year following *F. crenata* mast-fruiting. All individuals of *F. crenata*, including seedlings, various sizes of juveniles, and mature trees, found within or in near proximity to the transects were thoroughly mapped, assigned to one of six size classes according to their stem diameter, and identified by genotype (Table 19.1).

Extremely localized patterns of genetic substructurings in relation to the background environments were observed in both transects (Kitamura et al. 1997a). Notable evidence obtained by comparing genotypic, allelic, and zygotic frequency and several genetic parameters revealed that the genetic variation was quite high among the size classes. Regarding long-lived woody species like *F. crenata*, the genetic components of each size class can be considered as representative of that generation, and thus differences among size classes can be regarded as chronological changes in genetic structures. The temporal genetic substructure indicated by the differences in size class compositions of these two transects are obvious effects of past mating events. For example, the *Adh-a* allele in Transect 1 was shared only by juveniles of the intermediate size classes on the lower part of the slope (the right side of Transect 1, Fig. 19.2), and these *Adh-a*–carrying juveniles are likely to be the offspring of mother trees associated with past mating events. Somewhat similar situations were observed in several other loci. According to the demographic study using the matrix model for *F. crenata* (see Chapter 13), the elasticity of stasis is the highest among the four elasticity components of fecundity, stasis, progression, and retrogression. This result suggests that most juveniles stay in the same stage for a long period of time.

Table 19.1. Size-class discrimination and numbers of individuals (after Kitamura et al. 1997a)

	Size class	Transect 1	Transect 2	Age (years)[a]
Class 0	Seedlings	606	293	0
1	Juveniles (DGH[b]) < 5 mm	95	60	ca. 12
2	5 < Juveniles (DGH) < 10 mm	60	42	ca. 15–20
3	10 < Juveniles (DGH) < 25 mm	125	63	ca. 20–40
4	25 < Juveniles (DGH) < 60 mm	33	20	ca. 40
5	60 mm < Mature (DBH[c])	6	2	

[a]Age of suppressed juveniles was estimated by counting annual growth rings.

[b]Diameter at ground level.

[c]Diameter at breast height.

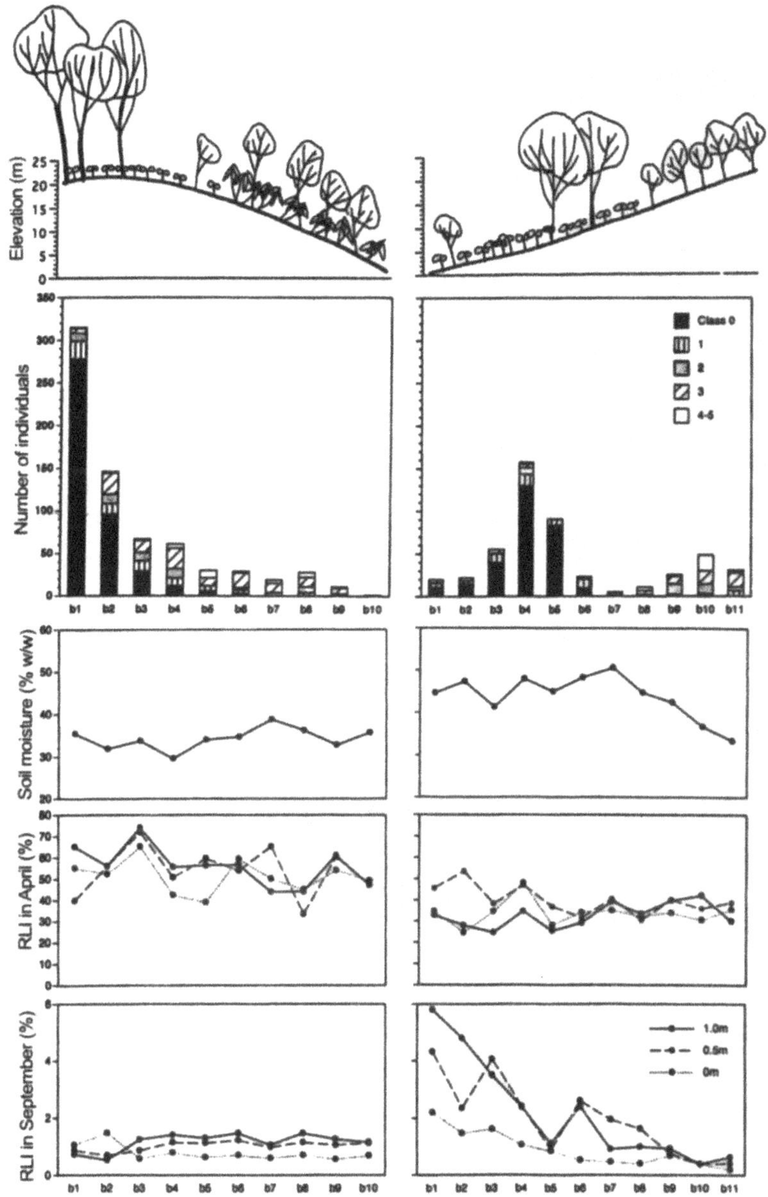

Fig. 19.1. Profiles of two *F. crenata* transects (after Kitamura et al. 1997a). *Left panels*, Transect 1, *Right panels*, Transect 2. The key in the *lower right panel* applies to the four panels depicting relative light intensity, and the distances represent height above ground level. *RLI*, relative light intensity

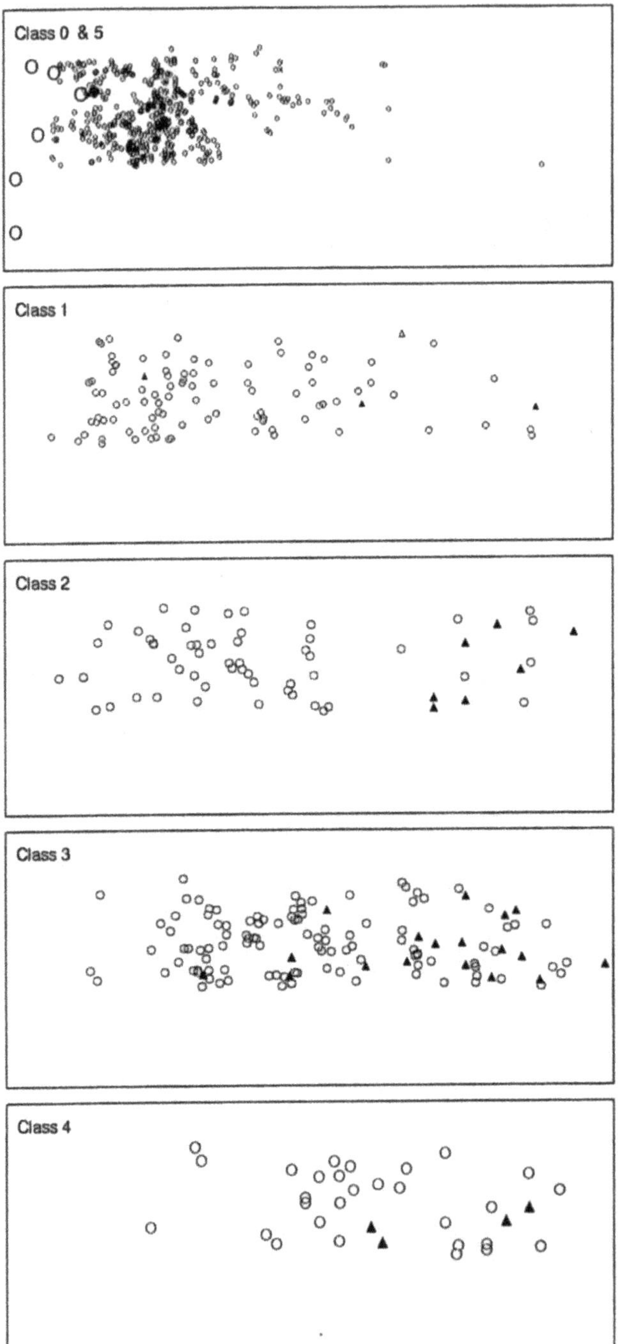

Fig. 19.2. Spatial distributions of *Adh* genotypes for each size class on Transect 1. The size of the rectangle is 25 × 60 m. In the *top panel*, *large circles* represent mother trees (class 5) and *small symbols* represent seedlings (class 0). Classes 0 to 5 correspond to the size classes in Table 19.1. *White triangles*, *black triangles*, *white circles*, and *crosses* represent genotype *aa*, *ab*, *bb*, and *bc*, respectively

Thus, from the genetic point of view, many genetic variations are accumulated and preserved during the juvenile stage, which is consistent with our result of genetic diversity within intermediate size classes. We also learned from the matrix analysis (see Chapter 13) that the death of an individual of mature stage is an important factor for the population persistence of *F. crenata*, and that these mature individuals affect the genetic component of the next generation via sexual reproduction.

The differences in genetic variability among size classes are also well manifested by the spatial genetic heterogeneity, since the spatial distribution of individuals differs among size classes. Most seedlings appear beneath the crowns of mother trees, reflecting extensive seed rains after mast-fruiting (cf. Chapter 13; Figs. 19.2 and 19.3), and most of the seedlings share genotypes that are very similar to those of the existing mother trees. In contrast, juveniles of intermediate size classes are established relatively far from current mother trees and they have most of the unique heterozygous genotypes (Figs. 19.2 and 19.3). Some of these juveniles are indeed considerably aged, being more than 30–70 years old, in spite of their small diameters. Such localized distributions of juveniles of the intermediate size classes with unique genotypes suggest the presence of gaps in the past, and that the juveniles are doubtless the offspring of mother trees that have died. It is apparent that new gaps have been colonized by the offspring of recent reproductive events, while older gaps are primarily composed of those trees that were produced and colonized by remote past reproductive events. We can recognize here typical regeneration dynamics in the forest community, called "gap dynamics" (see Chapter 12; Nakashizuka 1987; Yamamoto 1989). However, *F. crenata* is known to regenerate in small gaps, and small gaps are so frequently formed that the size of the genetic patchiness may thus reflect the small size of these regeneration gaps (Abe et al. 1995; see Chapters 7 and 12).

Limited seed dispersal, which is typical of a barochorous species such as *F. crenata* (Maeda 1988), is one of the factors that cause spatial heterogeneity. After seed dispersal, seedlings must establish themselves at safe sites for their subsequent successful growth to the next growth stage. The establishment process is greatly affected by various environmental factors of the habitat, including strong competitors with different growth forms on the forest floor such as dwarf bamboo, predation pressures by various animals, pest attacks, and so forth. However, the clustering of certain genotypes may be partly due to environmental heterogeneities of the habitats (see Chapter 3). The most significant feature is the localized spatial distribution of genotypic diversities, notably specific genotypes in juveniles of intermediate size classes. This is partly due to the availability of suitable safe sites for seedling establishment and further appropriate conditions for growth to the succeeding stage.

The presence of localized genetic structures may be evidence of the survival of reproductive individuals in the present or in the past. As has been noted, we have shown that the genetic diversity within local populations is not uniform, either chronologically or spatially. Such genetic substructuring is partly due to the limited gene flow, since wind-mediated pollen flow shows typical leptokurtic distributions (Levin and Kerster 1974; Hamrick 1982). Whether or not there is any allozyme-habitat association in the local populations of *F. crenata* in the OFR remains to be answered by future studies.

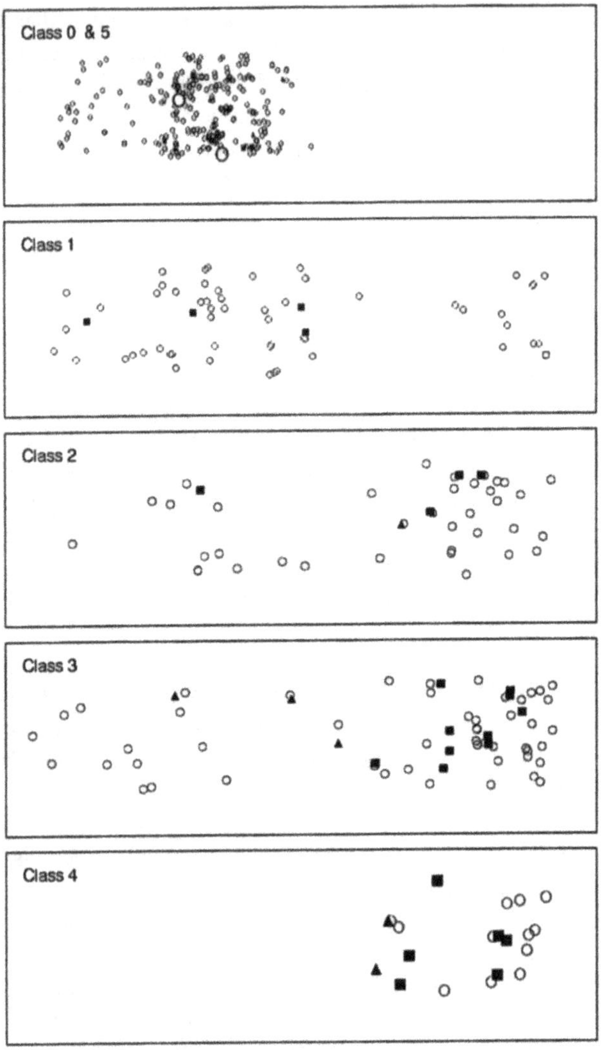

Fig. 19.3. Spatial distributions of *Got2* genotypes for each size class on Transect 2. The size of the rectangle is 20 × 60 m. In the *top panel*, *large circles* represent mother trees (class 5) and *small symbols* represent seedlings (class 0). Classes 0 to 5 correspond to the size classes in Table 19.1. *Black squares*, *white circles*, and *black triangles* represent genotype *ab*, *bb*, and *bc*, respectively

19.3 Development of Genetic Differentiation Within a Metapopulation

We mapped, sampled, and scored the genotypes of 138 mature *F. crenata* individuals in a 30-ha area, including the two intensively sampled transects (Fig. 19.4). Pa-

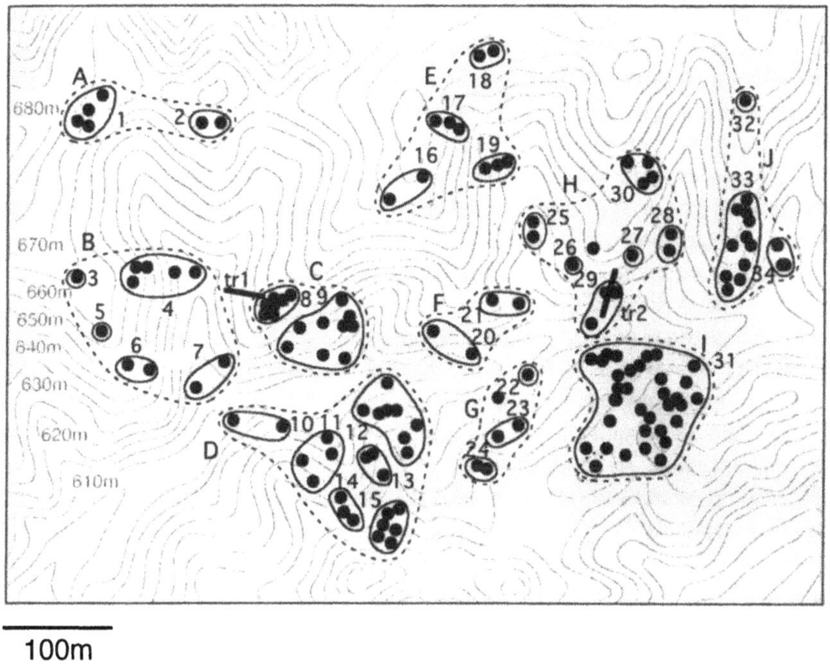

Fig. 19.4. Map of reproductive individuals of *F. crenata* in the 30-ha area, and groups of patch populations (*enclosed* by *solid lines* and *numbered*) and local populations (*enclosed* by *dotted lines* and represented by *letters*). North is to the *left*

rental candidates for seedlings and juveniles in Transects 1 and 2 were estimated using possible pairs of multilocus genotypes among mature trees. With regard to the distance between the seedlings or juveniles and the two parental candidates, the nearer mature trees were recognized as seed donors and the others as pollen donors. In both transects, 95% of the seedlings and juveniles were assumed to have seed donors within the transect. Pollen donors ranged from being the nearest neighbor to being ca. 500 m away, and most of them occurred within the same area, that is, in the same valley. The genotypes of about 3% of the seedlings and juveniles could not be explained by any combination of genotypes among the mature trees sampled. There are two possible reasons for the occurrence of unexplained genotypes: (i) some juveniles may have been recruited by parental trees that no longer exist today; and (ii) the gene flow may have come from outside of the sampled area.

In order to recognize the effective mating area for *F. crenata* in the OFR, the variance of distances of parental candidates from juveniles and seedlings that could explain 86.5% of their occurrence (Gliddon et al. 1987) was used to calculate the genetic neighborhood area and effective population size (Wright 1943, 1951; Crawford 1984; Silvertown and Lovett Doust 1993). Since *F. crenata* is assumed to be highly xenogamous, 0.6 to 1.0 outcrossing rates were used for the calculations. As a result, the gene flow distance was 11 to 13 m, the genetic neighborhood area was 3,000 to 4,000 m^2, and the effective population size was only 1.2 to 2.0 reproductive trees per neighborhood area. The effective population size is unexpectedly small, which may be due to the exceedingly patchy distribution of *F. crenata* in the OFR and which suggests the occurrence of hierarchical genetic substructurings.

The possible genetic substructurings can be analyzed in view of the metapopulation concept (Levins 1969). We considered the sampled 30-ha area, which could arbitrarily be divided into ten local populations and 34 patch populations (Fig. 19.4), as a metapopulation, with current structures (with respect both to size class and genetic substructure) doubtless reflecting changes due to natural "gap" formation and artificial disturbances in the area. The background concept of the patch population was based on (i) individuals occurring within the assumed genetic neighborhood area, ca. 50 m in diameter, (ii) the aggregated distribution of individuals, and (iii) the topography and community structures of the area where patchy distribution occurs. Patch populations were grouped into a local population on the basis of the following criteria: (i) adjacent groups of patch populations; (ii) estimated ranges of gene flow by means of pollen or seed distribution; and (iii) the topographic or community structure. The genetic substructurings were shown to be higher than expected by *Gst* (genetic differentiation among populations) values among both patch populations (*Gst* = 0.234) and local populations (*Gst* = 0.094).

In the OFR, strong artificial disturbances by selective cutting of canopy trees (mainly *Fagus* and *Quercus*) and grazing accompanied by forest fires in the recent past (see Chapter 4) may have caused considerable fragmentation of the *F. crenata* population. As a consequence, the OFR was split into a number of local and patch populations with different genetic components. Topographic complexity is another important background factor in the OFR. Topographically complex combinations of ridges and valleys, forming north-, south-, east-, or west-facing slopes, obviously hinder gene flow among local populations via pollen and seed, resulting in an exceedingly localized genetic substructuring in small patch and local populations.

The topographic complexity also makes the microenvironmental regimes very complex. For example, diurnal as well as seasonal distributions of sun flecks and light intensity on the forest floor (see Chapter 12) and various edaphic factors on a fine scale (see Chapters 3 and 8) strongly determine the fate of new colonizers with specific genetic characteristics. Seeds of *F. crenata* are spread primarily by barochory, but occasionally they are spread by small mammals such as mice. This is clearly shown by the remarkable caching of seedlings in the succeeding season after the mast-fruiting (Nakamura 1995; Miguchi 1996). The ranges of such secondary dispersal are reported to be approximately 4 m (Miguchi 1996), but propagule dispersal would also be affected extensively by the topography of the habitat.

Grazing was common in the Abukuma mountain range until the early 1980s (see Chapter 4). Due to severe grazing pressures, the dwarf bamboo thickets on the gentle slopes were eliminated in a relatively short time, and thus the dwarf bamboo layer is lacking in the OFR except on steep slopes. The dense coverage of dwarf bamboo thickets usually prevents the regeneration of other plants. Lack of these dense bamboo thickets may be one of the reasons why regeneration of *F. crenata* populations in the OFR was faster than in the forests of adjacent regions.

In addition, selective cutting of *F. crenata*, *F. japonica*, and other predominant oaks for mushroom cultivation (see Chapter 4) was obviously responsible for the fragmentation of the *F. crenata* populations; after selective cutting was done, deciduous oaks such as *Quercus serrata* and *Q. crispula* became predominant. This has no doubt further accelerated fragmentation of the *F. crenata* population in the OFR into many local and patch populations of various sizes.

Based on the results of spatial analyses, we found a genetically patchy distribution of *F. crenata* and thus a substructuring in a range of 10–20 m to be clearly present in the *F. crenata* population in the OFR. It can be assumed by the available evidence that local populations in the OFR already had localized genetic patterns before fragmentation of the population. If more fragmentation occurs within the OFR population, the genetic substructurings will occur to an even greater extent. Further fragmentation and isolation due to extinction will proceed locally, resulting in increasing genetic differentiation among patch populations and loss of genetic diversity.

References

Abe S, Masaki T, Nakashizuka T (1995) Factors influencing sapling composition in canopy gaps of a temperate deciduous forest. Vegetatio 120:21–32

Alvarez-Buylla ER, Garay AA (1994) Population genetic structure of *Cecropia obtusifolia*, a tropical pioneer tree species. Evolution 48:437–453

Argyres AZ, Schmitt J (1991) Microgeographic genetic structure of morphological and life history traits in a natural population of *Impatiens capensis*. Evolution 45:178–189

Clegg MT, Kahler AL, Allard RW (1978) Estimation of life cycle components of selection in an experimental plant population. Genetics 89:765–792

Crawford TJ (1984) The estimation of neighbourhood parameters for plant populations. Heredity 52:273–283

Dewey SE, Heywood JS (1988) Spatial genetic structure in a population of *Psychotria nervosa*. I. Distribution of genotype. Evolution 42:834–838

Epperson BK (1989) Spatial patterns of genetic variation within plant populations. In: Brown AHD, Clegg MT, Kahler AL, Wier BS (eds), Plant population genetics, breeding, and genetic resources. Sinauer, Sunderland, MA, pp 229–253

Epperson BK, Allard RW (1989) Spatial autocorrelation analysis of the distribution of genotypes within populations of lodgepole pine. Genetics 121:369–377

Epperson BK, Clegg MT (1986) Spatial-autocorrelation analysis of flower color polymorphisms within substructured populations of morning glory (*Ipomoea purpurea*). Am Nat 128:840–858

Geburek T, Tripp-Knowles P (1994) Genetic architecture in bur oak, *Quercus macrocarpa* (Fagaceae), inferred by means of spatial autocorrelation analysis. Plant Syst Evol 189:63–74

Gliddon C, Belhassen E, Gouyon P-H (1987) Genetic neighbourhoods in plants with diverse systems of mating and different patterns of growth. Heredity 59:29–32

Hamrick JL (1982) Plant population genetics and evolution. Am J Bot 69:1685–1693

Kawano S, Kitamura K (1997) Demographic genetics of the Japanese beech, *Fagus crenata*, at Ogawa Forest Preserve, Ibaraki, central Honshu, Japan. III. Population dynamics and genetic substructuring within a metapopulation. Plant Species Biol 12:157–177

Kitamura K, Shimada K, Nakashima K, Kawano S (1997a) Demographic genetics of the Japanese beech, *Fagus crenata*, at Ogawa Forest Preserve, Ibaraki, central Honshu, Japan. I. Spatial genetic substructuring in local populations. Plant Species Biol 12:107–135

Kitamura K, Shimada K, Nakashima K, Kawano S (1997b) Demographic genetics of the Japanese beech, *Fagus crenata*, at Ogawa Forest Preserve, Ibaraki, central Honshu, Japan. II. Genetic substructuring among size-classes in local populations. Plant Species Biol 12:137–155

Knowles P (1991) Spatial genetic structure within two natural stands of black spruce (*Picea mariana* (MILL.) B. S. P.). Silvae Genet 40:13–19

Leonardi S, Raddi S, Borghetti M (1996) Spatial autocorrelation of allozyme traits in a Norway spruce (*Picea abies*) population. Can J For Res 26:63–71

Levin DA, Kerster HW (1974) Gene flow in seed plants. Evol Biol 7: 139–220

Levins R (1969) Some demographic and genetic consequences of environmental heterogeneity for biological control. Bull Entomol Soc Am 15:237–240

Maeda T (1988) Studies on natural regeneration of beech (*Fagus crenata* BLUME) (in Japanese with English summary). Special Bull Agr Utsunomiya Univ 46:1–79

Miguchi H (1996) Dynamics of beech forest from the view point of rodents ecology—ecological interactions of the regeneration characteristics of *Fagus crenata* and rodents (in Japanese with English summary). Jpn J Ecol 46:185–189

Nakamura S (1995) Genetic variations and population structures of *Fagus crenata* in southwestern Japan. M.S. Thesis, Dept Botany, Graduate School of Science, Kyoto University

Nakashizuka T (1987) Regeneration dynamics of beech forests in Japan. Vegetatio 69:169–175

Perry DJ, Knowles P (1991) Spatial genetic structure within three sugar maple (*Acer saccharum* MARSH.) stands. Heredity 66:137–142

Schaal BA, Levin DA (1976) The demographic genetics of *Liatris cylindracea* MICHX. (Compositae). Am Nat 110:191–206

Schnabel A, Hamrick JL (1990) Organization of genetic diversity within and among populations of *Gleditsia triacanthos* (Leguminosae). Am J Bot 77:1060–1069

Schnabel A, Laushman RH, Hamrick JL (1991) Comparative genetic structure of two cooccurring tree species, *Maclura pomifera* (Moraceae) and *Gleditsia triacanthos* (Leguminosae). Heredity 67:357–364

Schoen EJ, Latta RG (1989) Spatial autocorrelation of genotypes in populations of *Impatiens pallida* and *Impatiens capensis*. Heredity 63:181–189

Shapcott A (1995) The spatial genetic structure in natural populations of the Australian temperate rainforest tree *Atherosperma moschatum* (LABILL.) (Monimiaceae). Heredity 74:28–38

Silvertown J, Lovett Doust J (1993) Introduction to plant population biology, 3rd ed. Blackwell, London

Williams CF (1994) Genetic consequences of seed dispersal in three sympatric forest herbs. II. Microspatial genetic structure within populations. Evolution 48:1959–1972

Wright S (1943) Isolation by distance. Genetics 28:114–138

Wright S (1951) The genetical structure of populations. Ann Eugenics 15:323–354

Xie CY, Knowles P (1991) Spatial genetic substructure within natural populations of jack pine (*Pinus banksiana*). Can J Bot 69:547–551

Yamamoto S (1989) Gap dynamics in climax *Fagus crenata* forests. Bot Mag Tokyo 102:93–114

20 Gene Flow Analysis of *Magnolia obovata* Thunb. Using Highly Variable Microsatellite Markers

YUJI ISAGI and TATSUO KANAZASHI

20.1 *Magnolia obovata*: A Subdominant Canopy Tree Species

In Ogawa Forest Reserve (OFR), the tree community consists of a few dominant tree species, such as *Quercus serrata, Fagus japonica,* and *F. crenata*, and many subdominant tree species, such as *Acer mono*, *Betula grossa*, *Carpinus laxiflora, Swida controversa*, *Ostrya japonica*, *Prunus verecunda*, *Sorbus japonica,* and *Magnolia obovata* (Masaki et al. 1992; also see Chapter 6). These latter species seldom occur as dominants in forest ecosystems, and their standing densities of adult individuals are low: a few to several trees per hectare. However, the populations of these tree species persist and have adequate regeneration, and thus contribute to the forest species diversity.

Magnolia obovata is a large deciduous tree of Japan and China that reaches 30 m in height. Although the species is common in mixed broad-leaved forests in Japan, it rarely accounts for a large proportion of the density in forest ecosystems, and adult densities of a few trees per hectare are typical. The flower of *M. obovata* is large (15–20 cm in diameter) and protogynous, and it closes its petals between the female and male periods. Although the anthesis of each flower continues for 3 to 4 days, the flowering period of individual trees continues for up to 40 days (Kikuzawa and Mizui 1990). It has been observed that *Magnolia* species have specialized pollination systems that rely on beetles (Thien 1974; Kikuzawa and Mizui 1990; Ishida 1996; Dieringer et al. 1999), which are thought to be less efficient pollinators than bees or butterflies (Thien 1974; Ramsey 1988).

Characterization of plant mating systems and the genetic structure of populations is essential for a thorough understanding of the factors that affect reproductive success, and also provides essential data for improving management and conservation of natural populations. Until recently, however, analyses of mating systems and gene flow by completely assigning parentage of the offspring have rarely used highly informative genetic markers. This is true for *M. obovata* as well as for most wild plant species.

Ecological Studies, Vol. 158
Nakashizuka, Matsumoto (eds.) Diversity and Interaction in a Temperate Forest Community: Ogawa Forest Reserve of Japan

20.2 Genetic Markers for Research on Plant Regeneration Processes

Because of the sessile nature of plants, gene transfer is conducted by the dispersal of pollen and seeds. The patterns and amounts of pollen and seed dispersal also affect regeneration processes and genetic structures of plant populations (Schaal 1980; Levin 1981; Hamrick and Murawski 1990; Ellstrand 1992; Hamrick et al. 1992). Gene flow estimates vary widely among plant species, and may be partially determined by differences in the dispersal characteristics of the species (Ellstrand 1992; Govindaraju 1988a, b; Hamrick et al. 1992). The type of pollen vector can strongly influence pollen dispersal, especially for living vectors with different foraging patterns (Schmitt 1980; Waser 1982). However, pollen-mediated gene flow has mostly been inferred using indirect methods; for instance, gene flow has been inferred by observing the flight distance of pollinators (Schmitt 1980) or by using some kind of dye as a pollen analogue (Campbell and Waser 1989). Based on the results obtained using these indirect methods, pollen movement by insect pollinators is generally considered to be quite leptokurtic; that is, most pollen is dispersed a short distance from the source (Levin and Kerster 1974; Campbell 1991; Ellstrand 1992).

Similarly, the range of seed dispersal distances by animal vectors or gravity dispersal have been reported to be relatively limited: mostly within several tens of meters from the source (Sork 1984; Jensen 1985; Yasuda et al. 1991; Rasmussen and Brødsgaad 1992; Iida 1996; Yumoto et al. 1998; also see Chapters 10 and 19).

Despite the fact that pollen and seed movement is commonly leptokurtic, several studies using allozyme markers revealed higher levels of gene flow than expected from measurements of pollen and seed dispersal (Karron et al. 1995; Campbell 1991). Allozyme markers are codominantly inherited, and they have been widely used for the measurement of genetic variation in populations, the genetic distance between populations, the estimation of self-pollination rates, and so on. However, the number of loci and alleles provided by allozyme markers is usually insufficient for direct measurements of gene flow or analysis of parentage. Therefore, in parentage-analysis studies in wild populations using allozyme markers, parentage has been only partially assigned (Krauss 1994; Smouse and Meagher 1994; Milligan and McMurray 1993) or some kind of statistical procedure has been applied to compensate for the insufficiency (Meagher 1986; Devlin et al. 1988; Roeder et al. 1989; Devlin and Ellstrand 1990; Smouse and Meagher 1994). In studies of gene flow that used allozyme markers, the estimates often have been made using, for example, G_{ST} statistics (genetic differentiation among populations, Nei 1973), with information on allele frequencies. Because these indirect methods assume evolutionary equilibrium between gene flow and genetic drift, the estimates of gene flow from this method tend to be higher than actual rates (Ellstrand 1992).

In order to investigate a mating system in detail, conduct parentage analysis of the offspring, precisely estimate gene flow, and learn what happens to seeds after their dispersal and germination, we need another class of genetic markers that is more precise and informative. Microsatellites, which are also called simple sequence repeats (SSRs), are tandem repeats of very short (one to six bases) motifs of nucle-

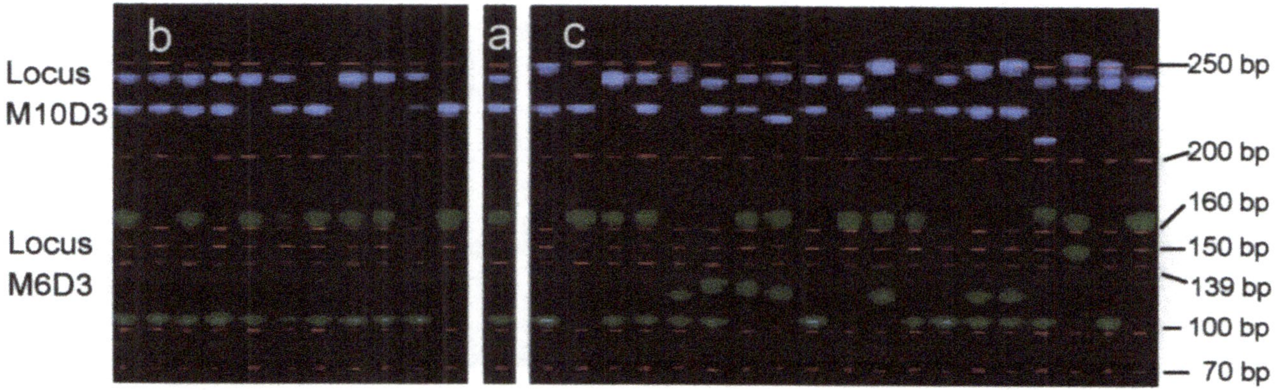

Fig. 20.1. Gel image of the microsatellite markers for a seed parent (tree no. #PD294) and its seeds. *Red, blue*, and *green bands* represent *size markers*, *locus M10D3*, and *locus M6D3*, respectively. **a** Genotype of the seed parent. **b** Genotypes of the seeds from an aggregate fruit. **c** Genotypes of seeds from another aggregate fruit. The genotypes of the seeds shown in panel **b** were all considered to be products of self-pollination because all alleles found at eight loci were common to those of the seed parent. In contrast, the genotypes of some seeds shown in panel **c** had alleles that were not present in the seed parent at one or more loci, and hence were considered to have originated from cross-pollination. Note that the pollination patterns of flowers collected in the same year from a single tree varies noticeably

otides, and are ideal for such studies because of their codominant inheritance, the presence of many highly polymorphic loci, and the existence of unequivocal alleles identified by differences in size (Fig. 20.1).

However, one drawback of using microsatellite markers is that the DNA sequences that flank the repeat structure must be known beforehand in order to design polymerase chain reaction (PCR) primers capable of amplifying the region. In comparison with using allozyme markers, much more investment in time and cost is necessary to develop microsatellite markers for the development of genomic libraries, screening, and DNA sequencing. Recently, microsatellite markers have been developed for a variety of wild plant species, and studies using microsatellite markers have revealed gene-flow phenomena in natural plant populations (e.g., Chase et al. 1996; Dow and Ashley 1996; Streiff et al. 1998; Miyazaki and Isagi 2000; Kameyama et al. 2001). We developed 11 microsatellite markers for *M. obovata* (Isagi et al. 1999) and applied some of them to studies of populations of this species in the OFR (Isagi et al. 2000).

20.3 Research Site and Development of Microsatellite Markers

Field research was conducted in two study plots (labeled A and B) in the OFR (Fig. 20.2a). Plot A was a 69-ha watershed (see Fig. 4.2, Chapter 4) in which all of the adult trees of *M. obovata* were mapped and their diameters at breast height (DBH) were measured. Plot B occupied an area of 6 ha (200 × 300 m) near the center of plot A (Fig. 20.2a). Leaf samples of *M. obovata* were obtained from all adult trees (83 trees) in plot A, and from 91 saplings in plot B, and were stored at –70 °C prior to DNA extraction.

For the construction of genomic libraries, crude genomic DNA was extracted using the CTAB method (Milligan 1992), and then digested with a restriction endonuclease, *Mbo*I. DNA fragments ranging in length from 400 to 600 bp were ligated to pUC19 that had been digested with *Bam*HI, and the plasmid DNA was introduced into JM109 *Escherichia coli* cells.

Recombinant colonies formed on agar plates were lifted with nylon membranes (Hybond-N^+, Amersham Pharmacia Biotech, Uppsala, Sweden), and screened using two synthetic oligonucleotides, $(GA)_{20}$ and $(CA)_{20}$, labeled by DIG oligonucleotide tailing according to the supplier's instructions (Roche Diagnostics, Tokyo, Japan). Plasmid DNA from positive colonies was sequenced with an ABI 377 autosequencer (Applied Biosystems, Foster, CA, USA), and PCR primers allowing amplification of the microsatellite were designed from the flanking DNA sequences.

PCR amplifications of the microsatellite loci were performed using a thermal cycler under the following conditions: initial denaturing at 94 °C for 9 min followed by 30 cycles of denaturation at 94 °C for 30 s, annealing for 30 s, and extension at 72 °C for 1 min, with a final incubation at 72 °C for 7 min after the last cycle. The volume of the reaction mixture was 10 µL, and the mixture contained 10 ng of DNA from *M. obovata*, 5 pmol of each primer labeled with fluorescent phosphoramidites

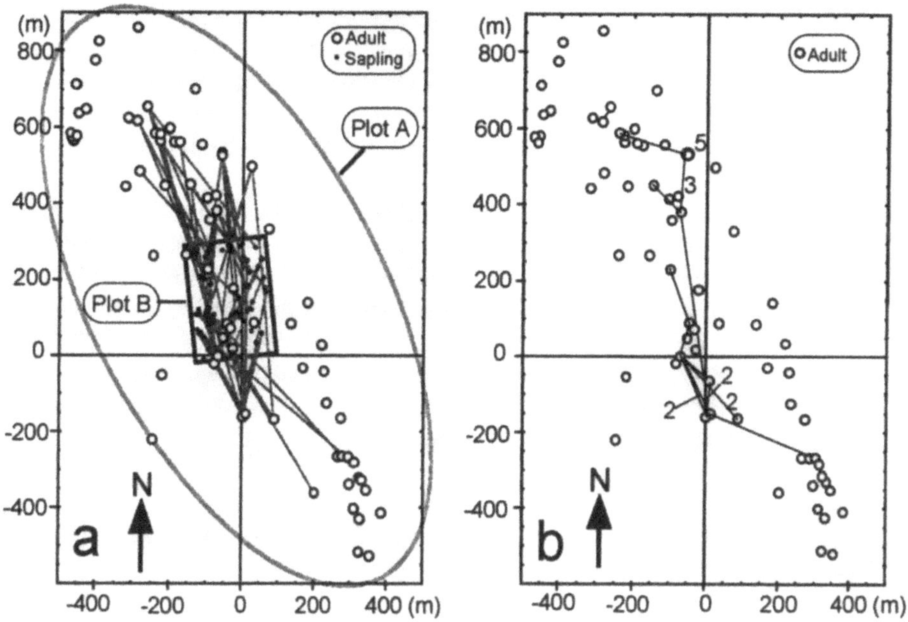

Fig. 20.2. Research plots used for parentage analysis of *Magnolia obovata*. **a** Locations of adults and saplings. Lines were drawn between parents in *plot A* and their offspring in *plot B*. **b** Locations of effective pollination estimated from the results of the parentage analysis using microsatellite markers. *Lines* were drawn to connect the two parents of each sapling in *plot B*, and represent the distance of effective pollination between pollen donors and seed parents. *Numbers* beside the *lines* indicate the number of events that occurred between the pairs

Table 20.1. Basic population statistics for eight microsatellite loci in the present population of *M. obovata*

Locus	Number of alleles	*He*[a]	Exclusionary power		Frequency of null alleles
			First parent	Second parent	
M6D1	30	0.95	0.792	0.883	–0.012
M6D3	29	0.95	0.806	0.893	–0.015
M6D4	36	0.91	0.695	0.819	–0.007
M10D3	21	0.91	0.685	0.812	–0.027
M10D6	8	0.84	0.495	0.667	–0.032
M10D8	21	0.91	0.674	0.805	–0.007
M15D5	6	0.68	0.256	0.423	0.012
M17D5	9	0.84	0.514	0.683	0.038
Mean	20.0	0.87			
Total	160		0.9998[b]	> 0.9999[b]	

[a]He, expected heterozygosity.

[b]Cumulative power.

(TET or 6-FAM), 0.25 U of *Taq* polymerase (Ampli TaqGold, Applied Biosystems), 200 μM of each dNTP, 1.5 mM of $MgCl_2$, 10 mM of Tris-HCl (pH 8.3), and 50 mM of KCl. PCR products were resolved on a 6% denaturing polyacrylamide gel, and the sizes were determined by automated fluorescent-scanning detection with the autosequencer (ABI 377, Applied Biosystems) and GeneScan analysis software (Applied Biosystems).

We used 8 out of the 11 polymorphic loci developed by Isagi et al. (1999). Each locus applied to the study population was highly variable; the number of alleles per locus ranged from 6 (locus M15D5) to 36 (locus M6D4), with a total of 160 alleles (Table 20.1). The average expected heterozygosity of these loci was 0.87, and the cumulative exclusion probabilities for the first and second parents were 0.9993 and more than 0.9999, respectively. The frequency of null alleles, which cannot be amplified by PCR because of mutations in primer regions, was considered low enough for each locus to permit a parentage analysis.

20.4 Parentage Analysis for *M. obovata* Using Microsatellite Markers

The parentage of saplings was assigned by comparing all possible haplotypes of each sapling with those of the 83 adult trees collected from plot A (Isagi et al. 2000). Of the 91 saplings found in plot B, 55 had at least one parent in plot A, whereas 36 had no parents within the plot (i.e., the watershed). Of the saplings with at least one parent in plot A, 24 had two parents with exact matches in the plot, versus 31 with only one parent in the plot. The data also revealed that no sapling in plot B was the product of self-pollination. Out of the 182 sapling genomes analyzed (= 91 saplings × 2), 103 (57%) were from outside plot A. This indicates a large amount of gene flow into the watershed.

The distances between parent trees and the offspring were large; these distances, which represent seed dispersal from maternal parents and pollen movement plus seed dispersal for paternal parents, ranged from 32 to 563 m, with an average of 264 m. The average value probably represents an underestimate of the true dispersal distance because 57% of the genomes originated outside plot A. Moreover, the distance between random pairs made up of an adult in plot A and a sapling in plot B was significantly longer (*U*-test, $P < 0.0001$) than the distance between offspring and adult trees (i.e., between nonrandom pairs). This indicates that near adults contribute more as gene sources to the saplings than do more distant adults.

The average DBH of adults having progeny in plot B was significantly larger (*U*-test, $P < 0.0001$) than that of adults not having their progeny in plot B, indicating that trees of larger size in plot A contributed more as pollen donors or seed parents to saplings in plot B.

It is notable that as much as 40% (= 36/91) of the saplings in plot B had no parent within the 69-ha research plot (plot A). One possibility is that saplings without parents in plot A originated from seeds pollinated by unknown parents outside plot A, and then dispersed into the plot from outside. The diaspores of *M. obovata* have a red, edible flesh, and the seeds are dispersed in the feces of birds. Generally, the distance of seed dispersal by birds has been considered to be quite long (up to 4 km, Johnson and Adkisson 1985), and the large ratio of saplings without a parent in plot A could certainly reflect active dispersal of the seeds of this species by birds. The range of seed dispersal by birds seems to reach more than several hundred meters, which is noticeably larger than the distances reported for seed dispersal by gravity or for mammals such as mice and monkeys; the latter dispersal ranges have been reported to be within 100 m from the source (Sork 1984; Jensen 1985; Iida 1996; Dow and Ashley 1996; Yumoto et al. 1998; Yasuda et al. 1991). Long-distance seed dispersal by birds was also implied for *Swida controversa* in Ogawa Forest Reserve; density of bird-dispersed fruits of *Swida controversa* decreased more slowly with increasing distance from the source than that of directly fallen fruits of the same species (see Chapter 10).

In one microsatellite analysis conducted for *Quercus macrocarpa*, Dow and Ashley (1996) estimated the proportion of saplings that had neither parent within the research plot (ca. 5 ha) to be 14%. Despite the much larger plot size (69 ha) for the present *M. obovata* population in the OFR, the proportion of the saplings that had no parent within the research plot was much larger (40%) than that determined for *Q. macrocarpa*. However, Kitamura et al. (see Chapter 19) found that about 97% of seedlings and saplings of *Fagus crenata* had their putative parents within the OFR. These contrasting differences among tree species in the amount of gene flow possibly reflects the difference between the seed dispersal mechanisms of each species: dispersal by gravity or small mammals (*Quercus macrocarpa* and *Fagus crenata*) and by birds (*M. obovata* and *Swida controversa*).

20.5 Pollen Movement

Although we found that 24 saplings had both their parents in plot A, it was not feasible to distinguish which parent was the pollen donor and which was the seed parent. Therefore, we could not determine the distance of seed dispersal, but we could infer the distance of pollen exchange, because the distance between the two parents represents the effective pollination distance (Fig. 20.2b). The distance of effective pollination ranged from 3.2 to 540 m, with an average of 131 m. The average distance between parents and their offspring (264 m; Fig. 20.2a) was significantly longer (U-test, $P < 0.0001$) than the effective pollination distance determined in the present study (Fig. 20.2b), because the former comprises both pollen movement and seed dispersal.

The average distance to the nearest neighbor for each adult tree of *M. obovata* was relatively short (44 m). The average distance of effective pollination was significantly larger (U-test, $P = 0.001$) than the distances between nearest neighbors for each adult tree, and significantly smaller (U-test, $P < 0.0001$) than that between random pairs of adult trees in plot A (562 m). This indicates that although pollination occurred between adults located closer together than the average distance between adult trees within the watershed, it did not always occur between nearest neighbors.

Although the dispersal of pollen from its source has been reported to show a leptokurtic distribution for some species (Levin and Kerster 1974; Harper 1977; Ellstrand 1992; Webb 1998), the frequency distribution for the distances of effective pollination (based on the distances between parents estimated using microsatellite markers) was not leptokurtic in the current study.

Using microsatellite markers, Chase et al. (1996) also reported that most pollination did not occur between nearest neighbors for a tropical canopy tree, *Pithecellobium elegans*. They inferred that this phenomenon was due to annual variations in flowering that occurred among adult trees: because flowering was episodic and varied among years and individuals, it is possible that only some of the trees in the population contributed to reproduction in a given year. Similarly, pollination of *M. obovata* did not always occur between nearest neighbors; in fact, more than 70% of the pollination was carried out between trees other than nearest neighbors. One possible reason for this pollination occurring between adults located far away from each other may be the difference in phenology of individual flowers. Although *M. obovata* has a long flowering period of up to 40 days (Kikuzawa and Mizui 1990), the longevity of each individual flower is only a few days, and its reproductive period consists of an initial female period followed by a male period. Therefore, each adult tree may switch between male, female, and bisexual phases, and not all of the trees in a population are likely to serve as pollen donors at the same time. This behavior would tend to increase the average distance between pollen donors and seed parents.

The pattern and distance of pollen dispersal are strongly affected by differences in pollen vectors and (for living vectors) their patterns of behavior (Govindaraju 1988a, b; Hamrick et al. 1992; Schmitt 1980; Waser 1982; Webb 1998). The pollinators of *M. obovata* are beetles (Kikuzawa and Mizui 1990), which are thought to be less efficient pollen vectors (Ramsey 1988) than bees and butterflies. However, the distance of pollen movement in the present stand was quite large (an average of 131 m, with a maximum of at least 540 m) (Fig. 20.2b). Despite the putative specialization of *Magnolia* for beetle pollination (Thien 1974; Kikuzawa and Mizui 1990; Ishida 1996; Dieringer et al. 1999), Tanaka and Yahara (1988) observed that various other insects, such as butterflies and bumblebees, were pollinators of *M. obovata* at a site in central Japan. Such effective pollinators might also explain the long-distance pollen movement estimated for the population of *M. obovata* in the present study.

About one third of the saplings (= 31/91) in plot B had only one of their two parents within plot A. These saplings probably grew from seeds fertilized with pollen from adults growing outside plot A. Hence, the average pollen movement distance estimated based on the distance between saplings with both parents within plot A must underestimate the maximum distance of pollen exchange, which may well be more than 540 m.

20.6 Factors that Increase the Amount of Gene Transfer for *M. obovata*

In natural plant populations, it has been often observed that the actual gene flow estimated using genetic markers is much larger than the expected rate based on an assumption of leptokurtic dispersal of pollen and seeds. There are several possible explanations for this difference: (1) the cumulative contribution of pollen dispersal may be leptokurtic, but with a long-tailed frequency distribution (Adams 1992; Ellstrand 1992); (2) underestimation of the extent of gene dispersal as a result of neglecting carry-over of pollen on vectors (Schaal 1980; Levin 1981); and (3) some kind of selective mechanism may be at work, by which trees preferentially choose pollen from distant sources over the more abundant pollen from nearer neighbors by means of pollen competition, selective abortion, embryo competition, etc. (Levin 1981; Waser 1993; Dow and Ashley 1996, 1998).

If a population is genetically structured, and inbreeding or outbreeding depression occurs based on the genetic relatedness of each adult tree, some kind of selection could be acting on pollen grains. It is known that *M. obovata* suffers greatly from inbreeding depression (Ishida and Nakamura 1997), and consequently, pollination between less-related (more spatially distant) trees might be favored and thereby counteract leptokurtic pollen dispersal. In addition to such genetic factors, long-distance seed dissemination by birds and limitations on the acceptance of pollen caused by temporal differences in anthesis among individual flowers also seem to favor a high degree of gene flow and provide chances to exchange genes between less related or distant trees. Although the population of *M. obovata* in the present study grew in a physically distinct landscape component (i.e., within a watershed), the amount of gene flow from outside the watershed was probably sufficient to prevent genetic differentiation by means of genetic drift because even a small amount of gene flow can prevent population differentiation (Hartl and Clark 1988; Hamrick et al. 1992).

20.7 Further Studies to Be Conducted Using Microsatellite Markers

We found that no sapling in plot B was the product of self-pollination by adult trees in plot A. However, genotype analysis of the seeds from an adult tree in plot A indicated that many seeds were the products of self-pollination (Fig. 20.1b). What kind of mechanism caused such genetic differences between the seed and sapling populations? Because the alleles of microsatellite loci were identified through PCR amplification, it is possible to score the genotype of each pollen grain on pollinators and use this information to infer the pollen donors. By comparing the genetic structure between cohorts of (1) pollen grains attached to pollinators found visiting pistils, (2) embryos, (3) seeds, and (4) saplings, we would be able to evaluate the effect of pollen competition, embryo abortion, and inbreeding depression on the genetic composition of each stage. This work would also shed some light on the mechanism by which pollen from far away can achieve effective pollination in spite of the supposedly leptokurtic nature of pollen dispersal.

Microsatellite analysis revealed a large degree of gene flow across the watershed and effective long-distance pollination (Fig. 20.2) in the present population of *M. obovata*. Chase et al. (1996) also analyzed the range of pollen dispersal for a tropical tree species, *Pithecellobium elegans*, and found that the average effective distance of pollen exchange was almost equivalent to that in the present study: 142 m, with a maximum of 350 m. As was the case for *M. obovata*, adult trees of *P. elegans* occur at low densities, and it is possible that pollination in tree species that are found at low densities occurs regularly over a wide range of distances. However, the universality of long-distance pollen exchange for such tree species remains to be tested for trees with different life-cycle characteristics, pollinator availability, stand age, tree density, effective population size, and mating systems, among other factors.

Several authors have reported that gene flow in natural plant populations is idiosyncratic; that is, it ranges from very low to high, varying not only among species but also among populations and individuals, and over the course of a season (Ellstrand and Marshall 1985; Slatkin 1987; Hamrick 1987; Ellstrand et al. 1989; Devlin and Ellstrand 1990). The range of such idiosyncratic within-species variation is sometimes more than an order of magnitude (Ellstrand 1992). Fig. 20.1 shows that the pattern of pollination differed between flowers on a single adult tree of *M. obovata* within a single year: the band patterns in Fig. 20.1b indicate self-pollination for entire seeds, whereas those in Fig. 20.1c indicate that at least some seeds were the result of outcrossing.

We also analyzed the paternity of seeds from several flowers of adult trees and found that individual flowers differed markedly in the rate of self-pollination, the distance of pollen movement, and the composition of the pollen donors (Y. Isagi et al. unpublished data). These results suggest that the pollination pattern of *M. obovata* is idiosyncratic and varies between flowers even on a single adult tree within a single year. Kikuzawa and Mizui (1990) observed the flowering and fruiting phenology of this species for 3 years and found that weather conditions affected the activity of pollinators, and that this in turn caused substantial annual differences in fruit production.

Species-specific characteristics might, of course, determine the pattern and amount of gene flow, but conditions such as annual weather fluctuations, small differences in the timing of anthesis of each flower, variable behavior of pollinators, the degree of inbreeding depression, and the degree of cross-incompatibility also affect the pattern and amount of gene flow dramatically. We lack sufficient information on the extent to which factors such as differences in genetic structure among growth stages, a high degree of gene flow across the watershed, and gene flow idiosyncratic to each flower, would affect the dynamics and maintenance of communities as a whole. In order to fully understand the processes of forest dynamics and maintenance, and to develop better management and conservation procedures, we need to monitor the forest system for a long time over a large area by examining highly informative and precise genetic markers such as microsatellites.

References

Adams WT (1992) Gene dispersal within forest tree populations. New For 6:217–240

Campbell DR (1991) Comparing pollen dispersal and gene flow in a natural population. Evolution 45:1965–1968

Campbell DR, Waser NM (1989) Variation in pollen flow within and among populations of *Ipomopsis aggregata*. Evolution 43:1444–1455

Chase MR, Moller C, Kesseli R, Bawa KS (1996) Distant gene flow in tropical trees. Nature 383:398–399

Devlin B, Ellstrand NC (1990) The development and application of a refined method for estimating gene flow from angiosperm paternity analysis. Evolution 44:248–259

Devlin B, Roeder K, Ellstrand NC (1988) Fractional paternity assignment: theoretical development and comparison to other methods. Theor Appl Genet 76:369–380

Dieringer G, Cabrera LR, Lara M, Loya L, Reyes-Castillo P (1999) Beetle pollination and floral thermogenicity in *Magnolia tamaulipana* (Magnoliaceae). Int J Plant Sci 160:64–71

Dow BD, Ashley MV (1996) Microsatellite analysis of seed dispersal and parentage of saplings in bur oak, *Quercus macrocarpa*. Mol Ecol 5:615–627

Dow BD, Ashley MV (1998) High levels of gene flow in bur oak revealed by paternity analysis using microsatellites. J Heredity 89:62–70

Ellstrand NC (1992) Gene flow among seed plant populations. New For 6:241–256

Ellstrand NC, Marshall DL (1985) Interpopulation gene flow by pollen in wild radish, *Raphanus sativus*. Am Nat 126:606–616

Ellstrand NC, Devlin B, Marshall DL (1989) Gene flow by pollen into small populations: data from experimental and natural stands of wild radish. Proc Nat Acad Sci USA 86:9044–9047

Govindaraju DR (1988a) Relationship between dispersal ability and levels of gene flow in plants. Oikos 52:31–35

Govindaraju DR (1988b) A note on the relationship between outcrossing rate and gene flow in plants. Heredity 61:401–404

Hamrick JL (1987) Gene flow and distribution of genetic variation in plant populations. In: Urbanska K (ed) Differentiation patterns in higher plants. Academic, New York, pp 53–67

Hamrick JL, Murawski DA (1990) The breeding structure of tropical tree populations. Plant Species Biol 5:157–165

Hamrick JL, Godt MJ, Sherman-Broyles SL (1992) Factors influencing levels of genetic diversity in woody plant species. New For 6:95–124

Harper JL (1977) Population biology of plants. Academic, London

Hartl DL, Clark AG (1988) Principles of population genetics, 2nd ed. Sinauer, Sunderland, MA

Iida S (1996) Quantitative analysis of acorn transportation by rodents using magnetic locator. Vegetatio 124:39–43

Isagi Y, Kanazashi T, Suzuki W, Tanaka H, Abe T (1999) Polymorphic DNA markers for *Magnolia obovata* Thumb. and their utility in the related species. Mol Ecol 8:698–700

Isagi Y, Kanazashi T, Suzuki W, Tanaka H, Abe T (2000) Microsatellite analysis of the regeneration process of *Magnolia obovata*. Heredity 84:143–151

Ishida K (1996) Beetle pollination of *Magnolia praecocissima* var. *borealis*. Plant Species Biol 11:199–206

Ishida K, Nakamura K (1997) Outcrossing rate and inbreeding depression in natural populations of *Magnolia hypoleuca*. Am J Bot Abstr 84:121

Jensen TS (1985) Seed-seed predator interactions of European beech, *Fagus silvatica,* and forest rodents, *Clethrionomys glareolus* and *Apodemus flavicollis*. Oikos 44:149–156

Johnson WC, Adkisson CA (1985) Dispersal of beech nuts by blue jays in fragmented landscapes. Am Midl Nat 113:319–324

Kameyama Y, Isagi Y, Nakagoshi N (2001) Patterns and levels of gene flow in *Rhododendron metternichii* var. *hondoense* revealed by microsatellite analysis. Mol Ecol 10:205–216

Karron JD, Tucker R, Thumser NN, Reinartz JA (1995) Comparison of pollinator flight movements and gene dispersal patterns in *Mimulus ringens*. Heredity 75:612–617

Kikuzawa K, Mizui N (1990) Flowering and fruiting phenology of *Magnolia hypoleuca*. Plant Species Biol 5:255–261

Krauss SL (1994) Restricted gene flow within the morphologically complex species *Persoonia mollis* (Proteaceae): contrasting evidence from the mating system and pollen dispersal. Heredity 73:142–154

Levin DA (1981) Dispersal versus gene flow in plants. Ann Mo Bot Gard 68:233–253

Levin DA, Kerster HW (1974) Gene flow in seed plants. Evol Biol 7:139–220

Masaki T, Suzuki W, Niiyama K, Iida S, Tanaka H, Nakashizuka T (1992) Community structure of a species-rich temperate forest, Ogawa Forest Reserve, central Japan. Vegetatio 98:97–111

Meagher TR (1986) Analysis of paternity within a natural population of *Chamaelirium luteum*. 1. Identification of most-likely male parents. Am Nat 128:199–215

Milligan B (1992) Plant DNA isolation. In: Hoezel, AR (ed) Molecular genetic analysis of populations: a practical approach. IRL Press, Oxford, pp 59–88

Milligan BG, McMurray CK (1993) Dominant vs codominant genetic markers in the estimation of male mating success. Mol Ecol 2:275–283

Miyazaki Y, Isagi Y (2000) Pollen flow and intrapopulation genetic structure of *Heloniopsis orientalis* on forest floor determined using microsatellite markers. Theor Appl Genet 101:718–723

Nei M (1973) Analysis of gene diversity in subdivided populations. Proc Nat Acad Sci USA 70:3321–3323

Ramsey MW (1988) Differences in pollinator effectiveness of birds and insects visiting *Banksia menziesii* (Proteaceae). Oecologia 76:119–124

Rasmussen IR, Brødsgaad B (1992) Gene flow inferred from seed dispersal and pollinator behavior compared to DNA analysis of restriction site variation in a patchy population of *Lotus corniculatus* L. Oecologia 89:277–283

Roeder K, Devlin B, Lindsay BG (1989) Application of maximum-likelihood methods to population genetic data for the estimation of individual fertilities. Biometrics 45:363–379

Schaal BA (1980) Measurement of gene flow in *Lupinus texensis*. Nature 284:450–451

Schmitt J (1980) Pollinator foraging behavior and gene dispersal in *Senecio* (Compositae). Evolution 34:934–943

Slatkin M (1987) Gene flow and the geographic structure of natural populations. Science 236:787–792

Smouse PE, Meagher TR (1994) Genetic analysis of male reproductive contributions in *Chamaelirium luteum* (L.) Gray (Liliaceae). Genetics 136:313–322

Sork VL (1984) Examination of seed dispersal and survival in red oak, *Quercus rubra* (Fagaceae), using metal-tagged acorns. Ecology 65:1020–1022

Streiff R, Labbe T, Bacielieri R, Steinkellner H, Glössl J, Kremer A (1998). Within population genetic structure in *Quercus robur* L. and *Quercus petraea* (Matt.) Liebl. Assessed within isozymes and microsatellites. Mol Ecol 7:317–328

Tanaka H, Yahara T (1988) The pollination of *Magnolia obovata*. In: Kawano S (ed) The world of plants II (in Japanese; title translated by the author). Kenkyusya, Tokyo, p 37

Thien LB (1974) Floral biology of *Magnolia*. Am J Bot 61:1037–1045

Waser NM (1982) A comparison of distances flown by different visitors to flowers of the same species. Oecologia 55:251–257

Waser NM (1993) Population structure, optimal outbreeding, and assortative mating in angiosperms. In: Thornhill NW (ed) The natural history of inbreeding and outbreeding. Univ of Chicago Press, Chicago, pp 173–199

Webb CL (1998) The selection of pollen and seed dispersal in plants. Plant Species Biol 13:57–67

Yasuda M, Nakagoshi N, Nehira K (1991) Examination of the spool-and-line method as a quantitative technique to investigate seed dispersal by rodents. Jpn J Ecol 41:257–262

Yumoto T, Noma N, Maruhashi T (1998) Cheek-pouch dispersal of seeds by Japanese monkeys (*Macaca fuscata yakui*) on Yakushima Island, Japan. Primates 39:325–338

Part 7

Interaction Between Plants and Other Organisms

21 Dynamics of Ectomycorrhizas and Actinorhizal Associations

HIROAKI OKABE

21.1 Introduction

A mycorrhiza is a symbiotic association between a plant root and a fungus. Mycorrhizas are classified into seven categories according to how the fungi penetrate the root cells, i.e., arbuscular mycorrhiza (forming a symbiotic association between more than 80% of vascular plants and zygomycetous fungi in the Glomales), ectomycorrhiza (an association between about 3% of seed plants and basidiomycetous fungi), ascomycetous and zygomycetous fungi, ericoid mycorrhiza, ectendomycorrhiza, arbutoid mycorrhiza, and monotropoid and orchidoid mycorrhiza (Smith and Read 1997; Raina et al. 2000). Mycorrhizal associations, both directly and indirectly, play a very important role in forest ecosystems: they enhance the uptake of water and plant nutrients, they promote plant growth, they increase resistance to drought, freezing, salinity, and root pathogens, they influence plant communities, and they play a role in the circulation of materials.

A group of mycorrhizal plants can form root nodules in the whole root system, e.g., a large number of legumes associated with nitrogen-fixing bacteria (*Rhizobium* and *Bradyrhizobium*) and nonlegumes (actinorhizal plants) such as *Alnus*, *Casuarina*, *Eraeagnus*, etc. with actinomycete *Frankia*. A mycorrhizal root system often consists of some ectomycorrhizal and some arbuscular mycorrhizal fungi, and the same mycorrhizal system might consist of only ectomycorrhizal fungi. These may be either short-term or long-term associations. These dynamic and complicated combinations with different root conditions may influence the establishment, persistence, and replacement of each symbiotic member in or on the root (Deacon and Fleming 1992). It is generally accepted that the role and function of ectomycorrhizas and actinorhizal root nodules in forests are deeply involved in tree growth and survival. However, the dynamics of symbiotic and fungal associates in forest soil are not well understood. Although nondestructive, long-term, and large-scale ecological methods in the field are necessary for such investigations, they have only been attempted for a few species and under limited conditions (Read 1992; Deacon and Fleming 1992).

The aim of this chapter is to show the results of using some ecological methods to investigate the dynamics of these symbionts in broad-leaved deciduous forests

Ecological Studies, Vol. 158
Nakashizuka, Matsumoto (eds.) Diversity and Interaction in a Temperate Forest Community: Ogawa Forest Reserve of Japan

(Okabe 1998). Specifically, it covers (1) how ectomycorrhizal fungi attack and become established on feeder roots after germination, (2) the characteristic distribution of mat-forming ectomycorrhizal fungi, (3) how to observe ectomycorrhizal fungi in forest soils, (4) the nodulation process on alder roots, (5) the spread of *Frankia* inoculants, and (6) the distribution of alder root nodules in a forest.

21.2 Ectomycorrhizal Associations

21.2.1 The Presence and Role of Black Mycorrhizas

The hornbeam, *Carpinus laxiflora,* is known to be associated with ectomycorrhizal fungi. Hornbeam seedlings with cotyledons or first leaves, which were probably less than 2 weeks old, and 1-year-old seedlings (Okabe 1994) were collected from mature deciduous forests in the Ogawa Forest Reserve (OFR) and at Nakoso, near OFR, They were grown on in 100-ml polypropylene film tubes packed with volcanic soil mixed with diatomite, under slightly dry conditions, for mycorrhizal assay over a period of 6 months. Table 21.1 lists the ectomycorrhizas obtained from the seedlings. These mycorrhizas were classified into two types according to color, i.e., black (or blackish) and not black. There were six types of melanized black ectomycorrhiza. The most typical (Cg, Fig. 21.1), *Cenococcum geophilum* Fr., formed black ectomycorrhizas and black sclerotia, with hyphae of the same color protecting the host from the changeable environment in the top layer of the soil (Mikola 1948; Pigott 1982; Unestam 1991).

Of the younger seedlings, 29.3% from OFR and 31.6% from Nakoso had ectomycorrhizas, while 22.4% and 0%, respectively, had Cg. In total, 27.6% and 31.6%, respectively, were the black type, and 1.7% and 0%, respectively, were the not-black type. There were very few types other than black.

In the younger seedlings the type of mycorrhiza was limited, and most seedlings had only one (Table 21.2). All the older seedlings were infected: the black type was

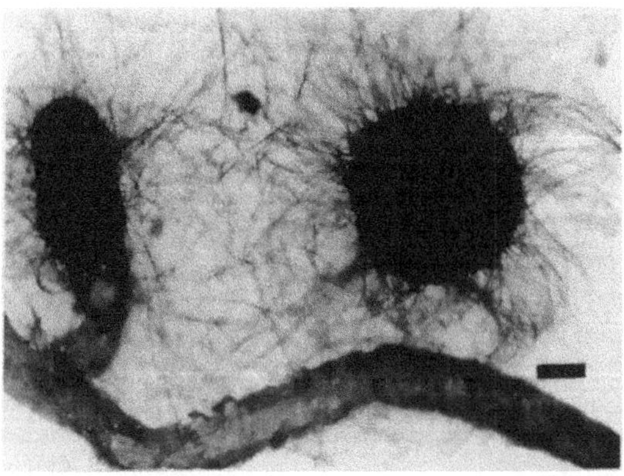

Fig. 21.1. Black ectomycorrhizal fungus, *Cenococcum geophilum*. Black mycorrhizas and spherical sclerotium with jet black hyphae. *Bar* 0.5 mm

Table 21.1 Types of ectomycorrhiza of *Carpinus laxiflora* seedlings (Okabe 1994)

Site	Seedling age	NO	Types of ectomycorrhiza												Total number of seedlings
			Cg	C1	C2	C3	C4	C5	M1	M2	M3	M4	M5	M6	
Ogawa	< 1 month	41	13	0	0	0	0	3	0	1	0	0	0	0	58
	1–2 years	0	32	1	1	1	1	0	18	6	2	1	4	0	67
Nakoso	< 1 month	13	0	0	0	0	0	6	0	0	0	0	0	0	19
	1–2 years	0	13	0	0	0	0	2	8	1	0	0	3	2	29

C(g,1–5), black mycorrhiza; M(1–6), not-black mycorrhiza; NO, not infected; Cg, *Cenococcum geophilum*; C1, black, without a rhizomorph from mycorrhiza, and a black rhizomorph between roots; C2, black, without a rhizomorph; C3, blacker than Cg, and easily blanched; C4, more dark brown than Cg, with a mantle with white–gray hyphae and a black rhizomorph between roots; C5, like Cg, but with a light-colored rhizomorph and very short; M1, long and monopodial, with a few emanating hyphae but an indistinct mantle; M2, short and monopodial, with dense emanating hyphae but an indistinct mantle; M3, pinnate, with a white mantle and a white rhizomorph; M4, light brown and monopodial, with dense hyphae but an indistinct mantle; M5, white and monopodial, without a rhizomorph and covered with rough hyphae; M6, white and monopodial, with dense hyphae.

Table 21.2 Combinations of ectomycorrhiza in *C. laxiflora* seedlings (Okabe 1994)

Site	Seedling age	Combinations of mycorrhiza
Ogawa	< 1 month	Cg(76), C5(18), M3(6)
	1–2 years	Cg+M1(31), Cg(27), M1(7), M2(5), Cg+C1(5), M5(5), M3(2), C3(2), Cg+M4+M2(2), Cg+C1+M1(2), Cg+M5+M2(2), Cg+M2(2), Cg+C4(2), Cg+M3(2), M1+M2(2), C2+M5(2)
Nakoso	< 1 month	C5(100)
	1–2 years	Cg+M1(25), Cg(25), Cg+M6(10), M5(10), M1(10), M2(5), C5(5), Cg+C5(5), Cg+M5(5)

Numbers in parentheses: (Number of infected seedlings/Total number of infected seedlings) × 100.

in 53.7% (Cg 48%) and 51.7% (Cg 45%) and the not-black type was in 46.3% and 48.3% from OFR and Nakoso, respectively. The black type of mycorrhiza in each seedling increased rapidly, and the number of each type also increased. However, thick mantles such as M6 were rare at this stage, since the taproots had not yet reached the deep soil layer, and other mantles (M1-5) were mostly indistinct except in the black types.

As mycorrhizal fungi associated with newly germinated seedlings have to develop in the litter layer, the environmental factors on the soil surface may strongly influence ectomycorrhizas and their fungus associates. Black mycorrhizas may play an important role in the first stage of the growth of seedlings, especially in terms of drought tolerance and ectomycorrhizal succession. The black hyphae of Cg, which are easily cut into fragments by any disturbance, may be able to form small propagation units. The positive behavior of Cg in the topsoil in any environment can provide a unique habitat. An association with Cg may guarantee the survival and growth of some seedlings.

However, while approximately 70% of the younger seedlings had bare roots and were exposed to hazardous conditions, the 1-year-old seedlings had a high rate of infection and a great diversity in types of mycorrhiza.

21.2.2 Long-Lived and Huge Mat-Forming Ectomycorrhizal Fungi

Little is known about the growth and development of ectomycorrhizal fungi in the field. In this study, a large mat-forming ectomycorrhizal fungus, *Rhodophyllus rhodopolius* (Fr.) Quel, was examined in a mixed *Quercus*, *Carpinus,* and *Fagus* forest on a high slope (Okabe 1994). Mats with a large volume and a strong hydrophobic or hydrophilic force may have an ecophysiological effect on soil organisms (Unestam 1991).

In an 8 m × 8 m plot, 495 fruiting bodies of this fungus occurred in 1990, 589 in 1991, and 0 in 1992. Their distribution was mapped by their presence or absence in a small quadrat (5 cm × 5 cm) within the plot (Fig. 21.2). In May 1993, the fungal mats were mapped on the basis of their thickness in another small quadrat (10 cm × 10 cm) within the same plot (Fig. 21.3). Although the mat extended to the upper A layer, it did not reach the mineral soil layer. The mats did not always cover the whole of the plot and were absent in places. Fresh mats were very tight, while old ones were tattered and crumbly.

In the absence of mats, fruiting bodies occurred in only 5.8% of quadrats. The fruiting bodies were irregularly distributed and found mostly on the mats. The mean thickness of the mats was 3.2 cm (maximum 6 cm). The sum of the mat thicknesses measured in each quadrat was 120 m. Its biomass and its influence on the soil environment were probably considerable.

Table 21.3 shows differences in mat thickness (which is a rough indication of fungal biomass) between adjacent quadrats. The difference was generally within 1 cm. There was little unevenness in the surface of the mats, and the thickness varied gradually. Although the forest floor was often rough and uneven, the surfaces of the

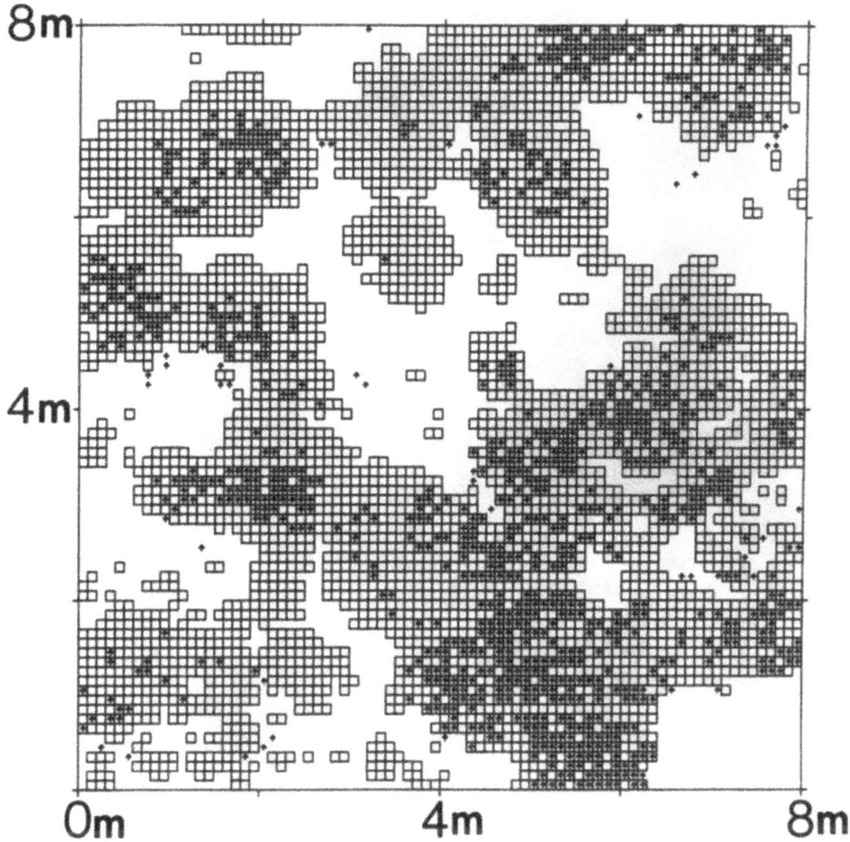

Fig. 21.2. Fruiting-body distribution of an ectomycorrhizal fungus, *Rhodophyllus rhodopolius*, and the mat extension (from Okabe 1994, with permission). *Squares*, quadrats (10 cm × 10 cm, total number 6,400) where the mat was present. *Arrows*, position of a fruiting-body in a small quadrat (5 cm × 5 cm, total number 25,600). Ectomycorrhizal trees in the plot (8 m × 8 m) were *Quercus serrata, Carpinus laxiflora*, *Fagus crenata,* and *Fagus japonica*

Table 21.3 Differences in the thickness of *R. rhodopolius* mats between adjacent quadrats (Okabe 1994)

Difference (cm)	Total	Both[a]	Either[b]
0	7104	2691	–
1	3401	2881	520
2	1281	953	328
3	569	196	373
4	231	23	208
5	45	2	43
6	9	0	0

[a]Mats formed in two adjacent quadrats.

[b]Mats formed in one of two adjacent quadrats.

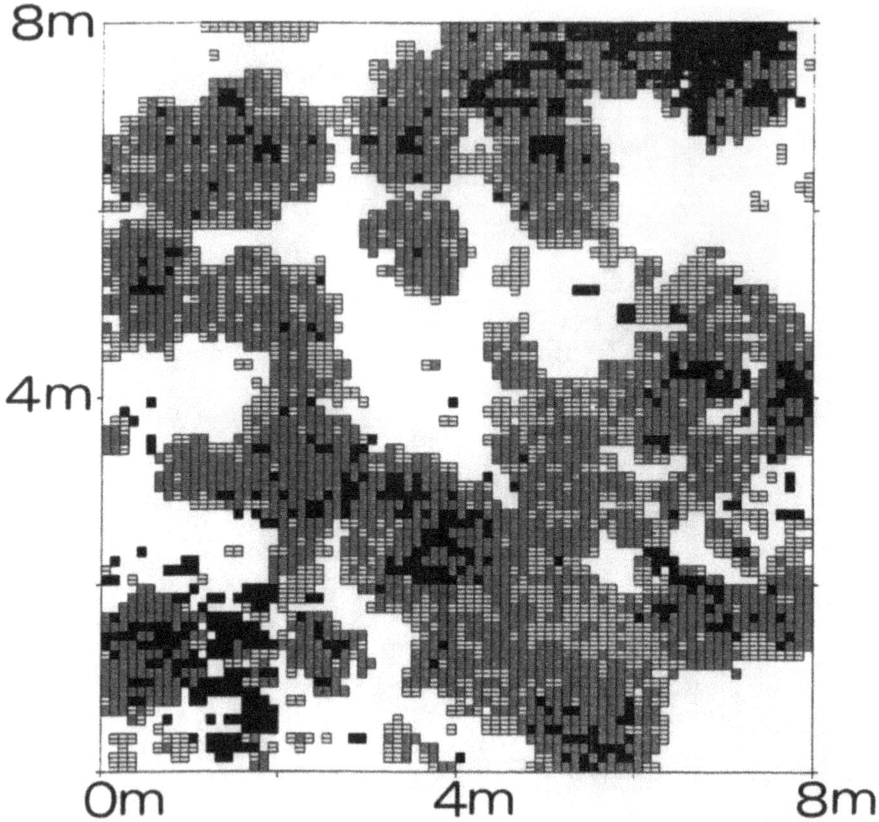

Fig. 21.3. Distribution of mats and their thickness (from Okabe 1994, with permission). *White squares,* quadrats (10 cm × 10 cm) where the mat was present. The thickness of the mat was classified into five stages from 1 cm to 5 cm at 1-cm intervals. The fifth stage (*black squares*) includes thicknesses of more than 5 cm

mats were flatter than the places without mats, because the fungi gradually made rough places flat by producing litter. It is believed that the mats grow horizontally and vertically with litter-fall every season. Periodical litter-fall may support the activity of the mats and the occurrence of fruiting bodies in the mats for a long time.

21.2.3 Spread of Ectomycorrhizas on the Topsoil

The mechanism of ectomycorrhizal development in soil is not well understood. A method has been developed to induce mycorrhizal development under nonwoven polyester cloth (mycorrhiza-developing (MD) method) set on the surface of the soil. Because water can pass easily through the cloth the appropriate physical conditions are maintained, although fresh falling litter cannot be supplied directly. The litter-fall was allowed to accumulate over the cloth naturally. In the growing season, my-

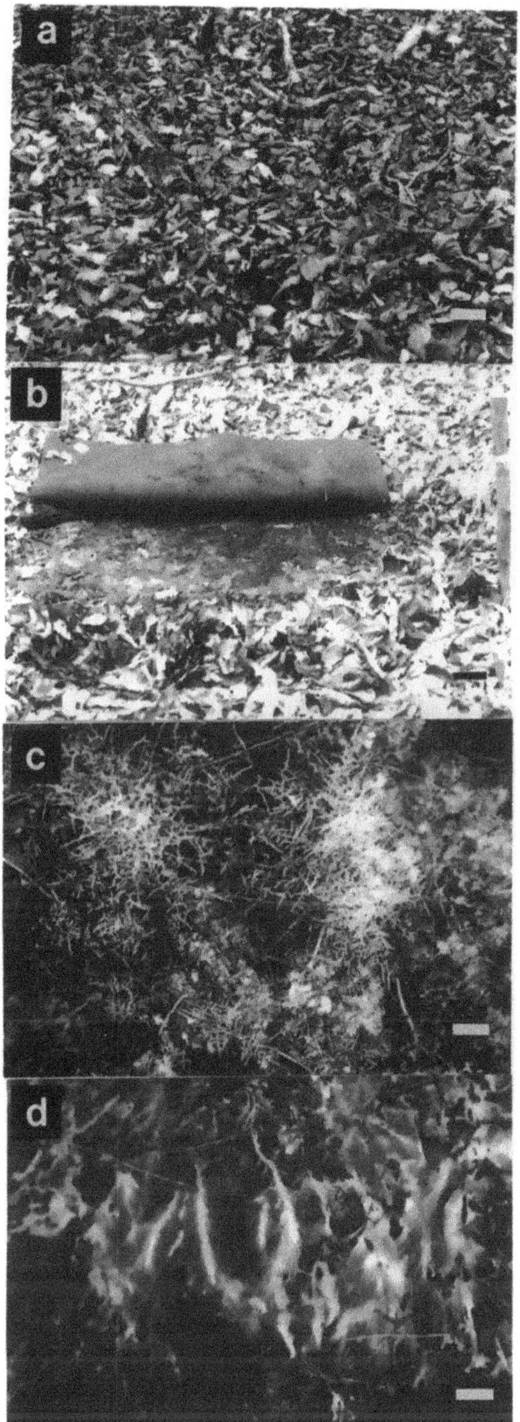

Fig.21.4. Investigation of ectomycorrhizas in the topsoil layer using the micorrhiza-developing (MD) method. This method uses the properties of root development. **a** A nondecomposing cloth was set into the humus layer, and **b** was torn off. Then the roots and associates developing immediately under the cloth were investigated. Hyphae passed through the cloth, which was made of nonwoven polyester, but roots did not. The ectomycorrhizas which developed under the cloth started branching out laterally within a month in the growing season, and it will be possible to observe them for some years. **c** Mixed mycorrhizas. **d** A pattern of *Cortinarius* mycelia. **a**, **b** *Bars* 5 cm; **c** *bar* 2 cm; **d** *bar* 5 mm

corrhizal extensions started within at least 1 month after setting, and the mycorrhizas could be observed for several years. Horizontal extensions of the mycorrhizas, obtained by applying the MD method to *Quercus* and *Carpinus* forests, could be drawn under the cloth (Fig. 21.4).

By using the method described above, the dynamics of mycorrhizas and fungal associates (i.e., extraradical hyphae, mycerial strands, rhizomorphs, mats, sclerotia) were observed in detail without soil contamination. Although it is well known that mycorrhizal fungal growth is very slow in vitro, fungal growth in the field may be very active because of the constant energy supply from the host trees, and the strong competition between microorganisms. The rapid extension of ectomycorrhizas means that nutrients in the topsoil become available for uptake, which may be a very effective strategy for overcoming limited resources.

21.3 Actinorhizal Root Nodules and Their Endophyte, *Frankia*

It is easy to find root nodules of nonleguminous trees such as alder. It is well known that the nitrogen-fixing function of actinorhizal root nodules is closely involved in the promotion of tree growth and survival, and nitrogen cycling in a forest ecosystem (Schwintzer and Tjepkema 1990). An alder, *Alnus hirsuta* var. *sibirica,* forms typical root nodules (= actinorhizas) in association with actinomycetous *Frankia.* Although actinorhizas that are perennial and occasionally grow over 10 cm in diameter are commonly found in deciduous forests, their mechanism of life is not well known. The development, growth, disappearance, and occupation of habitat of actinorhizas in forests were studied as described below.

21.3.1 Development of Nodules and Occupation of Host Roots

When the MD method was used under an alder tree, new nodules were observed when fine roots developed underneath the cloth (Fig. 21.5). The dynamics of the nodules was observed for 6 years. In the first year, many nodules developed. The next year, the ectomycorrhizas expanded and the fungi covered the nodules, but the ectomycorrhizas disappeared later. Ultimately, the nodules grew only on thick roots, and all those on fine roots disappeared. It seems that although thick roots received sufficient carbohydrates from the host, fine ones probably did not.

The distribution of nodules and ectomycorrhizas on fine roots changed rapidly, and they then disappeared every year. All nodules consume some carbohydrate, so the flow of energy to the outermost infected root may decrease. Endophytes (i.e., ectomycorhizal fungi, arbuscular mycorrhizal fungi, and *Frankia*) living in the outermost roots, particularly small roots, must receive a certain amount of carbohydrate in order to grow. If their nutrient supply fails, endophytes cannot grow in front of the nodules. The energy consumption of the nodules may be one of the reasons why the

Fig. 21.5. Formation and growth of root nodules. Numerous root nodules of alder, *Alnus hirsuta* var. *sibirica*, formed on the lateral roots, but not on main and fine roots. *Bar* 5 mm

ectomycorrhizas disappeared. The regulation of energy flow from host to endophyte may be related to competition between different symbionts.

21.3.2 Spread of *Frankia*

How do *Frankia*-forming nodules extend in soil? Although mature nodules are always good inocula, old and collapsed nodules also have effective inoculum potential. It was observed that soil and the dung of earthworms found around root nodules had the potential to form nodules on alder seedling roots. *Frankia* propagules containing hyphae, sporangia, and vesicles in one host cell may be an effective unit with high inoculum potential (Fig. 21.6a). The cell walls of the roots collapsed over time, and then the propagules gradually emerged from the host cell (Fig. 21.6b,c). Each unit may be carried by animals in the soil or dispersed by water and disturbance to form new symbionts.

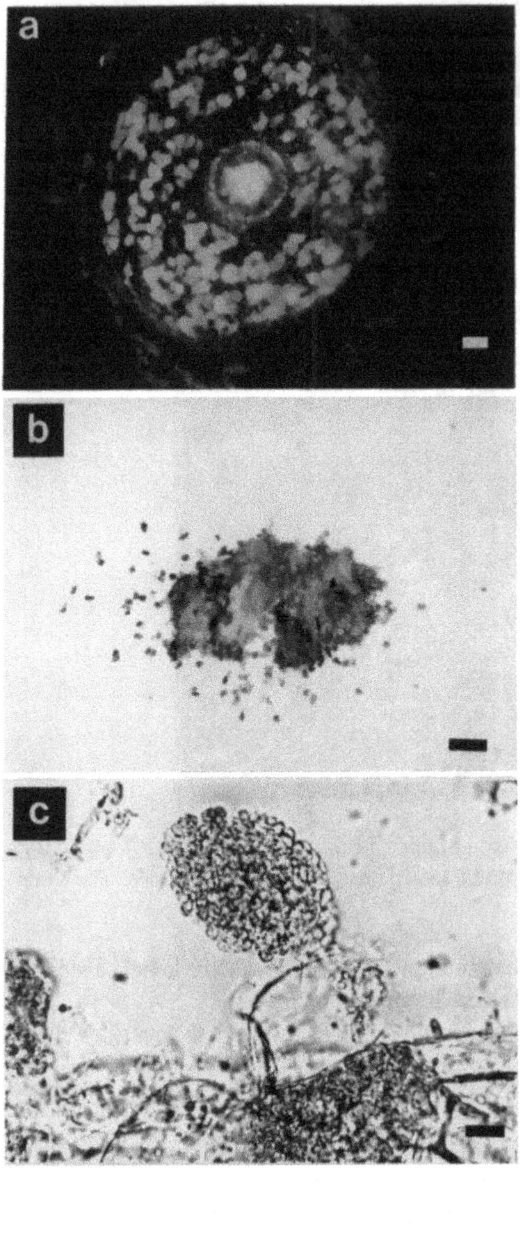

Fig. 21.6. Bodies of *Frankia* that get out of the host. **a** *Frankia* propagules extending into the root nodules of alder. **b** The cell walls of the host root are gradually decayed and degraded. **c** Ultimately the endophyte forms a unit outside the host cell. **a** *Bar* 0.1 mm; **b,c** *bars* 0.01 mm

21.3.3 Distribution of Nodules Around the Tree

The tree being investigated was situated in the middle of a plot (16 m × 16 m). The distribution of nodules (Fig. 21.7) in the plot was mapped by checking for the presence or absence of nodules in a minimum quadrat (20 cm × 20 cm), down to 20 cm depth, and all nodule diameters were measured.

The tree (*A. hirsuta* var. *sibirica*) was fully isolated from other trees and its di-

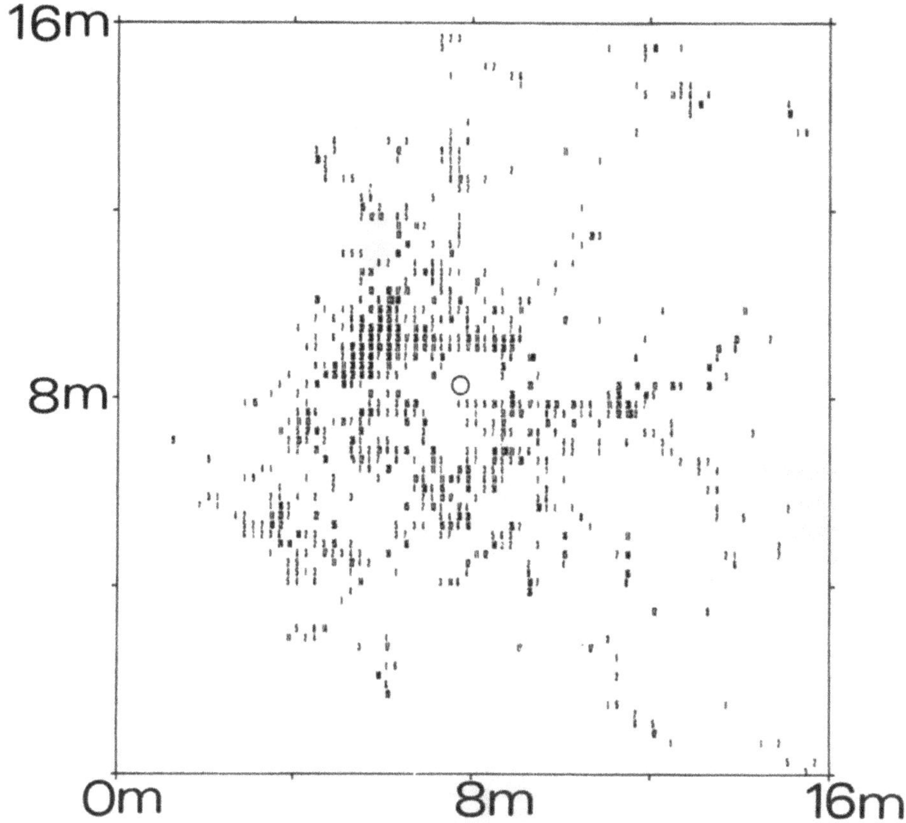

Fig. 21.7. Distribution of actinorhizal root nodules. The numbers show the number of nodules found in each quadrat (20 cm × 20 cm, a total of 6400 quadrats) in the plot (16 m × 16 m). The *circle* in the middle indicates the position of the big alder stem. The left side of the plot had deep soil, and the right had shallow soil. The nodules on the left were larger and found in greater density than those on the right. On the right, they were scattered for some distance along the roots

ameter at breast height was found to be 46 cm, which is close to the maximum size for this species. It seemed that the spread of the roots may have reached a maximum, and that the nodules were being formed in the same places. The diameter classes of the nodules showed a clear L-type distribution, and many small new nodules continued to be produced. Their production seemed to have proceeded smoothly.

The activity of the nodules was judged from their color. Active nodules were light brown, but nonactive ones were dark brown. Nodules were distributed in all directions, and were formed along thick roots running near the surface of the soil. There were few nodules below around 20 cm depth, but there were more and larger ones in deeper soil. There were approximately 7000 nodules, with a total volume of 41 l. This volume was calculated from the approximate diameters of nodules found under this tree only. The volume of active nodules was 23 l from 55% of the total number of nodules. The active nodules were generally small, and the nonactive ones were generally large. Active types were found in 80% of quadrats where nodules were identified, partially active ones in 60%, and nonactive ones in 36%. This association may mean that a high level of activity increases the tree's effective uptake of nutrients.

References

Deacon JW, Fleming LV (1992) Interactions of ectomycorrhizal fungi. In: Allen MF (ed) Mycorrhizal functioning. Chapman & Hall, New York, pp 249–300

Mikola P (1948) On the physiology and ecology of *Cenococcum graniforme* especially as a mycorrhizal fungus of birch. Commun Inst For Fenn 36:1–101

Okabe H (1994) Life-mode of ectomycorrhizal fungi (in Japanese). Soil Microorganisms 44:15–24

Okabe H (1998) Ecology of ectomycorrhizal and actinorhizal root nodules in broad-leaved deciduous forests. In: Proceedings of the 6th International Symposium of the Mycological Society of Japan, 1998. Japan–UK International Mycological Symposium, The Mycological Society of Japan, Chiba, pp 48–51

Pigott CD (1982) Survival of mycorrhiza formed by *Cenococcum geophilum* Fr. in dry soils. New Phytol 92:513–517

Raina S, Chamola BP, Mukerji KG (2000) Evolution of mycorrhiza. In: Mukerji KG, Chamola BP, Singh J (eds) Mycorrhizal biology. Kluwer Academic/Plenum, New York, pp 1–26

Read DJ (1992) The mycorrhizal mycelium. In: Allen MF (ed) Mycorrhizal functioning. Chapman & Hall, New York, pp 102–133

Schwintzer CR, Tjepkema JD (1990) The biology of *Frankia* and actinorhizal plants. Academic, San Diego

Smith SE, Read DJ (1997) Mycorrhizal symbiosis, 2nd ed. Academic. San Diego, pp 161–254, 377–452, 470–489

Unestam T (1991) Water repellency, mat formation, and leaf-stimulated growth of some ectomycorrhizal fungi. Mycorrhiza 1:13–20

22 Interactions Between Seeds of Family Fagaceae and Their Seed Predators

Akira Ueda

22.1 Introduction: Hypotheses of the Interaction Between Seeds and Seed Predators

Regarding the interaction between seeds and seed predators, Janzen (1971) proposed the predator satiation hypothesis, which states that synchronous and fluctuating seed production allows for the survival of seeds by alternately starving and satiating seed predators. This hypothesis depends on a strong relationship between the tree species and the seed predators specialized to prey on the seed of that species (that is, a one-to-one relationship). This sort of predator is called a typical specialist in the present study. However, if the seed predators can feed on the seeds of several tree species, predator satiation can occur only if all the prey species in a forest community fluctuate their seed production synchronously (Silvertown 1980). Otherwise, the populations of predators remain at high levels by feeding on the seeds of tree species that produce seeds asynchronously. If the seed predators can feed on other foods as well, then even synchronous seed production of several species in a forest community may be unable to control the predator populations.

For this sort of situation, Ims (1990a, b) pointed out that the degree of satiation caused by synchronous reproduction depends on the functional response of the seed predators. If the predators are generalists, and they switch from alternative food resources to the seeds of particular species only when they are abundant, synchronous seed production might lead to more intense predation. Thus, asynchronous seed production may be the best reproductive strategy when predators are generalists, which are defined in the present study as those species that can switch their food resources from seeds to alternative resources. If a tree (or a tree species) whose seeds are intensively damaged by these generalists cannot somehow forecast the intermittent seed production of other trees (or other tree species), it may produce relatively constant numbers of seeds every year in an attempt to divert the attention of the generalist predators to the abundant seeds of trees that produce intermittent large seed crops.

Ecological Studies, Vol. 158
Nakashizuka, Matsumoto (eds.) Diversity and Interaction in a Temperate Forest Community: Ogawa Forest Reserve of Japan

Five species of family Fagaceae, *Fagus crenata, F. japonica, Castanea crenata, Quercus serrata*, and *Q. crispula* found in the Ogawa Forest Reserve (OFR) include both the most fluctuating species (*Fagus* spp.) and the most constant species (*Quercus* spp.) with respect to seed production (see Fig. 9.3 in Chapter 9). This sort of variation in seed production among related species is suitable for testing Ims's (1990a, b) hypothesis. The key question is whether the reproductive habit of a tree can satiate specialist seed predators through synchronous, fluctuating seed production, or whether generalist predators can be diverted to other seeds or food resources through asynchronous, nonfluctuating seed production. To resolve this problem, insect species that prey on seeds of family Fagaceae were identified to determine whether they were generalist or specialist predators. Then the annual and individual variation in seed production of each species of tree was estimated to determine whether the mode of production matched an anti-specialist or an anti-generalist strategy. Finally, the interaction between seeds and seed predators was investigated by comparing the actual predation rates by specialist and generalist predators with the respective predation levels estimated from the characteristics of seed production.

22.2 Insects that Prey on Seeds of Family Fagaceae and Variations in Their Food Preferences

Insect predators of seeds of family Fagaceae in the OFR are shown in Table 22.1. It proved impossible to rear any of the moth larvae; thus, unequivocal species identification was not possible. However, 98.7% of the bored seeds of *F. crenata* may have been attacked by *Pseudopammene fagivora* Komai (Lepidoptera: Tortricidae) (Table 22.1, Fig. 22.1A) based on the species' distinctive feeding characteristics: damaged seeds had a large hole on the attached side of the two seeds within the cupule, and these were filled with the insect's frass (Fig. 22.1B) (Igarashi and Kamata 1997). This species of moth was unquestionably present in the study area, because the author collected two adults that had emerged from samples of the forest's organic layer (Ueda 2000a). This species is believed to be a specialist predator of the seeds of *F. crenata* (Komai 1980), and was the most important seed predator in studies by Igarashi and Kamata (1997). Eggs of this moth are laid on the cupule surface, and the insect overwinters in its pupal stage in the soil (Igarashi and Kamata 1997).

The species that feed on *F. japonica* have never been definitively studied. Because all the seeds of *F. japonica* that had been bored into during this study showed damage that resembled the pattern of attack by *P. fagivora*, it is reasonable to assume that *P. fagivora* is able to bore into the seeds of both *Fagus* species. If the moths that attacked *F. japonica* were not *P. fagivora*, it is likely that they were other predators specific to *F. japonica* or common to both *Fagus* species.

The timing of moth larvae boring into seeds differed between the *Fagus* species because the seed growth of *F. crenata* begins 2 months earlier than that of *F. japonica* (Ueda 2000a). Even if *P. fagivora* attacks both *Fagus* species, the insect must have

Table 22.1. Insect predators of seeds of Fagaceae in the OFR

Bore into		*Fagus crenata*[a]	*Castanea crenata*	*Quercus serrata* *Quercus crispula*
Predispersal seeds	Moths[b]	*Pseudopammene fagivora* ?[c]	*Neoblastobasis biceratala*	*Neoblastobasis biceratala* *Cydia danilevski* *Cydia glandicolana*[d]
	Weevils	None	*Curculio sikkimensis*	*Curculio distinguendus* *Curculio dentipes* *Curculio sikkimensis* *Kobuzo rectirostris*
	Gall midges	Family Cecidomyiidae[e]	none	none
Postdispersal seeds	Moths[b]	None	*Cryptaspasma marginifasciata*	*Cryptaspasma marginifasciata* *Cryptaspasma trigonana*

[a]Moths boring into seeds of *Fagus japonica* were not identified because of difficulties rearing them to the adult stage.

[b]All moths other than *Neoblastobasis biceratala* (Blastobasidae) belong to Tortricidae.

[c]This moth was not identified because of difficulties rearing it to the adult stage.

[d]This moth was confirmed only on *Q. crispula.*

[e]This insect was also seen on *F. japonica*. Two types, from one or two cocoons that suddenly appeared in empty seeds, and one or two larvae that infested and made cocoons in the pericarps of seeds of all categories, were found. The latter type did not destroy sound nuts. Thus the damage by this insect is not shown in Figs. 22.1 and 22.2, but their attack levels were low (Ueda, 2000a).

specialized on one of the two species because the period during which seed phenology is suitable for feeding by moth larvae is limited (Igarashi and Kamata 1997). Thus, the relationship between the trees and predators of *Fagus* species must be one to one (i.e., typical specialist).

Gall midges (family Cecidomyiidae) were seen on both *F. crenata* and *F. japonica* (Table 22.1). There were two types of damage. The first was characterized by the presence of one or two cocoons in empty seeds, and the second was characterized by the presence of one or two larvae that infested and made cocoons in the pericarp (seed coat) of the seeds (all categories of seed). The second type of damage did not destroy sound seeds. The attack levels by this insect are not shown in the present study, but they were low (Ueda 2000a).

Three species of moths were observed to have larvae that bored commonly into the predispersal seeds of both *Quercus* species: *Neoblastobasis biceratala* (Park), *Cydia danilevski* (Kuznezov), and *C. glandicolana* (Danilevsky) (Table 22.1, Fig. 22.1C, D). *N. biceratala* was also seen on *Castanea crenata* (Table 22.1). The eggs of *Cydia glandicolana* are laid singly on seeds and on the right side of leaves near the seeds starting in late August (Maeto 1993), but eggs of the other species have not been recognized. The larvae of the three moths bore into the seeds through the pericarps, which are partially covered by cupules, and they leave the seeds by making a lateral exit hole after finishing their feeding (Fig. 22.1E). Some of *N. biceratala* pupated in the seeds without leaving. *Kobuzo rectirostris* (Roelofs) was seen on both *Quercus* species (Table 22.1, Fig. 22.1F). The larva of *K. rectirostris* has a dark purple head, and the seeds bored by this insect have a large hole on the bottom of their pericarp both before and after the larva leaves the seed (Fig. 22.1G). Three species of *Curculio*, *C. distinguendus* (Roelofs), *C. dentipes* (Roelofs,), and *C. sikkimensis* (Heller) were observed in the predispersal seeds of both *Quercus* species (Table 22.1). *C. sikkimensis* was also seen on *Castanea crenata* (Table 22.1). The eggs of *C. sikkimensis* were buried at each end of a finger-shaped scar created by the weevil's long proboscis on the reverse side of the pericarp (Fig. 22.1H, I) starting in September, but eggs of the other species were not observed. The larvae of *Curculio* spp., which have orange heads, leave the seeds by making a lateral exit hole after finishing their feeding (Fig. 22.1J).

The larvae of two moth species, *Cryptaspasma marginifasciata* (Walsingham) and *C. trigonana* (Walsingham), were observed also as postdispersal seed predators for both *Quercus* species in the OFR (Table 22.1, Fig. 22.1K). *C. marginifasciata* was also seen on *Castanea crenata* (Table 22.1). These moths have been known to attack the seeds of various species of family Fagaceae, but not *Fagus* species (Maeto 1993; Ueda et al. 1993). The eggs of these species are laid in an egg mass on fallen leaves, and the larvae bore into the seeds through the part of the cotyledons that emerges from the pericarps during germination (Fig. 22.1L, M) (Ueda et al. 1993).

All seed predators of *Castanea crenata*, *Quercus serrata*, and *Q. crispula* had more than two host tree species (Table 22.1), and are known to be able to bore into the predispersal seeds of various *Quercus* species (e.g., Ueda et al. 1992). Since the phenology of seed growth in *C. crenata* and the two *Quercus* species is similar and overlapped in the OFR (A. Ueda unpublished data), their predators are able to attack more

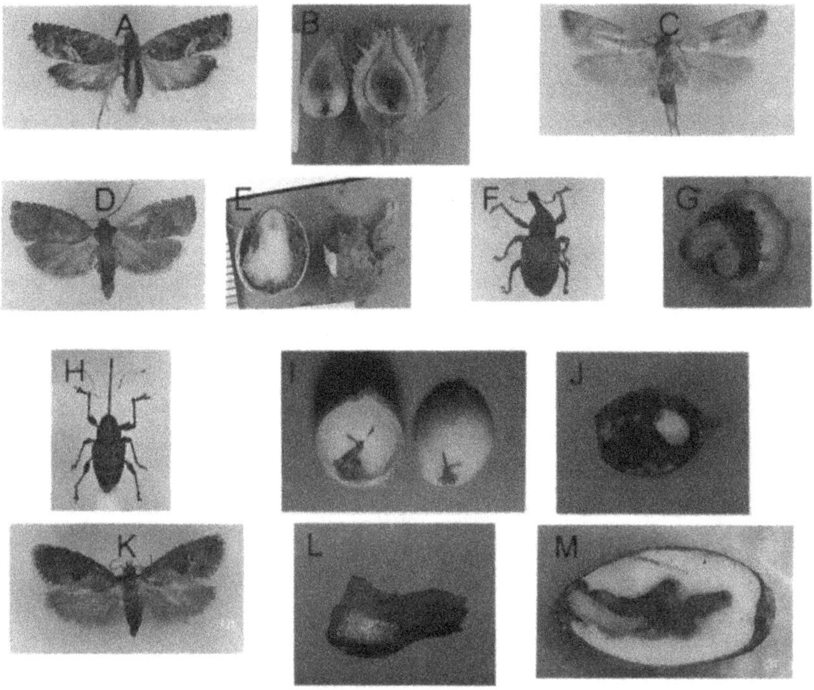

Fig. 22.1. Photographs of seed borers observed on the species of Fagaceae in this study. **A** *Pseudopammene fagivora*; **B** Feeding characteristics of *P. fagivora*; **C** *Neoblastobasis biceratala*; **D** *Cydia danilevski*; **E** a moth larva boring into the predispersal seed of *Quercus serrata*; **F–G** *Kobuzo rectirostris* and its larva; **H–J** *Curculio sikkimensis* and its eggs and larva; **K–M** *Cryptaspasma* spp., its hatched egg mass on a fallen leaf, and its larva

than two host species simultaneously. The food resources of these predators are limited to the seeds of a few co-occurring tree species in the forest, but, even so, the relationship between the predator and the tree species was not one to one. Thus, in the present study, they were classified as atypical specialists having several host species.

22.3 Determining Whether Seed Production Strategies Are Effective Against Specialist or Generalist Predators

To identify intraspecific synchrony of seed production and observe the levels of predispersal damage to seeds, two or three 0.5-m^2 seed traps (made from nylon cloth) were placed beneath each tree crown, within 5 m of the trunk, from 1992 to 1994. This arrangement of seed traps was different from the regular matrix of seed traps used by Shibata and Tanaka (see Chapter 9), who investigated the annual fluctuation and interspecific synchrony of reproduction.

Seed falls of *F. crenata* and *F. japonica* were concentrated in 1993 (Fig. 22.2), and, without exception, all trees examined exhibited highly synchronous seeding. Moreover, Shibata and Tanaka (see Chapter 9) observed large fluctuations in seed production for both species (see Fig. 9.3). These results suggest that *Fagus* species vary in their seed production to satiate specialist predators (Ims 1990a, b).

Since *C. crenata*, *Q. serrata,* and *Q. crispula* dispersed many small, immature seeds (probably aborted seeds), the seed production levels analyzed in this chapter deal only with the seeds larger than a defined minimum mature size for each species (Ueda 1996, 2000b). This minimum size was 1.0 cm^3 (length × width × thickness) for *Castanea crenata*, and 0.8 cm^3 (length × $width^2$) for *Quercus* species. The seed fall of *C. crenata* was concentrated in 1994 (Fig. 22.2), and seed production by all individuals examined was synchronous. Moreover, Shibata and Tanaka (see Chapter 9) observed large fluctuations in the levels of seed production of this species (see Fig. 9.3). These results suggest that *C. crenata* also has large variation in annual seed production to satiate specialist predators (Ims 1990a, b).

Seed falls of *Q. serrata* and *Q. crispula* were abundant in both 1993 and 1994

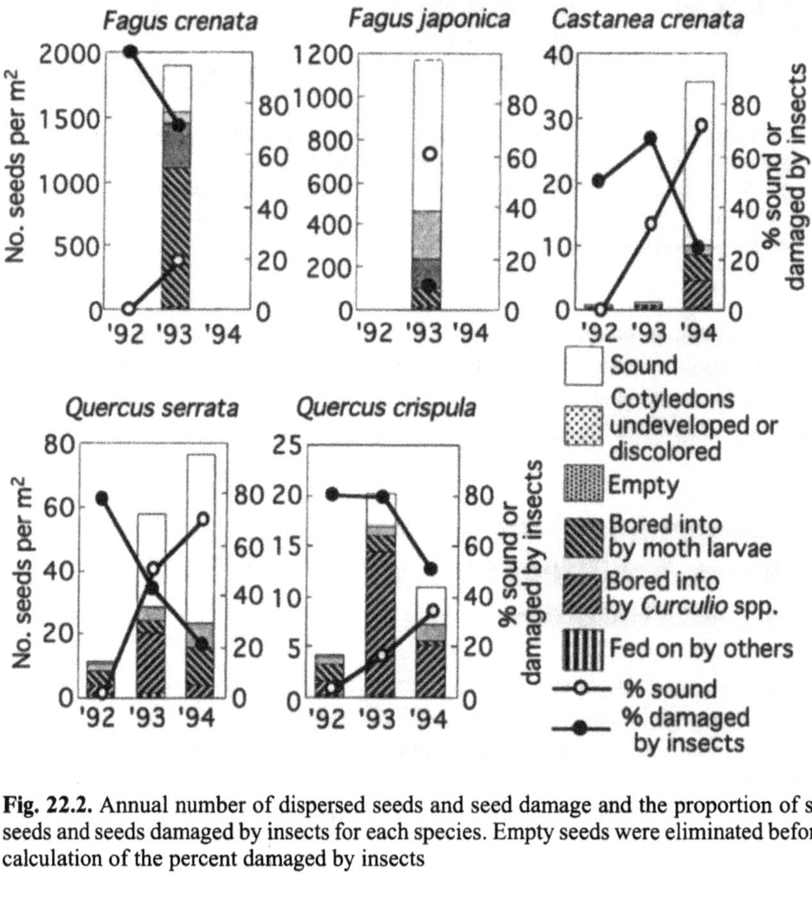

Fig. 22.2. Annual number of dispersed seeds and seed damage and the proportion of sound seeds and seeds damaged by insects for each species. Empty seeds were eliminated before the calculation of the percent damaged by insects

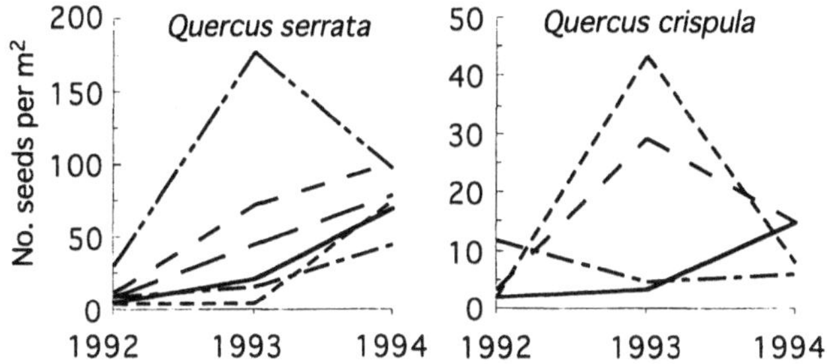

Fig. 22.3. Annual number of dispersed seeds for *Quercus serrata* (*left*) and *Q. crispula* (*right*)

(Fig. 22.2). Five of the six trees of *Q. serrata* examined in the study produced seeds synchronously, but seed production of *Q. crispula* was asynchronous among individuals (Fig. 22.3). Moreover, Shibata and Tanaka (Chapter 9) also observed weak fluctuations and continuous seed production from year to year in these two species (see Fig. 9.3). These results suggest that seed production by *Q. serrata* may both satiate specialist predators and avert damage by generalists, whereas seed production by *Q. crispula* may be more adapted to averting damage by generalists (Ims 1990a, b).

22.4 Predispersal Damage to Seeds

Predispersal insect damage levels of seeds of *F. crenata* and *F. japonica* in 1993 were 70.7% and 9.3%, respectively (Fig. 22.2). These levels show that *F. japonica* largely succeeded in escaping predispersal damage by insects, whereas *F. crenata* failed to escape such damage. However, the *F. crenata* trees in the present study could nonetheless be considered to have escaped attack to some extent because the observed damage levels were considerably lower than those observed for a *F. crenata* tree that produced seeds every year; in the case of the latter tree, more than 92% of the seeds had been bored into by moth larvae each year (Igarashi and Kamata 1991). On the basis of the results of the present study, the synchronous and fluctuating seeding habits of both *Fagus* species succeeded to some extent in reducing levels of predispersal seed damage by typical specialist insects.

The percentages of seeds of *C. crenata*, *Q. serrata,* and *Q. crispula* with predispersal insect damage were 20.7%, 20.4%, and 79.3%, respectively, in the year with the largest seed crop (Fig. 22.2). These figures indicate that, although *C. crenata* and *Q. serrata* succeeded in reducing predispersal damage by insects, *Q. crispula* failed to do so. Although synchronism between the various host species is required to successfully control the population of atypical specialist predators of these three tree species, their seed production was not strongly synchronized (see Fig. 9.3 in Chapter 9 and Fig. 22.2). Thus, the existence of alternative food resources (i.e., seeds of coexisting host tree species) for the atypical specialists makes it difficult to explain the results by means of the predator-satiation hypothesis.

This raises a question as to the cause of the low attack rate on the seeds of *C. crenata* and *Q. serrata*. The number of seeds bored into by *K. rectirostris* (the category labeled "fed on by others" in Fig. 22.2) was small every year for both *Quercus* species, and the number of seeds attacked by moths remained relatively stable even after the large crop year for both *Quercus* species (Fig. 22.2). Igarashi and Kamata (1993) also showed that the number of *Q. crispula* seeds attacked by moth larvae was uniform regardless of the fluctuations in seed production. These results suggest that factors other than the quantity of the food resource may keep the population levels of both *K. rectirostris* and the moths at low levels. Although it appears reasonable that the effects of heavy predation, parasitism, and disease on these insects may be additional factors affecting the observed results, this conclusion is not unequivocal.

In contrast, the number of seeds bored into by *Curculio* spp. increased along with seed production from 1992 to 1993 for *Q. serrata*, and then decreased, despite an increase in seed production from 1993 to 1994 (Fig. 22.2). Maeto (1993) showed that most larvae of *Curculio distinguendus*, *C. dentipes,* and *C. sikkimensis* (Table 22.1) overwintered at least two seasons before their emergence. In my study, *K. rectirostris* and all moths emerged in the year after seed dispersion, but only one individual of *C. distinguendus*, two of *C. dentipes,* and one of *C. sikkimensis* emerged in the year after seed dispersion; the others of these Curculio species overwintered at least two seasons in the soil. This long-term diapause could be an adaptation that helps the insect utilize a fluctuating food resource and makes it difficult for host trees to control the predator's population by intra- and interspecific synchronization of fluctuating levels of seed production.

Seed production of *Castanea crenata* and the two *Quercus* species in 1992 was synchronously low (Fig. 22.2), and the population of *Curculio* spp. seemed to be limited to a low level. Then, 2 years later (in 1994, when most adults of the suppressed 1992 population emerged), *Castanea crenata* and *Q. serrata* exhibited synchronized intraspecific seed production and were able to escape from predation by *Curculio* spp. However, the seed production of *Castanea crenata* and *Q. serrata* did not affect *Curculio* spp. populations positively. This is because unequivocal interspecific synchronism did not occur between *Castanea crenata* and the *Quercus* species; as well, succeeding lean years and fruit-bearing in alternate years were not observed for the *Quercus* species (Fig 9.3 in Chapter 9). It is likely that the wide host range and the long diapause of *Curculio* spp. makes it difficult for host trees to evolve a mechanism to produce seed synchronously interspecifically. However, only when these three tree species happened to have low seed production synchronously could they could escape predation by *Curculio* spp., with some time lags caused by the diapause of the predators. Thus, *Castanea crenata* and *Q. serrata* partially succeeded in escaping predispersal damage by *Curculio* spp.

22.5 Postdispersal Damage to Seeds

To observe postdispersal damage to seeds, four 0.5 m² quadrats surrounding each seed trap were established on the forest floor. For seeds that fell in 1993, all seeds were collected from one of the four quadrats on 16 March 1994. This procedure was repeated in unsampled quadrats on 1 May and 25 July 1994. For seeds that fell in 1994, the same procedure was carried out on 8 November 1994 and 27 March, 1 May, and 29 June 1995.

The numbers of seeds of *Fagus crenata*, *F. japonica*, and *Castanea crenata* collected from the forest floor in March were only 45.9%, 38.8%, and 30.2% , respectively, of the numbers caught in the seed traps the previous year (Fig. 22.4). This difference seemed to be caused mainly by seed caching by vertebrates, since the greatest reduction was observed for sound seeds, which are preferred by vertebrates (Shimano and Masuzawa 1995). Seeds on the forest floor may have been removed by some unidentified bird species and by any of three rodents, *Apodemus speciosus* (Temminck), *A. argenteus* (Temminck), and *Sciurus lis* Temminck (Yasuda, personal communication). These predators are generalists because they feed on a variety of foods (e.g., worms, insects, and seeds), and thus must be able to switch food resources in response to changes in the quantity and quality of each food resource. The differences between the number of seeds fallen in 1993 and the number on the forest floor in March 1994 were much larger for *Fagus* species and *C. crenata* than for *Quercus* species (Fig. 22.4); this suggests that *Fagus* species and *C. crenata* did not succeed in avoiding postdispersal damage by vertebrate generalists.

The numbers of seeds of *Quercus* species on the forest floor in March 1994 and 1995 were almost the same as the numbers caught by seed traps in the previous year (Fig. 22.4); this indicates that few seeds were removed by vertebrates, suggesting that *Quercus* species succeeded in avoiding postdispersal damage by vertebrate generalists. However, it is probable that many seeds of *Quercus* species would be removed by vertebrates in years in which only *Q. serrata* or *Q. crispula* produced large seed crops (i.e., when the two *Fagus* species and *C. crenata* contemporaneously produced small crops) because generalists will switch their food resource in response to changes in the quantity and quality of foods. Actually, in 1992 when all species of Fagaceae produced a small crop in the OFR, Iida (1996) showed that all 60 seeds of *Q. serrata* and *Q. crispula* that had been artificially placed on the forest floor in autumn disappeared within 2 weeks (also see Chapter 11, in which a high rate of disappearance of artificially dispersed *Q. serrata* seeds is described). This confirms that vertebrates remove seeds of *Quercus* species in years when other species of the Fagaceae produce small seed crops. In 1993, vertebrates removed the abundant seeds of *Fagus* species. In 1994, seeds of both *C. crenata* and *Q. serrata* were abundant, but only those of *C. crenata* were removed. If the seeds of both species are equivalent for vertebrates as a food resource, comparable amounts of the seed of these species should have been removed, but instead, many more seeds of *C. crenata* disappeared than those of *Q. serrata* (Fig. 22.4). This difference was probably caused by differences in the preferences of vertebrates for these seeds. Seeds of *Q. serrata* and *Q. crispula* have high tannin levels (4.8% and 6.7%, respectively) (Matsuyama 1982) that provide chemical protection against predation by vertebrates. The relatively constant seed production by *Quercus* species is disadvantageous in the years when other Fagaceae species produce few seed crops, because many *Quercus* seeds are then the only resource for vertebrates. In contrast, this pattern is advantageous in the years when at least one *Fagus* species or *C. crenata* produces large seed crops because their seeds are more attractive for vertebrates than those of *Quercus* species. Thus, *Q. serrata* and *Q. crispula* succeeded in avoiding attack by vertebrate generalists both by their relatively constant seed production and by means of chemical seed protection.

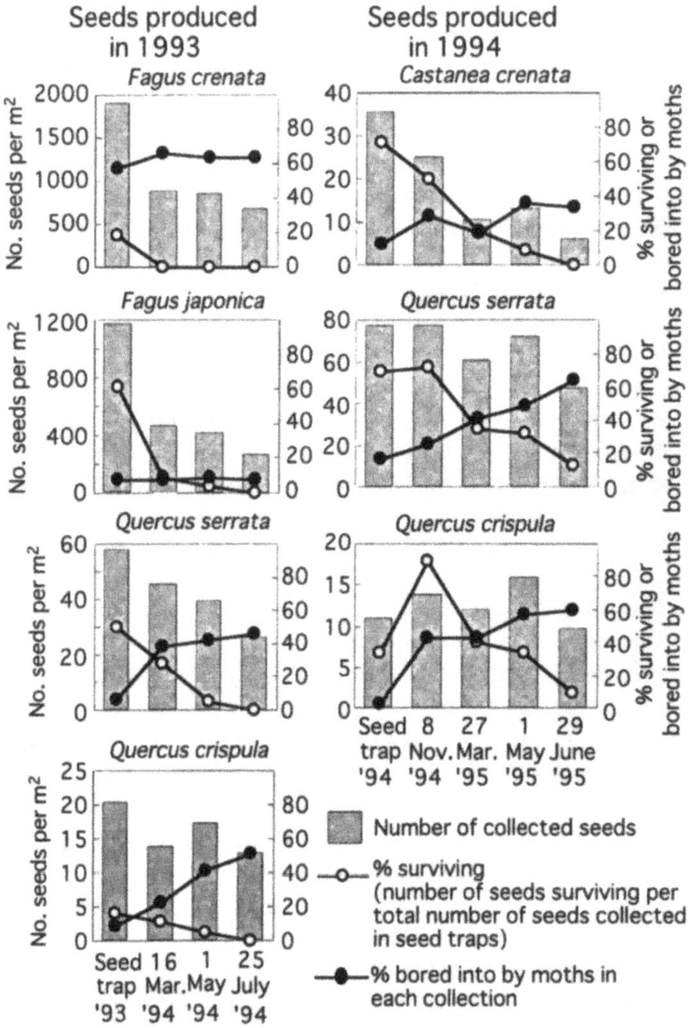

Fig. 22.4. Number of seeds collected from seed traps and quadrats, percent of these seeds bored into by moth larvae (*solid circles*), and percent surviving (*open circles*). *Left*: seeds produced in 1993. *Right*: seeds produced in 1994. The seed-trap data for percent surviving represents the percentage of seeds with no injuries; in quadrats, the percent surviving includes damaged seeds that were still alive, both those with still uninjured hypocotyls and those from which seedlings had begun to grow

It was impossible to distinguish predispersal damage by moth larvae from postdispersal damage solely on the basis of the larval feeding characteristics. For this reason, damage by moth larvae boring into postdispersal seeds was estimated by determining the increase in the proportion of seeds damaged by moth larvae between the seed traps and the quadrats. Since no insects bored into the postdispersal seeds of the *Fagus* species, the proportions of their seeds damaged by moth larvae did not increase (Fig. 22.4). Although larvae of *Cryptaspasma marginifasciata* bored into the postdispersal seeds of *Castanea crenata* (Table 22.1), the proportion of seeds damaged by moth larvae increased only slightly (Fig. 22.4). Since vertebrates had removed most dispersed sound seeds of *C. crenata*, few sound seeds remained on the forest floor for this moth's larvae.

The postdispersal seeds of both *Quercus* species were bored into by larvae of *Cryptaspasma marginifasciata* and *C. trigonana* (Table 22.1), and the proportions of seeds damaged by moth larvae increased greatly by March, and they continued to increase until at least the early summer (Fig. 22.4). The increasing proportion of seeds attacked by moth larvae was accompanied by a steadily decreasing proportion of surviving seeds and seedlings (Fig. 22.4). Since *Quercus* species use the cotyledons as the only source of nutrition in the pericarps after germination, attack on germinated seeds must affect the vitality of seedlings. Not all dead seedlings had been killed by the attacks of the moth larvae, but these larvae had damaged nearly all germinated seeds by summer. It is certain that almost all seedlings had been affected by moth larvae in their seeds, and that many of the seedlings weakened by the attacks were killed directly or indirectly. These results suggest that the two *Quercus* species failed to reduce postdispersal damage by moths, which are considered atypical specialists.

22.6 Conclusion

For the study of interactions between seeds and seed predators, the characteristics of seed production of five Fagaceae species co-occurring in the OFR and predation rates on their seeds were investigated in order to test the reproductive synchrony hypothesis of Ims (1990a, b). The results have been summarized in Table 22.2. *F. crenata* and *F. japonica* showed greatly fluctuating seed production that was synchronized within species, and this strategy succeeded in reducing the levels of attack by a monophagous species (or population) of moth classified as a typical specialist through predator satiation: however, it failed to avert attacks by generalist vertebrates capable of switching among alternative food resources.

Castanea crenata seed production also fluctuated greatly, with intraspecific synchronization, and this strategy succeeded in reducing predispersal attack levels. However, the observed low levels of predispersal attack by borers may be difficult to explain solely by the predator satiation hypothesis because the borers that attacked this species were atypical specialists that can bore into alternative hosts, such as the seeds of other species of the Fagaceae (excluding *Fagus* spp.). Moreover, interspecific synchronism of *C. crenata* with other host tree species was not observed. Like the two *Fagus* species, *C. crenata* also failed to avert postdispersal attack by vertebrates.

Table 22.2. Characteristics of seed production and pre- and postdispersal damage on each species of Fagaceae in the OFR

Tree species	*Fagus crenata*	*Fagus japonica*	*Castanea crenata*	*Quercus serrata*	*Quercus crispula*
Synchronism of seed production within years	Very high	Very high	High	High	Low
Fluctuation of seed production among years[a]	Very high	Very high	High	Low	Low
Predispersal insect damage in large crop years	Medium (Low)	Low (Low)	Low (Low)	Low (Medium)	High (High)
Postdispersal insect damage	Low[c] (Low)	Low[c] (Low)	Low (Low)	High (Medium)	High (High)
Disappearance of postdispersal seeds	Many (Many)	Many (Many)	Many (Many)	Few (Medium)	Few (Few)
Accumulation of tannin in the seed[b]	Low	Low	Low	High	High

Parentheses indicate predicted levels of damage from the hypothesis of Ims (1990a, b).

[a]See Shibata and Tanaka (Chapter 9).

[b]From Matsuyama (1982).

[c]No insects bored into seeds of *Fagus* spp.

Q. serrata exhibited weakly fluctuating seed production that was intraspecifically synchronized to some extent, and succeeded in reducing predispersal damage by borers. However, as in the case of *C. crenata*, this was difficult to explain solely by means of predator satiation because the borers that attacked this species were atypical specialists. This species also succeeded in reducing attack by vertebrate generalists because the seeds of *Quercus* species were not particularly abundant compared with those of the other species and had strong chemical protection (Matsuyama 1982); these factors were enough to divert the attention of vertebrate predators to the abundant seeds of other species of Fagaceae. However, the relatively constant seed production failed to deter postdispersal attack by moth larvae that were atypical specialists.

Quercus crispula exhibited weakly fluctuating seed production that was intraspecifically asynchronous and failed to reduce either pre- or postdispersal attack by borers that were atypical specialists. However, like *Q. serrata*, *Q. crispula* succeeded in reducing attack levels by vertebrates as a result of its relatively sparse seeds and chemical defenses.

These results coincide well with predictions based on the hypothesis of Ims (1990a, b) (Table 22.2). *Q. serrata* succeeded best of the five species of Fagaceae in the OFR in the early phases of reproduction under the crown of the parent tree, with many current year seedlings surviving (10.2 seedlings/m^2 in late June 1995) (Fig. 22.4). *Q. serrata*'s synchronous intraspecific seed production succeeded in reducing predispersal predation by *Curculio* spp. when low seed production by all host species and a delayed (two years later) large crop of this species occurred. *Q. serrata*'s weakly fluctuating seed production, combined with chemical protection of seeds, successfully diverted the attention of vertebrate predators to the abundant seeds produced by other species of Fagaceae. Thus, only the strategy of *Q. serrata* had the advantage of reducing attacks by both specialist predators of predispersal seeds and generalist predators of postdispersal seeds in a species-rich system such as the OFR.

The present study was designed to investigate damage to seeds in the early phases of reproduction under the crown of the parent tree, and one factor not yet discussed is the seeds once cached and forgotten by vertebrate predators. If a tree species relies on such cached and forgotten seeds for its reproduction, the most advantageous strategy would be to reduce attack by predispersal specialist predators and to increase removal of postdispersal seeds by vertebrate generalists, as was the case for *F. japonica*. To investigate this possibility, the number of seedlings that grow from seeds cached and forgotten by vertebrates should be investigated in the future (also see Chapter 11, where the same suggestion is discussed).

Acknowledgments

I wish to express my gratitude to Dr. Isamu Okochi, Dr. Masamichi Ito, Dr. Tadao Goto, Mr. Hiroshi Kitajima, and Mr. Yasunori Okamoto for their help in the field work; to Dr. Furumi Komai for his identification of moths; to Dr. Hajime Niiyama, Mr. Hitoshi Tojo, Ms. Mitsue Shibata, Mr. Takuya Shimada, Mr. Yutaka Igarashi, and Dr. Masatoshi Yasuda for their helpful suggestions; and to Dr. Simon Lawson and Dr. Kazuyuki Fujita for their reviews of the manuscript.

References

Igarashi Y, Kamata N (1991) Studies on seed insects of a beech, *Fagus crenata* (III) infestation of *Pseudopammene fagivora* Komai on beech nuts in case a beech tree produces many nuts every year (in Japanese). Trans Jpn For Soc 102:273–274

Igarashi Y, Kamata N (1993) Acorn insects of *Quercus mongolica* var. *grosseserrata* and their infestation (in Japanese). Trans Tohoku Branch Jpn For Soc 45:83–84

Igarashi Y, Kamata N (1997) Insect predation and seasonal seedfall of the Japanese beech, *Fagus crenata* Blume, in northern Japan. J Appl Entomol 121:65–69

Iida S (1996) Quantitative analysis of acorn transportation by rodents using magnetic locator. Vegetatio 124:39–43

Ims RA (1990a) The ecology and evolution of reproductive synchrony. Tree 5:135–140

Ims RA (1990b) On the adaptive value of reproductive synchrony as a predator-swamping strategy. Am Nat 136:485–498

Janzen DH (1971) Seed predation by animals. Annu Rev Ecol Syst 2:465–492

Komai F (1980) A new genus and species of Japanese Laspeyresiini infesting nuts of beech (Lepidoptera: Tortricidae). Tinea 11:1–7

Komai F, Ishikawa K (1987) Infestation of chestnut fruits in China with two species of the genus *Cydia* (Lepidoptera: Tortricidae) (in Japanese with English abstract). Jpn J Appl Entomol Zool 31:55–62

Maeto K (1993) Acorn insects of *Quercus mongolica* var. *grosseserrata* in Hitsujigaoka natural forest, Hokkaido—life-history of the principal species and their impacts on seed viability (in Japanese). Trans Hokkaido Branch Jpn For Soc 41:88–90

Matsuyama T (1982) Tree seeds (in Japanese)*. Hosei Univ Press, Tokyo. (*This English title is a tentative translation by the author of the original Japanese title.)

Silvertown JW (1980) The evolutionary ecology of mast seeding in trees. Biol J Linn Soc 14:235–250

Shimano K, Masuzawa T (1995) Comparison of seed preservation of *Fagus crenata* Blume under different snow conditions. J Jpn For Soc 77:79–82

Ueda A (1996) Insect infestation on acorns of Japanese chestnut, *Castanea crenata* Sieb. et Zucc., in natural forest (in Japanese). Trans Jpn For Soc 107:233–236

Ueda A (2000a) Pre- and post-dispersal damage to the nuts of two beech species (*Fagus crenata* Blume and *F. japonica* Maxim.) that masted simultaneously at the same site. J For Res 5:21–29

Ueda A (2000b) Pre- and post-dispersal damage to the acorns of two oak species (*Quercus serrata* Thumb and *Q. mongolica* Fischer) in a species-rich deciduous forest. J For Res 5:169–174

Ueda A, Igarashi M, Ito K (1992) Acorn infestation by insects on three evergreen oaks, *Quercus glauca*, *Q. myrsinaefolia* and *Pasania edulis* (I)—their invasion term and level before acorn dropping and comparison with the deciduous *Q. serrata* (in Japanese). Trans Jpn For Soc 103:529–532

Ueda A, Igarashi M, Ito K (1993) Acorn infestation by insects on three evergreen oaks, *Quercus glauca*, *Q. myrsinaefolia* and *Pasania edulis* (II)—their invasion term and level after acorn dropping (in Japanese). Trans Jpn For Soc 104:681–684

Part 8

Conclusion

23 General Conclusion: Forest Community Ecology and Applications

TOHRU NAKASHIZUKA, MORIYOSHI ISHIZUKA, and ISAMU OHKOCHI

23.1 Studies in Ogawa Forest Reserve

The overall purpose of the studies in Ogawa Forest Reserve (OFR) was to elucidate the mechanisms by which biological diversity is maintained, particularly analyzing the coexistence mechanisms of the tree species (see Chapter 1). The findings presented in this book from the 13 years of study also give us some suggestions for applications to forest management. We would like to summarize our studies on two points in this chapter: (1) mechanisms of tree coexistence, or mechanisms maintaining tree diversity in the OFR; and (2) the implications of these findings for forest management applications.

23.2 Coexistence Mechanisms

The studies in OFR demonstrated that several mechanisms may enable different tree species to coexist, or maintain tree species diversity, in this temperate forest community. We first summarize the relationship between the coexistence mechanisms and the life stages during which each mechanism operates. We propose some points for consideration when evaluating the importance of these coexistence mechanisms.

23.2.1 Stage-Specific Mechanisms

One of the implications of the results obtained in the OFR is that the stages at which the various coexistence mechanisms operate are different, as summarized in Fig. 23.1. Horizontal heterogeneity, forest architecture, disturbances, and interactions may enhance tree coexistence at different life stages (Nakashizuka 1996, 2001).

Ecological Studies, Vol. 158
Nakashizuka, Matsumoto (eds.) Diversity and Interaction in a Temperate Forest Community: Ogawa Forest Reserve of Japan

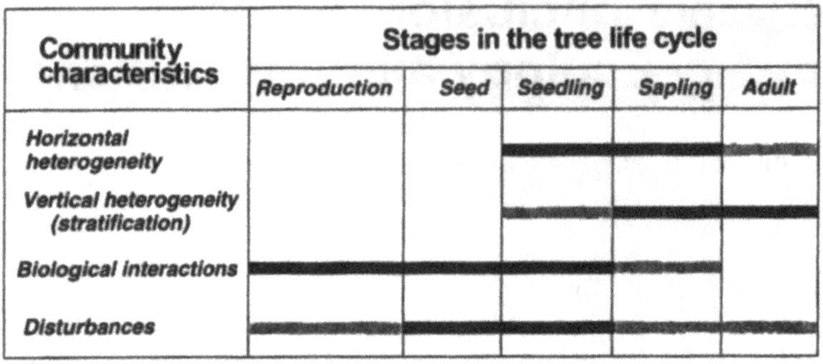

Fig. 23.1. The mechanisms that work to maintain species diversity and the life stages in which those mechanisms mainly operate. The stages during which the effects of the mechanism are strong are shaded *black*, and those during which the effects are weaker are shaded *gray*

Habitat Heterogeneity

The distributions of some trees were highly dependent on topography (see Chapters 6 and 8). Among species in the same genus, species specialized to ridge or valley locations (for example, *Carpinus*), or were distributed more generally, as was typical for the genus *Acer* in the OFR (Masaki et al. 1992). The processes leading to distributional patterns with respect to topography were conspicuous in seedling demography (see Chapter 17; Shibata and Nakashizuka 1995) and ecophysiology (see Chapter 17). The stages after establishment may also show some topographic preference in growth and survival; however, their contribution to the whole-life demography would be smaller, since the cohorts have already selected their favorable habitat through the preceding demographic processes. Kobe (1996) also suggested that juvenile survivorship might be sufficient to explain community composition. These facts suggest that horizontal habitat heterogeneity (heterogeneous distribution with respect to the ground environment and resources such as water and nutrients) is particularly important in the seed and seedling stages before establishment.

Forest Architecture

In contrast, the vertical gradient of the light resource, which is created by the forest architecture (Kohyama 1994), is important in the life stages after seedling establishment. Each tree species has reached its maximum size in this forest community, suggesting that the strata to which the species belongs when it matures can be determined by examining the forest structure. This type of coexistence requires a trade-off between the maximum size and the per capita recruitment rate among the species when we do not consider horizontal heterogeneity (Kohyama 1994). Among the species in the OFR, this trade-off is suggested to exist only for populations of established saplings and older trees, but not for populations of seedlings and seeds (Nakashizuka et al. 1992, 1995). The competition for light is the main mechanism regulating this type of coexistence, and it becomes important after the cohorts have experienced the high mortality of the young stages (Nakashizuka 2001).

Interactive Processes Between Trees and Other Organisms

Interactions between the trees and other organisms are particularly important from reproduction to the seedling and sapling stages. Several studies have demonstrated that seeds and seedlings usually suffer severe herbivory (see Chapter 11; Masaki et al. 1998; Tanaka 1995; Shibata and Nakashizuka 1995). In particular, the largest mortality among seeds and seedlings was from herbivory (or predation) by animals (see Chapter 11). The annual fluctuations in seed production of some species were clearly useful for avoiding high rates of seed predation (see Chapter 9; Shibata et al. 1998). Trees in the seed and seedling stages are small enough that individuals may die from heavy herbivory. Older trees are constructed with less consumable, and thus less attractive, substances such as lignin, making them resistant to herbivory. After the tree has attained a certain size, herbivory seldom causes individual death.

In spite of the high level of sacrifice, however, some dyszoochorous seeds take advantage of herbivores for their dispersal. *Quercus* seeds are distributed to areas significantly larger than their seed shadow by caching by rodents (see Chapters 10 and 22; Iida 1996). Also, the role of birds in seed dispersal has been evaluated quantitatively in the OFR (see Chapter 10; Masaki et al. 1994; Iida and Nakashizuka 1998). Whether seed dispersal by birds is an example of directed dispersal is still in question, although that it increases the chance that the seeds reach safe sites is apparent (Nakashizuka et al. 1995).

Other interactive features that were not intensively studied in the OFR may play important roles in the stages from reproduction to sapling. Some evidence for pollen transportation was quantified by genetic markers (Part 6) and could be extended to studies of interactions between trees and pollinators. The interactions between trees and fungi have just begun to be studied (see Chapter 21). Both the positive (mutualistic exchange with mycorrhizas) and negative interactions (pathogen) with fungi also seem to be more effective in these young stages (Allen 1991).

Disturbances

Disturbances are a potentially important species-coexistence mechanism affecting a wide range of life stages. The trade-off between seed size and number was fundamentally related to safe-site frequency for a species (see Chapter 10; Nakashizuka et al. 1995), which is dependent on the disturbance regime. Seed-dispersal mechanisms were also related to this trade-off (see Chapter 9; Tanaka et al. 1998).

Seedlings and saplings behaved differently with regard to leaf ecophysiology (see Chapters 15 and 16) and demography (see Chapter 12) according to the light level, suggesting adaptations to the heterogeneous light climate caused by canopy disturbances. There were actually variations among species in sapling density according to canopy gap size (see Chapter 12; Abe et al. 1995), suggesting some phenomenon such as gap size partitioning (Sipe and Bazzaz 1995).

Seed bank formation is a strategy that increases the chance of finding safe sites, which are formed unpredictably by disturbances. Thus, both seed dormancy and seed dispersal can be adaptations to disturbance: temporal and spatial enhancement of the chance and steadiness of finding a safe sites. Fresh pulp around seeds may work as a two-way strategy; it may reward the dispersal agent (frugivorous birds) and may inhibit germination if it is not consumed by the agent, as shown in *Kalopanax pictus* (Iida and Nakashizuka 1998).

Even among the adult trees, resistance to disturbance differs (Foster and Boose 1992). Also among the species in the OFR, the composition of gap makers was not proportional to the composition of canopy trees, but depended both on tree size and species (Abe et al. 1995). Although the contributions of these variations in demographic parameters to the maintenance of the species population have not been quantified in detail, the traits of the seed and seedling stages have larger variation among species than do those of the adult stage.

23.2.2 Which Mechanisms Are Important at What Life Stages?

Are all the parameters of the various life stages equally important? This question can be broken down into two questions: (1) Which life stages of a tree species are important for maintaining its population? and (2) Which mechanisms dominate the assembly rules in this tree community?

What Are the Critical Life Stages of a Tree Species?

Throughout the demographic changes of a tree's life cycle, the most rapid change in population density occurs in the stages from seed to seedling. The seed produced by *Betula grossa* was reduced to only 0.03% in the first year (Nakashizuka et al. 1995). Therefore, some researchers think these stages are critical to the tree population. In contrast, mortality resulting from competition for light in the stages after establishment is usually very small, usually as low as 1% yr^{-1} or so for trees more than 10 cm in diameter (Nakashizuka et al. 1992). However, for some species, mortality from this cause may occur after an individual tree has lived for hundreds of years, so a very small difference in annual mortality results in a large difference in fitness over a long period. Thus, it may be difficult to say which is the more critical stage for maintaining the population of a tree species simply by comparing annual mortality rates.

The elasticity analyses of the population matrix model generally show that the survival of reproductive individuals is usually more important for tree species (Silvertown et al. 1993). However, elasticity analysis assumes that the population growth rates lead to the transition probabilities changing by the same proportion as the original values. This assumption may not be valid, since transition probabilities for the young stages have much greater variability in relation to the environment than those for older stages. In the OFR, the survival rate of a first-year seedling varies from 0.1% to 90% yr^{-1}, while that of adult trees is very stable at around 95%–99% yr^{-1}. A 20% change in the former value occurs very frequently, depending on the environment (e.g., in a gap or not), while the latter value seldom changes. Thus, it is still problematic to say which stage is more critical for the maintenance of tree populations.

What Are the Important Mechanisms Determining Tree Coexistence?

The important mechanisms determining tree coexistence are site dependent, and it thus is not possible to answer this question from results obtained in only one forest community. However, some quantitative evaluation of the contribution of the assembly rules to tree species diversity is possible.

There are some congeneric species that in the OFR display different habitat preferences with respect to topography (*Acer, Carpinus*; Masaki et al. 1992), suggesting some niche separation, probably in terms of nutrient or water utilization. Among the species occupying the same topographic habitats, there are some differences in maximum size, suggesting that coexistence may depend on forest architecture (see Chapter 6). These characteristics will lead to multidimensional niche partitioning among constituent species. However, there still seems to be some redundancy among species if we try to explain community composition with habitat heterogeneity as the only assembly rule.

Many demographic parameters suggest that some features are adaptive to disturbances. Several of these mechanisms may be correlated or independent, such as those features characteristic of pioneer versus late-successional species (Ricklefs 1990). Some parameters that are both positively and negatively correlated, suggesting such groupings, could be observed in the OFR. For some wind-dispersed species, seed size is positively correlated with the frequency of safe sites but negatively with the area of the seed shadow (see Chapter 10). The size at which reproduction begins is positively correlated with the maximum size of the species (see Chapter 9). But not all relationships that have been documented in textbooks were observed, suggesting that a framework such as pioneer versus climax species is too simple. Some of these reported correlations might result from phylogenetic constraints that are not apparent in this forest at this moment. It is necessary to evaluate the contribution of such parameters to population maintenance in some way to avoid errors from this kind of omission.

The relative importance of coexistence mechanisms seems to vary among communities and ecosystems. Habitat heterogeneity is important in communities where there are large topographic variations, while disturbances are more important in a community with complicated disturbance regimes but without large environmental heterogeneity. In species-rich systems such as tropical rain forests, the interaction between organisms is important, as are chance effects (Brokaw and Busing 2000). Quantitative analyses of the contributions of each coexistence mechanism will become possible when in the future we have accurate measurements of all parameters and their variations.

23.2.3 Integration Tools

As the above summary shows, we have yet to answer completely the question regarding the mechanisms that make coexistence possible. We need some tools to answer the question, particularly in terms of evaluating the contributions of each parameter to population maintenance or the relative importance of each coexistence mechanism. The matrix model and the individual-based model are two tools introduced in this book.

The matrix model is a well-established system to describe life-history traits as a whole (Caswell 1989). It is particularly effective for identifying key stages in the tree life cycle. However, as seen in the earlier modeling attempt by Silvertown et al. (1993), most long-lived trees tend to be given greater weight and importance in the adult stage compared with short-live species. Thus, such a model may be more applicable to a discussion of the comparative ecology of short-lived herbaceous plants, which have a small range of variation in life span, than to a discussion of the coexistence of tree species.

Also, a matrix model cannot fully utilize the information obtained in field studies. For example, information on growth rates and their variation with respect to certain environmental gradients is not really utilized in a transition probability matrix. In the OFR, such data were originally collected with high accuracy, although that may not be reflected in the matrix.

The individual-based model (IBM) model is another advanced system for analyzing a community (Pacala and Rees 1998). This model is applicable to a wide range of topics, from the description of a tree's life history to the prediction of global warming effects. An IBM model can be used for a sensitivity analysis to quantitatively evaluate coexistence mechanisms. The model developed in the OFR (see Chapter 14) will give us useful insights into these mechanisms.

This model can utilize almost all of the parameters we collected in the field studies; in fact, it requires even more. Even after 13 years of parameterization, we have not obtained some parameters at the desirable level of accuracy. On the other hand, by including many parameters, it might be difficult to isolate the contribution of a single parameter to the whole, because some parameters are correlated or in a trade-off relationship with others.

23.3 Forest Management Maintaining Biodiversity

In recent forestry practices, the purpose of forest management is not the maximization of timber products or effective regeneration alone. We discuss the application of the results of our studies in the OFR to forest operations or management practices with the goal of maintaining high biodiversity (see Chapter 1). As discussed above, the maintenance of biodiversity can be accomplished by controlling the assembly rules. Such controls should be based on knowledge of the response of each species to every factor affecting the community condition.

23.3.1 Evaluating Habitat Heterogeneity

To understand tree diversity arising from habitat heterogeneity (see Chapter 3), ecophysiological knowledge of each tree species is specifically important. Trees seem to be favored by the environment of the habitat particularly in their young stages, and nursery techniques have been developed because we knew this empirically.

Although habitat heterogeneity may determine tree diversity to some extent (see Chapter 8), its relative contribution to the diversity observed may vary from site to site. A site with large habitat heterogeneity may have many species that prefer specific microenvironments, while a site with a uniform habitat may have few species. For this reason forest reserves are chosen such that they include as many different habitats as possible to keep biodiversity high (Spellerberg 1991).

23.3.2 Controlling Disturbances and Forest Structure

Most human activities constitute some kind of disturbance for a ecosystem. The control of the disturbance regime is one of the more tractable methods of forest management, and it has been frequently used, although there are not many successful examples of sustainable management by this means. Traditional techniques of silviculture can be regarded as disturbance control practices designed to maximize a crop or the fitness of the target species. For sustainable management, however, we must modify forestry practices to maximize other criteria, such as biodiversity (Johns 1997).

Creating artificial canopy gaps in a forest with a relatively homogeneous structure (such as tree farms, coppice forests, etc.) is one recent approach to the maintenance of high biodiversity among trees and other plants and their associated organisms (Peterken 1991; Spies et al. 1994; Kobayashi and Kamitani 2000; Yoshida and Kamitani 1998). A test of this approach has been conducted in the vicinity of the OFR (see Chapter 18). In this case, however, the aim of the artificial gap creation was to compromise the maximization of biodiversity by an increase in the growth or biomass of the target species.

The regeneration of Japanese beech forests has been studied for a long time by Japanese foresters (Kataoka 1991). Until the early 1960s, logging by clear-cutting was usual in the National Forests of Japan. Since then, a shelterwood logging system has been in use. There have been many reports on this method's effectiveness as a regeneration system. Because of the large amount of understory plants growing on the forest floor, typically dwarf bamboo, unsuccessful regeneration after logging has been documented many times in the Japanese literature (Kataoka 1991; Maeda 1988). Most sites failed to have sufficient regeneration unless shrubs were cut. Recently, this system was evaluated from the view point of the diversity of the forest floor plants (Nagaike et al. 1999) and was found to cause relatively little decrease in the diversity of the forest floor flora. However, its effects on other organisms are still unknown.

Not only the disturbance intensity, but also the nature of the disturbance is important. Exposing mineral soil and retaining old or fallen logs have been considered effective methods for maintaining the biodiversity of organisms other than plants so that the interactions among all the organisms could be maintained (Franklin 1989; Kirby and Patterson 1992).

The importance of large-scale, but rare disturbances should also not be overlooked. *Quercus* spp., *Carpinus* spp., and some other tree species in the OFR are apparently dependent on large-scale disturbances such as forest fire (see Chapters 4 and 7). Since fire is strictly controlled under present forest policy, the population of these species will decrease if we keep this policy. A similar situation has also been reported for North American *Quercus* populations (Abrams 1992; Runkle 1990). The use of such large-scale disturbances for forest management is currently a controversial topic, particularly in areas such as Japan, where forest reserves and residential areas are often in close proximity.

23.3.3 Controlling Interactions

Controlling interactions between trees and other organisms is the most difficult technique to develop at this time. Many indirect effects are possible; therefore, controlling one process might cause other unexpected results.

Seed predation and seed dispersal are well-documented interactions in the OFR (see Chapters 10 and 22). Mast seeding seems to effectively increase fitness in many species, particularly by helping the tree escape predispersal predation by relatively host-specific insects (see Chapters 9 and 22). However, predation by generalists such as rodents is apparently rather complicated. There are five dominant large-seeded trees of the family Fagaceae in the OFR, and their fruiting tended to be asynchronous. For this reason, population fluctuations of rodents, which depend largely on these seeds in winter, are very unpredictable. Thus, some species are able to avoid postdispersal mortality from this cause, but others are not (see Chapter 22). Such a complex system needs more (probably long-term) study before it can be applied to forest management.

The survival of seedlings is also influenced by other organisms, such as herbivorous insects that fall from adult tree crowns. The negative effect of host-specific plant parasites and predators on seedlings declines with increasing distance from the parent and other adult trees (Janzen 1970). This mechanism was first reported from a tropical forest (Janzen 1971), but it has since been reported in a northern forest in which insects are major agents of seedling mortality (Fukuyama 1997). We can expect such interactions to also be present in the OFR.

Defoliating insects sometimes damage adult trees. The beech caterpillar, *Syntypistis* (= *Quadricalcarifera*) *punctatella* (Motschulsky), is the most well-known defoliator of the beech trees *Fagus japonica* and *F. crenata*. Outbreaks occur cyclically at 8–11 year intervals, especially in northern Japan (Liebhold et al. 1996). Delayed density-dependant factors such as the entomopathogenic fungus *Cordyceps militaris*, may affect its dynamics (Kamata et al. 1997). Beech caterpillars are also present in the OFR, but we have no records of outbreaks. It would be interesting to know whether such variation in insect populations among sites relates to differences in biodiversity among the sites.

Studies on natural enemies and their roles in the pest population dynamics of a natural forest such as the OFR are important and more should be undertaken. The use of introduced natural enemies to a pest as a biological control agent has been known to cause the crash of a population of a nontarget endemic species (Murray et al. 1988). Importing commercial exotic natural enemies as an easy but dangerous solution that should be replaced by an effort to identify endemic natural enemies. A natural primary forest such as the OFR may offer opportunities for this kind of research while maintaining a high biological diversity with complex interactions among organisms. Future directions of biological control may be based on endemic natural enemies after productive research on their biology and ecology in a natural forest.

23.4 Future Directions

Overall, we propose several directions for future studies: (1) interactions between plants and other organisms; and (2) landscape-level processes should be studied with long-term perspectives.

23.4.1 Interactions Among Organisms

Interactions between trees and other organisms have not been studied enough in the OFR, although even those studies that have been carried out suggest the complexity of the effects of interactions. A small number of key species may play very important roles in a community. The principles and techniques necessary to maintain a diverse interactive network must be studied as the next step for the proper conservation and management of forests.

More collaborative studies between plant ecologists and scientists specializing on other organisms are essential. In addition, in the area of studies on plant responses to other organisms, the allocations of resources to the various organs and substances involved in interactions with other organisms have not yet been studied in detail. To ensure effective pollination, trees invest in rewards for the pollinating agents (Pyke 1991). Allocations to defensive substances might be in a trade-off relationship with allocations to growth (Coley et al. 1985). Similarly, we do not know how large a proportion of net production is allocated to symbiotic fungi (Allen 1991).

23.4.2 Landscape-Level Processes

Considering the status of forest ecosystems today, the conservation of biodiversity should be discussed beyond a single ecosystem. There are some key species which have large home ranges extending across many different ecosystems. Most forest ecosystems and habitats have been extensively fragmented, and thus some species have confronted obstacles to the maintenance of genetic diversity and population size. Thus, large-scale designs are required to take advantage of desirable ecosystem services.

To address these problems, manipulative experiments of landscape-level interactions are truly needed. Some pioneer studies have already been begun in several areas (Lovejoy et al. 1986). We hope that the studies in the OFR can be expanded to address such problems in the future, building upon the core model that has already been established, as described in this book.

References

Abe S, Masaki T, Nakashizuka T (1995) Factors influencing sapling composition in canopy gaps of a temperate deciduous forest. Vegetatio 120:21–32

Abrams MD (1992) Fire and the development of oak forests. Bioscience 42:346–353

Allen MF (1991) The ecology of mycorrhizae. Cambridge University Press, New York

Brokaw N, Busing R (2000) Niche versus chance and tree diversity in forest gaps. Trends Ecol Evol 15:183–188

Caswell H (1989) Matrix population models—construction, analysis, and interpretation. Sinauer, Sunderland, MA

Coley PD, Bryant JP, Chapin III FS (1985) Resource availability and plant antiherbivore defense. Science 230:895–899

Foster DR, Boose ER (1992) Patterns of forest damage resulting from catastrophic wind in central New England, USA. J Ecol 80:79–98

Franklin J (1989) Toward a New Forestry. American Forests, November/December, pp 7–44

Fukuyama K (1997) Effects of mature host trees on distribution of *Prociphilus oriens* Mordvilko (Homoptera: Aphididae) on *Fraxinus lanuginosa* Koidz. Saplings (in Japanese with English summary). Jpn J Appl Entomol Zool 41:105–107

Iida S (1996) Quantitative analysis of acorn transportation by rodents using magnetic locator. Vegetatio 124:39–43

Iida S, Nakashizuka T (1998) Spatial and temporal dispersal of *Kalopanax pictus* seeds in a temperate deciduous forest, central Japan. Plant Ecol 135:243–248

Janzen DH (1970) Herbivores and the number of tree species in tropical forests. Am Nat 104:501–528

Janzen DH (1971) Escape of juvenile *Dioclea megacarpa* (Leguminosae) vines from predators in a deciduous tropical forest. Am Nat 105:97–112

Johns AG (1997) Timber production and biodiversity conservation in tropical rainforests. Cambridge University Press, Cambridge, UK

Kamata N, Sato H, Shimazu M (1997) Seasonal changes in the infection of pupae of the beech caterpillar, *Quadricalcarifera punctatella* (Motsch.) (Lep., Notodontidae), by *Cordyceps militalis* Link (Clavicipitales, Clavicipitaceae) in the soil of the Japanese beech forest. J Appl Entomol 121:17–21

Kataoka H (1991) Desirable treatment for beech regeneration (in Japanese). In: Murai H, Yamaya K, Kataoka H, Yui M (eds) Natural environment and conservation of beech forests. Soft Science, Tokyo, pp 351–394

Kirby K J and Patterson G (1992) Ecology and management of semi-natural tree species mixture. In: Cannell MGR, Malcolm DC, Robertson PA (eds) The ecology of mixed-species stands of trees. Blackwell, London, pp 189–209

Kobayashi M, Kamitani T (2000) Effect of surface disturbance and light level on seedling emergence in a Japanese secondary deciduous forest. J Veg Sci 11:93–100

Kobe RK (1996) Intraspecific variation in sapling mortality and growth predicts geographic variation in forest composition. Ecol Monogr 66:181–201

Kohyama T (1994) Size-structure–based models of forest dynamics to interpret population- and community-level mechanisms. J Plant Res 107:107–116

Liebhold A, Kamata N, Jacob T (1996) Cyclicity and synchrony of historical outbreaks of the beech caterpillar, *Quadricalcarifera punctatella* (Motschulsky) in Japan. Res Popul Ecol 38:87–94

Lovejoy TE, Bierregaad RO Jr, Rynolds AB, Mallcolm JR, Quintela CE, Harper LH, Brown KSR Jr, Powell GVN, Schubart HOR, Hays MB (1986) Edge and other effects of isolation on Amazon forest fragments. In: Soule ME (ed) Conservation biology. Sinauer, Sunderland, MA, pp 257–285

Maeda T (1988) Studies on natural regeneration of beech (*Fagus crenata* Blume) (in Japanese with English summary). Special Bull Agr Utsunomiya Univ 46:1–79

Masaki T, Suzuki W, Niiyama K, Iida S, Tanaka H, Nakashizuka T (1992) Community structure of a species-rich temperate forest, Ogawa Forest Reserve, central Japan. Vegetatio 98:97–111

Masaki T, Kominami Y, Nakashizuka T (1994) Spatial and seasonal patterns of seed dissemination of *Cornus controversa* in a temperate forest. Ecology 75:1903–1910

Masaki T, Tanaka H, Shibata M, Nakashizuka T (1998) The seed bank dynamics of *Cornus controversa* and their role in regeneration. Seed Sci Res 8:53–63

Murray J, Murray E, Johnson M, Clarke B (1988) The Extinction of *Partula* on Moorea. Pacific Sci 42:150–153

Nagaike T, Kamitani T, Nakashizuka T (1999) The effect of shelterwood logging on the diversity of plant species in a beech (*Fagus crenata*) forest in Japan. For Ecol Manag 118:161–171

Nakashizuka T (1996) Factors maintaining forest tree diversity. In: Turner IM, Diong CH, Lim SSL, Ng PKL (eds) Biodiversity and the dynamics of ecosystems. The international network for DIVERSITAS in Western Pacific and Asia (DIWPA), pp 51–62

Nakashizuka T (2001) Species coexistence research in temperate, mixed deciduous forests. Trends Ecol Evol 16:205–210

Nakashizuka T, Iida S, Tanaka H, Shibata M, Abe S, Masaki T, Niiyama K (1992) Community dynamics of Ogawa Forest Reserve, a species-rich deciduous forest, central Japan. Vegetatio 103:105–112

Nakashizuka T, Iida S, Masaki T, Shibata M, Tanaka H (1995) Evaluating increased fitness through dispersal: a comparative study on tree populations in a temperate forest, Japan. Ecoscience 2:245–251

Pacala SW, Rees M (1998) Models suggesting field experiments to test two hypotheses explaining successional diversity. Am Nat 152:729–737

Peterken GF (1991) Ecological issues in the management of woodland nature reserves. In: Spellerberg IF, Goldsmith FB, Morris MG (eds) The scientific management of temperate communities for conservation. Blackwell, London, pp 245–272

Pyke GH (1991) What does it cost a plant to produce floral nectar? Nature 350:58–59

Ricklefs RE (1990) Ecology, 3 rd edn. WH Freeman, New York

Runkle JR (1990) Gap dynamics in an Ohio *Acer-Fagus* forest and speculations on the geography of disturbance. Can J For Res 20:632–641

Shibata M, Nakashizuka T (1995) Seed and seedling demography of four co-occurring *Carpinus* species in a temperate deciduous forest. Ecology 76:1099–1108

Shibata M, Tanaka H, Nakashizuka T (1998) Causes and consequences of mast seed production of four co-occurring *Carpinus* species in Japan. Ecology 79:54–64

Silvertown J, Franco M, Pisanty I, Mendoza A (1993) Comparative plant demography—relative importance of life-cycle components to finite rate of increase in woody and herbaceous perennials. J Ecol 81:465–476

Sipe TW, Bazzaz FA (1995) Gap partitioning among maples (*Acer*) in central New England: survival and growth. Ecology 76:1587–1602

Spellerberg IF (1991) Biogeographical basis of conservation. In: Spellerberg IF, Goldsmith FB, Morris MG (eds) The scientific management of temperate communities for conservation. Blackwell, London, pp 293–322

Spies TA, Ripple WJ, Bradshaw GA (1994) Dynamics and pattern of a managed coniferous forest landscape in Oregon. Ecol Appl 4:555–568

Tanaka H (1995) Seed demography of three co-occurring *Acer* species in a Japanese temperate deciduous forest. J Veg Sci 6:887–896

Tanaka H, Shibata M, Nakashizuka T (1998) A mechanistic approach for evaluating the role of wind dispersal in tree population dynamics. J Sust For 6:155–174

Yoshida T, Kamitani T (1998) Effects of crown release on basal area growth rates of some broad-leaved tree species with different shade-tolerance. J For Res 3:181–184

Subject Index

A

B

C

D

E

Q

R

S

Ecological Studies

Volumes published since 1996

Volume 120
Landscape Function and Disturbance in Arctic Tundra (1996)
J.F. Reynolds and J.D. Tenhunen (Eds.)

Volume 121
Biodiversity and Savanna Ecosystem Processes. A Global Perspective (1996)
O.T. Solbrig, E. Medina, and J.F. Silva (Eds.)

Volume 122
Biodiversity and Ecosystem Processes in Tropical Forests (1996)
G.H. Orians, R. Dirzo, and J.H. Cushman (Eds.)

Volume 123
Marine Benthic Vegetation. Recent Changes and the Effects of Eutrophication (1996)
W. Schramm and P.H. Nienhuis (Eds.)

Volume 124
Global Change and Arctic Terrestrial Ecosystems (1997)
W.C. Oechel et al. (Eds.)

Volume 125
Ecology and Conservation of Great Plains Vertebrates (1997)
F.L. Knopf and F.B. Samson (Eds.)

Volume 126
The Central Amazon Floodplain: Ecology of a Pulsing System (1997)
W.J. Junk (Ed.)

Volume 127
Forest Decline and Ozone: A Comparison of Controlled Chamber and Field Experiments (1997)
H. Sandermann, A.R. Wellburn, and R.L. Heath (Eds.)

Volume 128
The Productivity and Sustainability of Southern Forest Ecosystems in a Changing Environment (1998)
R.A. Mickler and S. Fox (Eds.)

Volume 129
Pelagic Nutrient Cycles: Herbivores as Sources and Sinks (1997)
T. Andersen

Volume 130
Vertical Food Web Interactions: Evolutionary Patterns and Driving Forces (1997)
K. Dettner, G. Bauer, and W. Völkl (Eds.)

Volume 131
The Structuring Role of Submerged Macrophytes in Lakes (1998)
E. Jeppesen et al. (Eds.)

Volume 132
Vegetation of the Tropical Pacific Islands (1998)
D. Mueller-Dombois and F.R. Fosberg

Volume 133
Aquatic Humic Substances: Ecology and Biogeochemistry (1998)
D.O. Hessen and L.J. Tranvik (Eds.)

Volume 134
Oxidant Air Pollution Impacts in the Montane Forests of Southern California (1999)
P.R. Miller and J.R. McBride (Eds.)

Volume 135
Predation in Vertebrate Communities: The Bialowieża Primeval Forest as a Case Study (1998)
B. Jedrzejewska and W. Jedrzejewski

Volume 136
Landscape Disturbance and Biodiversity in Mediterranean-Type Ecosystems (1998)
P.W. Rundel, G. Montenegro, and F.M. Jaksic (Eds.)

Volume 137
Ecology of Mediterranean Evergreen Oak Forests (1999)
F. Rodà et al. (Eds.)

Volume 138
Fire, Climate Change and Carbon Cycling in the North American Boreal Forest (2000)
E.S. Kasischke and B. Stocks (Eds.)

Volume 139
Responses of Northern U.S. Forests to Environmental Change (2000)
R. Mickler, R.A. Birdsey, and J. Hom (Eds.)

Volume 140
Rainforest Ecosystems of East Kalimantan: El Niño, Drought, Fire and Human Impacts (2000)
E. Guhardja et al. (Eds.)

Volume 141
Activity Patterns in Small Mammals: An Ecological Approach (2000)
S. Halle and N.C. Stenseth (Eds.)

Volume 142
Carbon and Nitrogen Cycling in European Forest Ecosystems (2000)
E.D. Schulze (Ed.)

Volume 143
Global Climate Change and Human Impacts on Forest Ecosystems: Postglacial Development, Present Situation and Future Trends in Central Europe (2001)
J. Puhe and B. Ulrich

Volume 144
Coastal Marine Ecosystems of Latin America (2001)
U. Seeliger and B. Kjerfve (Eds.)

Volume 145
Ecology and Evolution of the Freshwater Mussels Unionoida (2001)
G. Bauer and K. Wächtler (Eds.)

Volume 146
Inselbergs: Biotic Diversity of Isolated Rock Outcrops in Tropical and Temperate Regions (2000)
S. Porembski and W. Barthlott (Eds.)

Volume 147
Ecosystem Approaches to Landscape Management in Central Europe (2001)
J.D. Tenhunen, R. Lenz, and R. Hantschel (Eds.)

Volume 148
A Systems Analysis of the Baltic Sea (2001)
F.V. Wulff, L.A. Rahm, and P. Larsson (Eds.)

Volume 149
Banded Vegetation Patterning in Arid and Semiarid Environments (2001)
D. Tongway and J. Seghieri (Eds.)

Volume 150
Biological Soil Crusts: Structure, Function, and Management (2001)
J. Belnap and O.L. Lange (Eds.)

Volume 151
Ecological Comparisons of Sedimentary Shores (2001)
K. Reise (Ed.)

Volume 152
Global Biodiversity in a Changing Environment: Scenarios for the 21st Century (2001)
F.S. Chapin, O. Sala, and E. Huber-Sannwald (Eds.)

Volume 153
UV-Radiation and Arctic Ecosystems (2002)
D.O. Hessen (Ed.)

Volume 154
Geoecology of Antarctic Ice-Free Coastal Landscapes (2002)
L. Beyer and M. Bölter (Eds.)

Volume 155
Conserving Biodiversity in East African Forests: A Study of the Eastern Arc Mountains (2002)
W.D. Newmark

Volume 156
Urban Air Pollution and Forests: Resources at Risk in the Mexico City Air Basin (2002)
M.E. Fenn, M. de Lourdes de la Isla de Bauer, and T. Hernández-Tejeda

Volume 157
Mycorrhizal Ecology (2002)
M.G.A. van der Heijden and I.R. Sanders

Volume 158
Diversity and Interaction in a Temperate Forest Community: Ogawa Forest Reserve of Japan (2002)
T. Nakashizuka and Y. Matsumoto (Eds.)

www.ingramcontent.com/pod-product-compliance
Lightning Source LLC
Chambersburg PA
CBHW070821150926
51501CB00043B/937
* 9 7 8 4 4 3 1 6 8 0 0 0 0 *